Glass Physics

Glass Physics

Edited by
Douglas Suarez

Larsen & Keller
www.larsen-keller.com

Glass Physics
Edited by Douglas Suarez
ISBN: 978-1-63549-136-4 (Hardback)

Larsen & Keller
Published by Larsen and Keller Education,
5 Penn Plaza,
19th Floor,
New York, NY 10001, USA

Cataloging-in-Publication Data

Glass physics / edited by Douglas Suarez.
p. cm.
Includes bibliographical references and index.
ISBN 978-1-63549-136-4
1. Glass. 2. Glass--Standards. 3. Glass craft. 4. Physics.
I. Suarez, Douglas.
TA450 .G53 2017
620.144--dc23

Printed and bound in China.

For more information regarding Larsen and Keller Education and its products, please visit the publisher's website www.larsen-keller.com

Table of Contents

Preface

The book aims to shed light on some of the unexplored aspects of glass physics. It talks in detail about the various techniques and methods used in this medium of study. Glass physics is the study of the properties and characteristics of glass. This text is a compilation of chapters that discuss the most vital concepts in the field of glass physics. It will serve as a valuable source of reference for those interested in this field. The topics included in it are of utmost significance and are bound to provide incredible insights to readers. For all those who are interested in glass physics, this textbook can prove to be an essential guide.

A foreword of all Chapters of the book is provided below:

Chapter 1 - Glass is used in windowpanes, tableware and optoelectronics. The oldest form of glass is known as silicate glass. Glass transmits, reflects and refracts light. The chapter on glass offers an insightful focus, keeping in mind the subject matter; **Chapter 2** - The physical properties of glass are viscosity, low emissivity, corrosion, glass transition, thermal expansion, liquidus, heat capacity and surface tension among others. Glass transition is the transition of a material from a hard to a relatively brittle state. The brittle state is the "glassy" state; as the temperature increases it turns into a rubbery state. This section is an overview of the subject matter incorporating all the physical properties of glass; **Chapter 3** - The glass that is used as building material is known as architectural glass. Glass is usually used for internal partitions within a building. Glass bricks, annealing, laminated glass, heatable glass and self-cleaning glass are some of the examples given in this chapter. This section will provide an integrated understanding of architectural glass; **Chapter 4** - Libyan desert glass is a substance that is mostly found in the Libyan Desert. Darwin glass, just like Libyan Desert glass is found naturally. Darwin glass is found in the south of Queenstown. The other glasses mentioned in the following section are tektite, volcanic glass, tachylite, obsidian, Pele's tears, pumice etc. This chapter strategically incorporates the natural occurrences of glass; **Chapter 5** - Glass production has mainly two main methods; the float glass process and glassblowing. The most famous technique of forming glass is glassblowing; it involves inflating molten glass and converting them into a bubble with the help of blowpipe. The aspects that have been explained in the chapter are cylinder blown sheet, machine drawn cylinder sheet, silicon dioxide, polished plate glass and glazier. The aspects elucidated in this section are of vital importance and provide a better understanding of the production of glass; **Chapter 6** - The applications of glass that have been discussed in this section are glass tile, glass-bottom boat, fiberglass, windshield, storm glass etc. Glass-bottom boats are boats that have some portion that has a section of glass. This section helps the passengers in observing the underwater environment from within the boat. The chapter serves as a source to understand the applications of glass; **Chapter 7** - Glass recycling is the conversion of waste glass into products that can be used. Recycled glass countertops are entirely made up of glass that is 100% recyclable. Glass wool and abrasive are also explained in the chapter. Recycling of glass is best understood in confluence with the major topics listed in the following section.

I would like to thank the entire editorial team who made sincere efforts for this book and my family who supported me in my efforts of working on this book. I take this opportunity to thank all those who have been a guiding force throughout my life.

Editor

1

Introduction to Glass

Glass is used in windowpanes, tableware and optoelectronics. The oldest form of glass is known as silicate glass. Glass transmits, reflects and refracts light. The chapter on glass offers an insightful focus, keeping in mind the subject matter.

The joining of two tubes made of lead glass during glass welding

Glass is a non-crystalline amorphous solid that is often transparent and has widespread practical, technological, and decorative usage in, for example, window panes, tableware, and optoelectronics. Scientifically, the term "glass" is often defined in a broader sense, encompassing every solid that possesses a non-crystalline (that is, amorphous) structure at the atomic scale and that exhibits a glass transition when heated towards the liquid state.

The most familiar, and historically the oldest, types of glass are "silicate glasses" based on the chemical compound silica (silicon dioxide, or quartz), the primary constituent of sand. The term *glass*, in popular usage, is often used to refer only to this type of material, which is familiar from use as window glass and in glass bottles. Of the many silica-based glasses that exist, ordinary glazing and container glass is formed from a specific type called soda-lime glass, composed of approximately 75% silicon dioxide (SiO_2), sodium oxide (Na_2O) from sodium carbonate (Na_2CO_3), calcium oxide, also called lime (CaO), and several minor additives. A very clear and durable quartz glass can be made from pure silica, but the high melting point and very narrow glass transition of quartz make glassblowing and hot working difficult. In glasses like soda lime, the compounds added to quartz are used to lower the melting temperature and improve workability, at a cost in the toughness, thermal stability, and optical transmittance.

Many applications of silicate glasses derive from their optical transparency, which gives rise to one of silicate glasses' primary uses as window panes. Glass will transmit, reflect and refract light; these qualities can be enhanced by cutting and polishing to make optical lenses, prisms, fine glassware, and optical fibers for high speed data transmission by light. Glass can be colored by adding metallic salts, and can also be painted and printed with vitreous enamels. These qualities have led to the extensive use of glass in the manufacture of art objects and in particular, stained glass windows. Although brittle, silicate glass is extremely durable, and many examples of glass fragments exist from early glass-making cultures. Because glass can be formed or molded into any shape, and also because it is a sterile product, it has been traditionally used for vessels: bowls, vases, bottles, jars and drinking glasses. In its most solid forms it has also been used for paperweights, marbles, and beads. When extruded as glass fiber and matted as glass wool in a way to trap air, it becomes a thermal insulating material, and when these glass fibers are embedded into an organic polymer plastic, they are a key structural reinforcement part of the composite material fiberglass. Some objects historically were so commonly made of silicate glass that they are simply called by the name of the material, such as drinking glasses and reading glasses.

In science, porcelains and many polymer thermoplastics familiar from everyday use are glasses too. These sorts of glasses can be made of quite different kinds of materials than silica: metallic alloys, ionic melts, aqueous solutions, molecular liquids, and polymers. For many applications, like glass bottles or eyewear, polymer glasses (acrylic glass, polycarbonate or polyethylene terephthalate) are a lighter alternative than traditional glass.

Silicate Glass

Ingredients

Silica (the chemical compound SiO_2) is a common fundamental constituent of glass. In nature, vitrification of quartz occurs when lightning strikes sand, forming hollow, branching rootlike structures called fulgurite.

Fused quartz is a glass made from chemically-pure SiO_2 (silica). It has excellent thermal shock characteristics, being able to survive immersion in water while red hot. However, its high melting-temperature (1723 °C) and viscosity make it difficult to work with. Normally, other substances are added to simplify processing. One is sodium carbonate (Na_2CO_3, "soda"), which lowers the glass transition temperature. The soda makes the glass water-soluble, which is usually undesirable, so lime (calcium oxide [CaO], generally obtained from limestone), some magnesium oxide (MgO) and aluminium oxide (Al_2O_3) are added to provide for a better chemical durability. The resulting glass contains about 70 to 74% silica by weight and is called a soda-lime glass. Soda-lime glasses account for about 90% of manufactured glass.

Most common glass contains other ingredients to change its properties. Lead glass or flint glass is more 'brilliant' because the increased refractive index causes noticeably more specular reflection and increased optical dispersion. Adding barium also increases the refractive index. Thorium oxide gives glass a high refractive index and low dispersion and was formerly used in producing high-quality lenses, but due to its radioactivity has been replaced by lanthanum oxide in modern eyeglasses. Iron can be incorporated into glass to absorb infrared energy, for example in heat absorbing filters for movie projectors, while cerium(IV) oxide can be used for glass that absorbs UV wavelengths.

The following is a list of the more common types of silicate glasses, and their ingredients, properties, and applications:

1. Fused quartz, also called fused silica glass, vitreous silica glass, is silica (SiO_2) in vitreous or glass form (i.e., its molecules are disordered and random, without crystalline structure). It has very low thermal expansion, is very hard, and resists high temperatures (1000–1500 °C). It is also the most resistant against weathering (caused in other glasses by alkali ions leaching out of the glass, while staining it). Fused quartz is used for high temperature applications such as furnace tubes, lighting tubes, melting crucibles, etc.

2. Soda-lime-silica glass, window glass: silica 72% + sodium oxide (Na_2O) 14.2% + lime (CaO) 10.0% + magnesia (MgO) 2.5% + alumina (Al_2O_3) 0.6%. Is transparent, easily formed and most suitable for window glass. It has a high thermal expansion and poor resistance to heat (500–600 °C). It is used for windows, some low temperature incandescent light bulbs, and tableware. Container glass is a soda-lime glass that is a slight variation on flat glass, which uses more alumina and calcium, and less sodium and magnesium which are more water-soluble. This makes it less susceptible to water erosion.

3. Sodium borosilicate glass, Pyrex: silica 81% + boric oxide (B_2O_3) 12% + soda (Na_2O) 4.5% + alumina (Al_2O_3) 2.0%. Stands heat expansion much better than window glass. Used for chemical glassware, cooking glass, car head lamps, etc. Borosilicate glasses (e.g. Pyrex) have as main constituents silica and boron oxide. They have fairly low coefficients of thermal expansion (7740 Pyrex CTE is 3.25×10^{-6}/°C as compared to about 9×10^{-6}/°C for a typical soda-lime glass), making them more dimensionally stable. The lower coefficient of thermal expansion (CTE) also makes them less subject to stress caused by thermal expansion, thus less vulnerable to cracking from thermal shock. They are commonly used for reagent bottles, optical components and household cookware.

4. Lead-oxide glass, crystal glass: silica 59% + lead oxide (PbO) 25% + potassium oxide (K_2O) 12% + soda (Na_2O) 2.0% + zinc oxide (ZnO) 1.5% + alumina 0.4%. Because of its high density (resulting in a high electron density) it has a high refractive index, making the look of glassware more brilliant (called "crystal", though of course it is a glass and not a crystal). It also has a high elasticity, making glassware "ring". It is also more workable in the factory, but cannot stand heating very well.

5. Aluminosilicate glass: silica 57% + alumina 16% + lime 10% + magnesia 7.0% + barium oxide (BaO) 6.0% + boric oxide (B_2O_3) 4.0%. Extensively used for fiberglass, used for making glass-reinforced plastics (boats, fishing rods, etc.) and for halogen bulb glass.

6. Oxide glass: alumina 90% + germanium oxide (GeO_2) 10%. Extremely clear glass, used for fiber-optic waveguides in communication networks. Light loses only 5% of its intensity through 1 km of glass fiber. Most optical fiber is based on silica, as are all the glasses above.

Another common glass ingredient is crushed alkali glass or "cullet" ready for recycled glass. The recycled glass saves on raw materials and energy. Impurities in the cullet can lead to product and equipment failure. Fining agents such as sodium sulfate, sodium chloride, or antimony oxide may be added to reduce the number of air bubbles in the glass mixture. Glass batch calculation is the method by which the correct raw material mixture is determined to achieve the desired glass composition.

Moldavite, a natural glass formed by meteorite impact, from Besednice, Bohemia

Tube fulgurites

Quartz sand (silica) is the main raw material in commercial glass production

Trinitite, a glass made by the Trinity nuclear weapon test

Physical Properties

Optical Properties

Glass is in widespread use largely due to the production of glass compositions that are transparent to visible light. In contrast, polycrystalline materials do not generally transmit visible light. The individual crystallites may be transparent, but their facets (grain boundaries) reflect or scatter light resulting in diffuse reflection. Glass does not contain the internal subdivisions associated with grain boundaries in polycrystals and hence does not scatter light in the same manner as a polycrystalline material. The surface of a glass is often smooth since during glass formation the molecules of the supercooled liquid are not forced to dispose in rigid crystal geometries and can follow surface tension, which imposes a microscopically smooth surface. These properties, which give glass its clearness, can be retained even if glass is partially light-absorbing—i.e., colored.

Glass has the ability to refract, reflect, and transmit light following geometrical optics, without scattering it. It is used in the manufacture of lenses and windows. Common glass has a refraction index around 1.5. This may be modified by adding low-density materials such as boron, which lowers the index of refraction, or increased (to as much as 1.8) with high-density materials such as (classically) lead oxide, or in modern uses, less toxic oxides of zirconium, titanium, or barium. These high-index glasses (inaccurately known as "crystal" when used in glass vessels) cause more chromatic dispersion of light, and are prized for their diamond-like optical properties.

According to Fresnel equations, the reflectivity of a sheet of glass is about 4% per surface (at normal incidence in air), and the transmissivity of one element (two surfaces) is about 90%. Glass with high germanium oxide content also finds application in optoelectronics—e.g., for light-transmitting optical fibers.

A wine glass

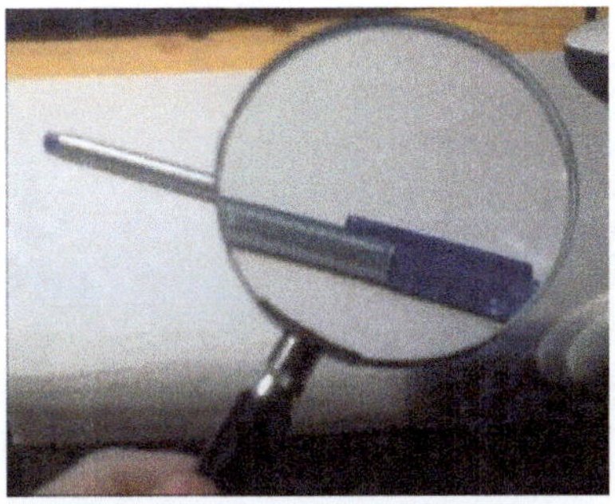

Simple optical device: the magnifying glass

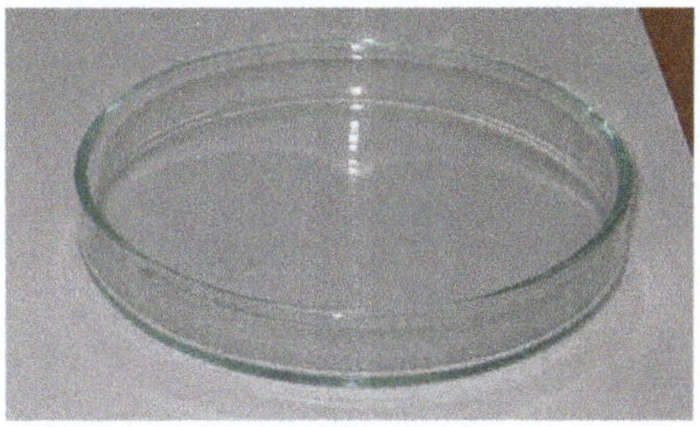

Glass petri dish

Other Properties

In the process of manufacture, silicate glass can be poured, formed, extruded and molded into forms ranging from flat sheets to highly intricate shapes. The finished product is brittle and will fracture, unless laminated or specially treated, but is extremely durable under most conditions. It erodes very slowly and can withstand the action of water. It is resilient to chemical attack and is an ideal material for the manufacture of containers for foodstuffs and most chemicals.

Contemporary Production

Following the glass batch preparation and mixing, the raw materials are transported to the furnace. Soda-lime glass for mass production is melted in gas fired units. Smaller scale furnaces for specialty glasses include electric melters, pot furnaces, and day tanks. After melting, homogenization and refining (removal of bubbles), the glass is formed. Flat glass for windows and similar applications is formed by the float glass process, developed between 1953 and 1957 by Sir Alastair Pilkington and Kenneth Bickerstaff of the UK's Pilkington Brothers, who created a continuous ribbon of glass using a molten tin bath on which the molten glass flows unhindered under the influence of gravity. The top surface of the glass is subjected to nitrogen under pressure to obtain a polished finish. Container glass for common bottles and jars is formed by blowing and pressing methods. This glass is often slightly modified chemically (with more alumina and calcium oxide) for greater water resistance. Further glass forming techniques are summarized in the table Glass forming techniques.

Once the desired form is obtained, glass is usually annealed for the removal of stresses. Surface treatments, coatings or lamination may follow to improve the chemical durability (glass container coatings, glass container internal treatment), strength (toughened glass, bulletproof glass, windshields), or optical properties (insulated glazing, anti-reflective coating).

Impurities give the glass its color

Some of the many color possibilities of glass

Transparent and opaque examples

Glass can be blown into an infinite number of shapes

Color

Color in glass may be obtained by addition of electrically charged ions (or color centers) that are homogeneously distributed, and by precipitation of finely dispersed particles (such as in photochromic glasses). Ordinary soda-lime glass appears colorless to the naked eye when it is thin, although iron(II) oxide (FeO) impurities of up to 0.1 wt% produce a green tint, which can be viewed in thick pieces or with the aid of scientific instruments. Further FeO and Cr_2O_3 additions may be used for the production of green bottles. Sulfur, together with carbon and iron salts, is used to form iron polysulfides and produce amber glass ranging from yellowish to almost black. A glass melt can also acquire an amber color from a reducing combustion atmosphere. Manganese dioxide can be added in small amounts to remove the green tint given by iron(II) oxide. When used in art glass or studio glass is colored using closely guarded recipes that involve specific combinations of metal oxides, melting temperatures and "cook" times. Most colored glass used in the art market is manufactured in volume by vendors who serve this market, although there are some glassmakers with the ability to make their own color from raw materials.

History of Silicate Glass

Bohemian flashed and engraved ruby glass (19th-century)

Wine goblet, mid-19th century. Qajar dynasty. Brooklyn Museum.

Roman cage cup from the 4th century CE

Studio glass. Multiple colors within a single object increase the difficulty of production, as glasses of different colors have different chemical and physical properties when molten.

Naturally occurring glass, especially the volcanic glass obsidian, has been used by many Stone Age societies across the globe for the production of sharp cutting tools and, due to its limited source areas, was extensively traded. But in general, archaeological evidence suggests that the first true glass was made in coastal north Syria, Mesopotamia or ancient Egypt. The earliest known glass objects, of the mid third millennium BCE, were beads, perhaps initially created as accidental by-products of metal-working (slags) or during the production of faience, a pre-glass vitreous material made by a process similar to glazing.

Glass remained a luxury material, and the disasters that overtook Late Bronze Age civilizations seem to have brought glass-making to a halt. Indigenous development of glass technology in South Asia may have begun in 1730 BCE. In ancient China, though, glassmaking seems to have a late start, compared to ceramics and metal work. The term *glass* developed in the late Roman Empire. It was in the Roman glassmaking center at Trier, now in modern Germany, that the late-Latin term *glesum* originated, probably from a Germanic word for a transparent, lustrous substance. Glass objects have been recovered across the Roman empire in domestic, industrial and funerary contexts.

Glass was used extensively during the Middle Ages. Anglo-Saxon glass has been found across England during archaeological excavations of both settlement and cemetery sites. Glass in the Anglo-Saxon period was used in the manufacture of a range of objects including vessels, beads, windows and was also used in jewelry. From the 10th-century onwards, glass was employed in stained glass windows of churches and cathedrals, with famous examples at Chartres Cathedral and the Basilica of Saint Denis. By the 14th-century, architects were designing buildings with walls of stained glass such as Sainte-Chapelle, Paris, (1203–1248) and the East end of Gloucester Cathedral. Stained glass had a major revival with Gothic Revival architecture in the 19th-century. With the Renaissance, and a change in architectural style, the use of large stained glass windows became less prevalent. The use of domestic stained glass increased until most substantial houses had glass windows. These were initially small panes leaded together, but with the changes in technology, glass could be manufactured relatively cheaply in increasingly larger sheets. This led to larger window panes, and, in the 20th-century, to much larger windows in ordinary domestic and commercial buildings.

In the 20th century, new types of glass such as laminated glass, reinforced glass and glass bricks have increased the use of glass as a building material and resulted in new applications of glass. Multi-storey buildings are frequently constructed with curtain walls made almost entirely of glass. Similarly, laminated glass has been widely applied to vehicles for windscreens. While glass containers have always been used for storage and are valued for their hygienic properties, glass has been utilized increasingly in industry. Optical glass for spectacles has been used since the late Middle Ages. The production of lenses has become increasingly proficient, aiding astronomers as well as having other application in medicine and science. Glass is also employed as the aperture cover in many solar energy systems.

From the 19th century, there was a revival in many ancient glass-making techniques including cameo glass, achieved for the first time since the Roman Empire and initially mostly used for pieces in a neo-classical style. The Art Nouveau movement made great use of glass, with René Lalique, Émile Gallé, and Daum of Nancy producing colored vases and similar pieces, often in cameo glass, and also using luster techniques. Louis Comfort Tiffany in America specialized in stained glass, both secular and religious, and his famous lamps. The early 20th-century saw the large-scale factory production of glass art by firms such as Waterford and Lalique. From about 1960 onwards there have been an increasing number of small studios hand-producing glass artworks, and glass artists began to class themselves as in effect sculptors working in glass, and their works as part fine arts.

In the 21st century, scientists observing the properties of ancient stained glass windows, in which suspended nanoparticles prevent UV light from causing chemical reactions that change image colors, are developing photographic techniques that use similar stained glass to capture true color images of Mars for the 2019 ESA Mars Rover mission.

Chronology of Advances in Architectural Glass

- 1226: "Broad Sheet" first produced in Sussex.

- 1330: "Crown glass" for art work and vessels first produced in Rouen, France. "Broad Sheet" also produced. Both were also supplied for export.
- 1500s: A method of making mirrors out of plate glass was developed by Venetian glassmakers on the island of Murano, who covered the back of the glass with a mercury-tin amalgam, obtaining near-perfect and undistorted reflection.
- 1620: "Blown Plate" first produced in London. Used for mirrors and coach plates.
- 1678: "Crown Glass" first produced in London. This process dominated until the 19th century.
- 1843: An early form of "Float Glass" invented by Henry Bessemer, pouring glass onto liquid tin. Expensive and not a commercial success.
- 1874: Tempered glass is developed by Francois Barthelemy Alfred Royer de la Bastie (1830-1901) of Paris, France by quenching almost molten glass in a heated bath of oil or grease.
- 1888: "Machine Rolled" glass introduced allowing patterns to be introduced.
- 1898: "Wired Cast" glass invented by Pilkington for use where safety or security was an issue.
- 1959: "Float Glass" launched in UK. Invented by Sir Alastair Pilkington.

Mouth-blown window-glass in Sweden Kosta Glasbruk, (1742) with a pontil mark from the glassblower's pipe

A building in Canterbury, England, which displays its long history in different building styles and glazing of every century from the 16th to the 20th included.

Windows in the choir of the Basilica of Saint Denis, one of the earliest uses of extensive areas of glass. (early 13th-century architecture with restored glass of the 19th-century)

"Hardwick Hall, more glass than wall". (late 16th-century)

Windows at Österreichische Postsparkasse, Vienna, (early 20th-century)

Westin Bonaventure Hotel, USA, show the extensive use of glass as a building material in the 20th-21st centuries

Other Types of Glass

New chemical glass compositions or new treatment techniques can be initially investigated in small-scale laboratory experiments. The raw materials for laboratory-scale glass melts are often different from those used in mass production because the cost factor has a low priority. In the laboratory mostly pure chemicals are used. Care must be taken that the raw materials have not reacted with moisture or other chemicals in the environment (such as alkali or alkaline earth metal oxides and hydroxides, or boron oxide), or that the impurities are quantified (loss on ignition). Evaporation losses during glass melting should be considered during the selection of the raw materials, e.g., sodium selenite may be preferred over easily evaporating SeO_2. Also, more readily reacting raw materials may be preferred over relatively inert ones, such as $Al(OH)_3$ over Al_2O_3. Usually, the melts are carried out in platinum crucibles to reduce contamination from the crucible material. Glass homogeneity is achieved by homogenizing the raw materials mixture (glass batch), by stirring the melt, and by crushing and re-melting the first melt. The obtained glass is usually annealed to prevent breakage during processing.

To make glass from materials with poor glass forming tendencies, novel techniques are used to increase cooling rate, or reduce crystal nucleation triggers. Examples of these techniques include aerodynamic levitation (cooling the melt whilst it floats on a gas stream), splat quenching (pressing the melt between two metal anvils) and roller quenching (pouring the melt through rollers).

Network Glasses

A CD-RW (CD). Chalcogenide glasses form the basis of rewritable CD and DVD solid-state memory technology.

Some glasses that do not include silica as a major constituent may have physico-chemical properties useful for their application in fiber optics and other specialized technical applications. These include fluoride glasses, aluminosilicates, phosphate glasses, borate glasses, and chalcogenide glasses.

There are three classes of components for oxide glasses: network formers, intermediates, and modifiers. The network formers (silicon, boron, germanium) form a highly cross-linked network of chemical bonds. The intermediates (titanium, aluminium, zirconium, beryllium, magnesium, zinc) can act as both network formers and modifiers, according to the glass composition. The modifiers (calcium, lead, lithium, sodium, potassium) alter the network structure; they are usually present as ions, compensated by nearby non-bridging oxygen atoms, bound by one covalent bond to the glass network and holding one negative charge to compensate for the positive ion nearby. Some elements can play multiple roles; e.g. lead can act both as a network former (Pb^{4+} replacing Si^{4+}), or as a modifier.

The presence of non-bridging oxygens lowers the relative number of strong bonds in the material and disrupts the network, decreasing the viscosity of the melt and lowering the melting temperature.

The alkali metal ions are small and mobile; their presence in glass allows a degree of electrical conductivity, especially in molten state or at high temperature. Their mobility decreases the chemical resistance of the glass, allowing leaching by water and facilitating corrosion. Alkaline earth ions, with their two positive charges and requirement for two non-bridging oxygen ions to compensate for their charge, are much less mobile themselves and also hinder diffusion of other ions, especially the alkalis. The most common commercial glasses contain both alkali and alkaline earth ions (usually sodium and calcium), for easier processing and satisfying corrosion resistance. Corrosion resistance of glass can be achieved by dealkalization, removal of the alkali ions from the glass surface by reaction with e.g. sulfur or fluorine compounds. Presence of alkaline metal ions has also detrimental effect to the loss tangent of the glass, and to its electrical resistance; glasses for electronics (sealing, vacuum tubes, lamps...) have to take this in account.

Addition of lead(II) oxide lowers melting point, lowers viscosity of the melt, and increases refractive index. Lead oxide also facilitates solubility of other metal oxides and is used in colored glasses. The viscosity decrease of lead glass melt is very significant (roughly 100 times in comparison with soda glasses); this allows easier removal of bubbles and working at lower temperatures, hence its frequent use as an additive in vitreous enamels and glass solders. The high ionic radius of the Pb^{2+} ion renders it highly immobile in the matrix and hinders the movement of other ions; lead glasses therefore have high electrical resistance, about two orders of magnitude higher than soda-lime glass ($10^{8.5}$ vs $10^{6.5}$ Ohm·cm, DC at 250 °C).

Addition of fluorine lowers the dielectric constant of glass. Fluorine is highly electronegative and attracts the electrons in the lattice, lowering the polarizability of the material. Such silicon dioxide-fluoride is used in manufacture of integrated circuits as an insulator. High levels of fluorine doping lead to formation of volatile SiF_2O and such glass is then thermally unstable. Stable layers were achieved with dielectric constant down to about 3.5–3.7.

Amorphous Metals

Samples of amorphous metal, with millimeter scale

In the past, small batches of amorphous metals with high surface area configurations (ribbons, wires, films, etc.) have been produced through the implementation of extremely rapid rates of cooling. This was initially termed "splat cooling" by doctoral student W. Klement at Caltech, who showed that cooling rates on the order of millions of degrees per second is sufficient to impede the formation of crystals, and the metallic atoms become "locked into" a glassy state. Amorphous metal wires have been produced by sputtering molten metal onto a spinning metal disk. More recently a number of alloys have been produced in layers with thickness exceeding 1 millimeter. These are known as bulk metallic glasses (BMG). Liquidmetal Technologies sell a number of zirconium-based BMGs. Batches of amorphous steel have also been produced that demonstrate mechanical properties far exceeding those found in conventional steel alloys.

In 2004, NIST researchers presented evidence that an isotropic non-crystalline metallic phase (dubbed "q-glass") could be grown from the melt. This phase is the first phase, or "primary phase", to form in the Al-Fe-Si system during rapid cooling. Interestingly, experimental evidence indicates that this phase forms by a *first-order transition*. Transmission electron microscopy (TEM) images show that the q-glass nucleates from the melt as discrete particles, which grow spherically with a uniform growth rate in all directions. The diffraction pattern shows it to be an isotropic glassy phase. Yet there is a nucleation barrier, which implies an interfacial discontinuity (or internal surface) between the glass and the melt.

Electrolytes

Electrolytes or molten salts are mixtures of different ions. In a mixture of three or more ionic species of dissimilar size and shape, crystallization can be so difficult that the liquid can easily be supercooled into a glass. The best-studied example is $Ca_{0.4}K_{0.6}(NO_3)_{1.4}$.

Aqueous Solutions

Some aqueous solutions can be supercooled into a glassy state, for instance LiCl:RH_2O in the composition range $4<R<8$.

Molecular Liquids

A *molecular liquid* is composed of molecules that do not form a covalent network but interact only through weak van der Waals forces or through transient hydrogen bonds. Many molecular liquids can be supercooled into a glass; some are excellent glass formers that normally do not crystallize.

A widely known example is sugar glass.

Under extremes of pressure and temperature solids may exhibit large structural and physical changes that can lead to polyamorphic phase transitions. In 2006 Italian scientists created an amorphous phase of carbon dioxide using extreme pressure. The substance was named amorphous carbonia(a-CO_2) and exhibits an atomic structure resembling that of silica.

Polymers

Important polymer glasses include amorphous and glassy pharmaceutical compounds. These are useful because the solubility of the compound is greatly increased when it is amorphous compared to the same crystalline composition. Many emerging pharmaceuticals are practically insoluble in their crystalline forms.

Colloidal Glasses

Concentrated colloidal suspensions may exhibit a distinct glass transition as function of particle concentration or density.

In cell biology there is recent evidence suggesting that the cytoplasm behaves like a colloidal glass approaching the liquid-glass transition. During periods of low metabolic activity, as in dormancy, the cytoplasm vitrifies and prohibits the movement to larger cytoplasmic particles while allowing the diffusion of smaller ones throughout the cell.

Glass-ceramics

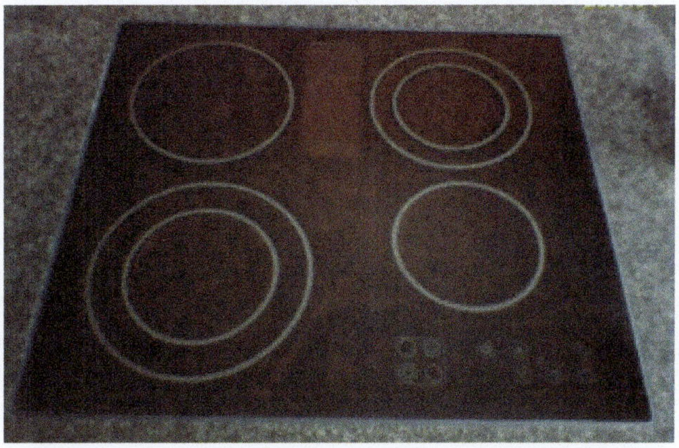

A high-strength glass-ceramic cooktop with negligible thermal expansion.

Glass-ceramic materials share many properties with both non-crystalline glass and crystalline ceramics. They are formed as a glass, and then partially crystallized by heat treatment. For example, the microstructure of whiteware ceramics frequently contains both amorphous and crystalline phases. Crystalline grains are often embedded within a non-crystalline intergranular phase of grain boundaries. When applied to whiteware ceramics, vitreous means the material has an extremely low permeability to liquids, often but not always water, when determined by a specified test regime.

The term mainly refers to a mix of lithium and aluminosilicates that yields an array of materials with interesting thermomechanical properties. The most commercially important of these have the distinction of being impervious to thermal shock. Thus, glass-ceramics have become extremely useful for countertop cooking. The negative thermal expansion coefficient (CTE) of the crystalline ceramic phase can be balanced with the positive CTE of the glassy phase. At a certain point (~70% crystalline) the glass-ceramic has a net CTE near zero. This type of glass-ceramic exhibits excellent mechanical properties and can sustain repeated and quick temperature changes up to 1000 °C.

Structure

As in other amorphous solids, the atomic structure of a glass lacks any long-range translational periodicity. Due to chemical bonding characteristics glasses do possess a high degree of short-range order with respect to local atomic polyhedra.

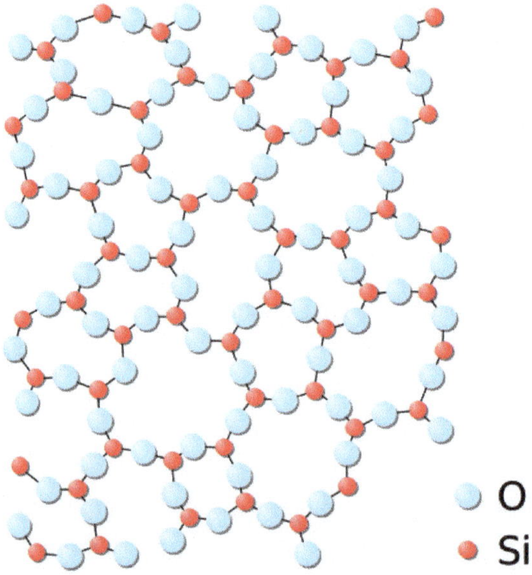

The amorphous structure of glassy silica (SiO_2) in two dimensions. No long-range order is present, although there is local ordering with respect to the tetrahedral arrangement of oxygen (O) atoms around the silicon (Si) atoms.

Formation from a Supercooled Liquid

In physics, the standard definition of a glass (or vitreous solid) is a solid formed by rapid melt quenching. The term glass is often used to describe any amorphous solid that exhibits a glass transition temperature T_g. If the cooling is sufficiently rapid (relative to the characteristic crystallization time) then crystallization is prevented and instead the disordered atomic configuration of the supercooled liquid is frozen into the solid state at T_g. The tendency for a material to form a glass while quenched is called glass-forming ability. This ability can be predicted by the rigidity theory. Generally, the structure of a glass exists in a metastable state with respect to its crystalline form, although in certain circumstances, for example in atactic polymers, there is no crystalline analogue of the amorphous phase.

Some people consider glass to be a liquid due to its lack of a first-order phase transition where certain thermodynamic variables such as volume, entropy and enthalpy are discontinuous through the glass transition range. The glass transition may be described as analogous to a second-order phase transition where the intensive thermodynamic variables such as the thermal expansivity and heat capacity are discontinuous. Nonetheless, the equilibrium theory of phase transformations does not entirely hold for glass, and hence the glass transition cannot be classed as one of the classical equilibrium phase transformations in solids.

Glass is an amorphous solid. It exhibits an atomic structure close to that observed in the supercooled liquid phase but displays all the mechanical properties of a solid. The notion that glass flows to an appreciable extent over extended periods of time is not supported by empirical research or theoretical analysis. Laboratory measurements of room temperature glass flow do show a motion consistent with a material viscosity on the order of 10^{17}–10^{18} Pa s.

Although the atomic structure of glass shares characteristics of the structure in a supercooled liquid, glass tends to behave as a solid below its glass transition temperature. A supercooled liquid behaves as a liquid, but it is below the freezing point of the material, and in some cases will crystallize almost instantly if a crystal is added as a core. The change in heat capacity at a glass transition and a melting transition of comparable materials are typically of the same order of magnitude, indicating that the change in active degrees of freedom is comparable as well. Both in a glass and in a crystal it is mostly only the vibrational degrees of freedom that remain active, whereas rotational and translational motion is arrested. This helps to explain why both crystalline and non-crystalline solids exhibit rigidity on most experimental time scales.

Behavior of Antique Glass

The observation that old windows are sometimes found to be thicker at the bottom than at the top is often offered as supporting evidence for the view that glass flows over a timescale of centuries, the assumption being that the glass has exhibited the liquid property of flowing from one shape to another. This assumption is incorrect, as once solidified, glass stops flowing. The reason for the observation is that in the past, when panes of glass were commonly made by glassblowers, the technique used was to spin molten glass so as to create a round, mostly flat and even plate (the crown glass process, described above). This plate was then cut to fit a window. The pieces were not absolutely flat; the edges of the disk became a different thickness as the glass spun. When installed in a window frame, the glass would be placed with the thicker side down both for the sake of stability and to prevent water accumulating in the lead cames at the bottom of the window. Occasionally such glass has been found installed with the thicker side at the top, left or right.

Mass production of glass window panes in the early twentieth century caused a similar effect. In glass factories, molten glass was poured onto a large cooling table and allowed to spread. The resulting glass is thicker at the location of the pour, located at the center of the large sheet. These sheets were cut into smaller window panes with nonuniform thickness, typically with the location of the pour centered in one of the panes (known as "bull's-eyes") for decorative effect. Modern glass intended for windows is produced as float glass and is very uniform in thickness.

Several other points can be considered that contradict the "cathedral glass flow" theory:

- Writing in the *American Journal of Physics*, the materials engineer Edgar D. Zanotto states "...the predicted relaxation time for GeO_2 at room temperature is 10^{32} years. Hence, the relaxation period (characteristic flow time) of cathedral glasses would be even longer." (10^{32} years is many times longer than the estimated age of the universe.)
- If medieval glass has flowed perceptibly, then ancient Roman and Egyptian objects should have flowed proportionately more—but this is not observed. Similarly, prehistoric obsidian blades should have lost their edge; this is not observed either (although obsidian may have a different viscosity from window glass).
- If glass flows at a rate that allows changes to be seen with the naked eye after centuries, then the effect should be noticeable in antique telescopes. Any slight deformation in the antique telescopic lenses would lead to a dramatic decrease in optical performance, a phenomenon that is not observed.

- There are many examples of centuries-old glass shelving that has not bent, even though it is under much higher stress from gravitational loads than vertical window glass.

References

- B. H. W. S. de Jong, "Glass"; in "Ullmann's Encyclopedia of Industrial Chemistry"; 5th edition, vol. A12, VCH Publishers, Weinheim, Germany, 1989, ISBN 978-3-527-20112-9, pp. 365–432.
- Pfaender, Heinz G. (1996). Schott guide to glass. Springer. pp. 135, 186. ISBN 978-0-412-62060-7. Retrieved 8 February 2011.
- Donald R. Uhlmann; Norbert J. Kreidl, eds. (1991). Optical properties of glass. Westerville, OH: American Ceramic Society. ISBN 0-944904-35-1.
- Vogel, Werner (1994). Glass Chemistry (2 ed.). Springer-Verlag Berlin and Heidelberg GmbH & Co. K. ISBN 3-540-57572-3.
- Thomas P. Seward, ed. (2005). High temperature glass melt property database for process modeling. Westerville, Ohio: American Ceramic Society. ISBN 1-57498-225-7.
- Shackelford, James F.; Doremus, Robert H. (2008). Ceramic and Glass Materials: Structure, Properties and Processing. Springer. p. 158. ISBN 0-387-73361-2.
- Doering, Robert; Nishi, Yoshio (2007). Handbook of semiconductor manufacturing technology. CRC Press. pp. 12–3. ISBN 1-57444-675-4.
- Kingery, W.D.; Uhlmann, H.K. Bowen, D.R. (1976). Introduction to ceramics (2 ed.). New York: Wiley. ISBN 978-0471478607.
- Richerson, David W. (1992). Modern ceramic engineering : properties, processing and use in design (2 ed.). New York, NY: Dekker. ISBN 0-8247-8634-3.
- Cusack, N. E. (1987). The physics of structurally disordered matter: an introduction. Adam Hilger in association with the University of Sussex press. ISBN 0-85274-829-9.
- Munguira, Ignacio (9 February 2016). "Glasslike Membrane Protein Diffusion in a Crowded Membrane". ACS Nano. doi:10.1021/acsnano.5b07595. Retrieved 1 March 2016.
- Vannoni, M.; Sordini, A.; Molesini, G. (2011). "Relaxation time and viscosity of fused silica glass at room temperature". Eur. Phys. J. E. 34: 9–14. doi:10.1140/epje/i2011-11092-9.

2

Physical Properties of Glass

The physical properties of glass are viscosity, low emissivity, corrosion, glass transition, thermal expansion, liquidus, heat capacity and surface tension among others. Glass transition is the transition of a material from a hard to a relatively brittle state. The brittle state is the "glassy" state; as the temperature increases it turns into a rubbery state. This section is an overview of the subject matter incorporating all the physical properties of glass.

Viscosity

The viscosity of a fluid is a measure of its resistance to gradual deformation by shear stress or tensile stress. For liquids, it corresponds to the informal concept of "thickness"; for example, honey has a much higher viscosity than water.

Viscosity is a property of the fluid which opposes the relative motion between the two surfaces of the fluid .in a fluid that are moving at different velocities. When the fluid is forced through a tube, the particles which compose the fluid generally move more quickly near the tube's axis and more slowly near its walls; therefore some stress (such as a pressure difference between the two ends of the tube) is needed to overcome the friction between particle layers to keep the fluid moving. For a given velocity pattern, the stress required is proportional to the fluid's viscosity.

A fluid that has no resistance to shear stress is known as an *ideal* or *inviscid* fluid. Zero viscosity is observed only at very low temperatures in superfluids. Otherwise, all fluids have positive viscosity, and are technically said to be viscous or viscid. In common parlance, however, a liquid is said to be *viscous* if its viscosity is substantially greater than that of water, and may be described as *mobile* if the viscosity is noticeably less than water. A fluid with a relatively high viscosity, such as pitch, may appear to be a solid.

Etymology

The word "viscosity" is derived from the Latin "*viscum*", meaning mistletoe and also a viscous glue made from mistletoe berries.

Definition

Dynamic (Shear) Viscosity

The dynamic (shear) viscosity of a fluid expresses its resistance to shearing flows, where adjacent layers move parallel to each other with different speeds. It can be defined through the idealized situation known as a Couette flow, where a layer of fluid is trapped between two horizontal plates, one fixed and one moving horizontally at constant speed . This fluid has to be homogeneous in the layer and at different shear stresses. (The plates are assumed to be very large, so that one need not consider what happens near their edges.)

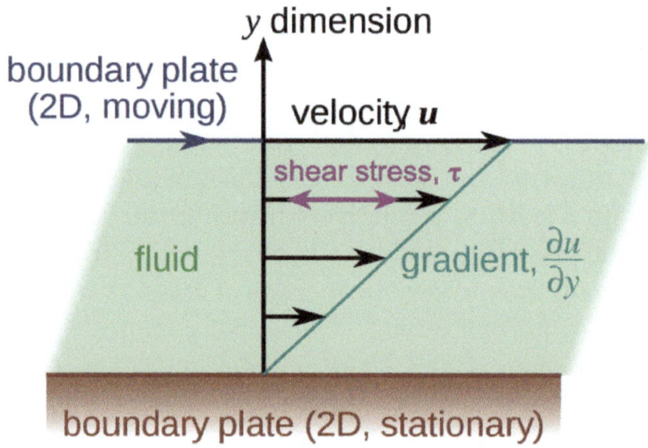

Laminar shear of fluid between two plates. Friction between the fluid and the moving boundaries/plates causes the fluid to shear. The force required for this action is a measure of the fluid's viscosity.

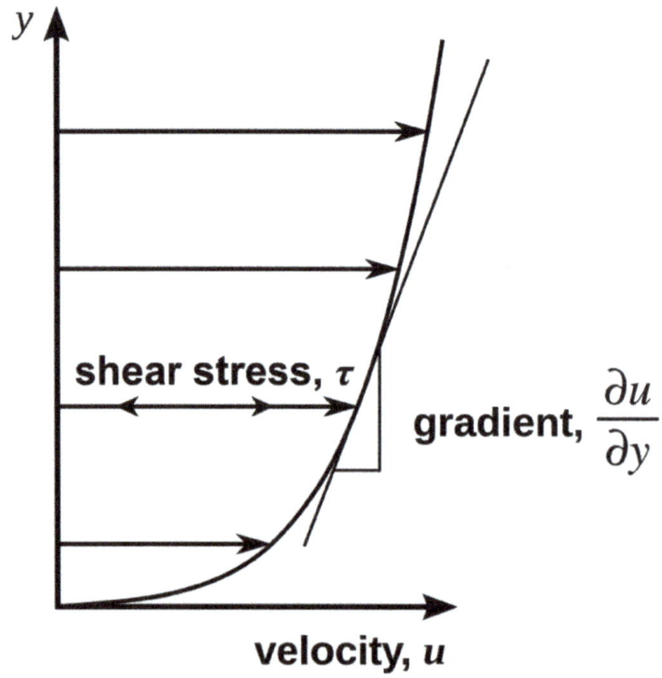

In a general parallel flow (such as could occur in a straight pipe), the shear stress is proportional to the gradient of the velocity

If the speed of the top plate is small enough, the fluid particles will move parallel to it, and their speed will vary linearly from zero at the bottom to u at the top. Each layer of fluid will move faster than the one just below it, and friction between them will give rise to a force resisting their relative motion. In particular, the fluid will apply on the top plate a force in the direction opposite to its motion, and an equal but opposite one to the bottom plate. An external force is therefore required in order to keep the top plate moving at constant speed.

The magnitude F of this force is found to be proportional to the speed u and the area A of each plate, and inversely proportional to their separation y:

$$F = \mu A \frac{u}{y}.$$

The proportionality factor μ in this formula is the viscosity (specifically, the *dynamic viscosity*) of the fluid.

The ratio u/y is called the *rate of shear deformation* or *shear velocity*, and is the derivative of the fluid speed in the direction perpendicular to the plates. Isaac Newton expressed the viscous forces by the differential equation

$$\tau = \mu \frac{\partial u}{\partial y},$$

where $\tau = F/A$, and $\partial u/\partial y$ is the local shear velocity. This formula assumes that the flow is moving along parallel lines and the y-axis, perpendicular to the flow, points in the direction of maximum shear velocity. This equation can be used where the velocity does not vary linearly with y, such as in fluid flowing through a pipe.

Use of the Greek letter mu (μ) for the dynamic stress viscosity is common among mechanical and chemical engineers, as well as physicists. However, the Greek letter eta (η) is also used by chemists, physicists, and the IUPAC.

Kinematic Viscosity

The kinematic viscosity (also called "momentum diffusivity") is the ratio of the dynamic viscosity μ to the density of the fluid ρ. It is usually denoted by the Greek letter nu (ν).

$$\nu = \frac{\mu}{\rho}$$

It is a convenient concept when analyzing the Reynolds number, which expresses the ratio of the inertial forces to the viscous forces:

$$\mathrm{Re} = \frac{\rho u L}{\mu} = \frac{uL}{\nu},$$

where L is a typical length scale in the system.

Bulk Viscosity

When a compressible fluid is compressed or expanded evenly, without shear, it may still exhibit a form of internal friction that resists its flow. These forces are related to the rate of compression or expansion by a factor called the volume viscosity, bulk viscosity or second viscosity.

The bulk viscosity is important only when the fluid is being rapidly compressed or expanded, such as in sound and shock waves. Bulk viscosity explains the loss of energy in those waves, as described by Stokes' law of sound attenuation.

Viscosity Tensor

In general, the stresses within a flow can be attributed partly to the deformation of the material from some rest state (elastic stress), and partly to the rate of change of the deformation over time (viscous stress). In a fluid, by definition, the elastic stress includes only the hydrostatic pressure.

In very general terms, the fluid's viscosity is the relation between the strain rate and the viscous stress. In the Newtonian fluid model, the relationship is by definition a linear map, described by a viscosity tensor that, multiplied by the strain rate tensor (which is the gradient of the flow's velocity), gives the viscous stress tensor.

The viscosity tensor has nine independent degrees of freedom in general. For isotropic Newtonian fluids, these can be reduced to two independent parameters. The most usual decomposition yields the stress viscosity μ and the bulk viscosity σ.

Newtonian and Non-newtonian Fluids

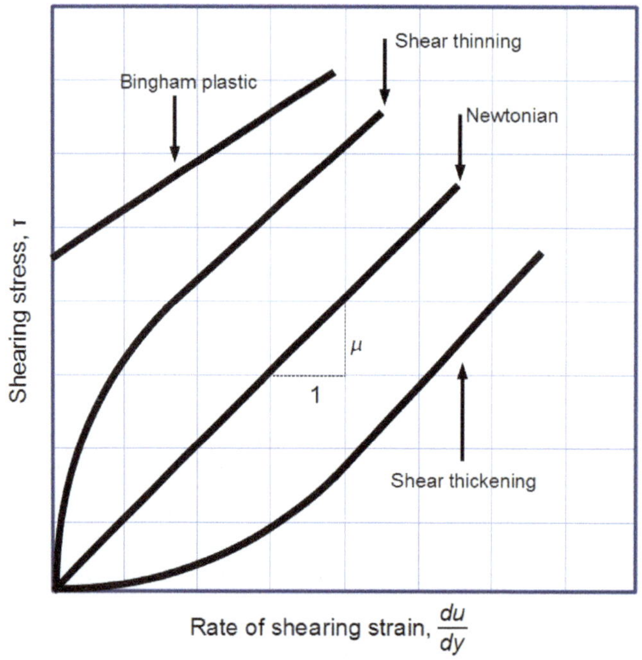

Viscosity, the slope of each line, varies among materials

Newton's law of viscosity is a constitutive equation (like Hooke's law, Fick's law, Ohm's law): it is not a fundamental law of nature but an approximation that holds in some materials and fails in others.

A fluid that behaves according to Newton's law, with a viscosity μ that is independent of the stress, is said to be Newtonian. Gases, water, and many common liquids can be considered Newtonian in ordinary conditions and contexts. There are many non-Newtonian fluids that significantly deviate from that law in some way or other. For example:

- Shear-thickening liquids, whose viscosity increases with the rate of shear strain.
- Shear-thinning liquids, whose viscosity decreases with the rate of shear strain.
- Thixotropic liquids, that become less viscous over time when shaken, agitated, or otherwise stressed.
- Rheopectic liquids, that become more viscous over time when shaken, agitated, or otherwise stressed.
- Bingham plastics that behave as a solid at low stresses but flow as a viscous fluid at high stresses.

Shear-thinning liquids are very commonly, but misleadingly, described as thixotropic.

Even for a Newtonian fluid, the viscosity usually depends on its composition and temperature. For gases and other compressible fluids, it depends on temperature and varies very slowly with pressure.

The viscosity of some fluids may depend on other factors. A magnetorheological fluid, for example, becomes thicker when subjected to a magnetic field, possibly to the point of behaving like a solid.

Viscosity in Solids

The viscous forces that arise during fluid flow must not be confused with the elastic forces that arise in a solid in response to shear, compression or extension stresses. While in the latter the stress is proportional to the *amount* of shear deformation, in a fluid it is proportional to the *rate* of deformation over time. (For this reason, Maxwell used the term *fugitive elasticity* for fluid viscosity.)

However, many liquids (including water) will briefly react like elastic solids when subjected to sudden stress. Conversely, many "solids" (even granite) will flow like liquids, albeit very slowly, even under arbitrarily small stress. Such materials are therefore best described as possessing both elasticity (reaction to deformation) and viscosity (reaction to rate of deformation); that is, being viscoelastic.

Indeed, some authors have claimed that amorphous solids, such as glass and many polymers, are actually liquids with a very high viscosity (greater than 10^{12} Pa·s). However, other authors dispute this hypothesis, claiming instead that there is some threshold for the stress, below which most solids will not flow at all, and that alleged instances of glass flow in window panes of old buildings are due to the crude manufacturing process of older eras rather than to the viscosity of glass.

Viscoelastic solids may exhibit both shear viscosity and bulk viscosity. The extensional viscosity is a linear combination of the shear and bulk viscosities that describes the reaction of a solid elastic material to elongation. It is widely used for characterizing polymers.

In geology, earth materials that exhibit viscous deformation at least three orders of magnitude greater than their elastic deformation are sometimes called rheids.

Viscosity Measurement

Viscosity is measured with various types of viscometers and rheometers. A rheometer is used for those fluids that cannot be defined by a single value of viscosity and therefore require more parameters to be set and measured than is the case for a viscometer. Close temperature control of the fluid is essential to acquire accurate measurements, particularly in materials like lubricants, whose viscosity can double with a change of only 5 °C.

For some fluids, viscosity is a constant over a wide range of shear rates (Newtonian fluids). The fluids without a constant viscosity (non-Newtonian fluids) cannot be described by a single number. Non-Newtonian fluids exhibit a variety of different correlations between shear stress and shear rate.

One of the most common instruments for measuring kinematic viscosity is the glass capillary viscometer.

In coating industries, viscosity may be measured with a cup in which the efflux time is measured. There are several sorts of cup – such as the Zahn cup and the Ford viscosity cup – with usage of each type varying mainly according to the industry. The efflux time can also be converted to kinematic viscosities (centistokes, cSt) through the conversion equations.

Also used in coatings, a Stormer viscometer uses load-based rotation in order to determine viscosity. The viscosity is reported in Krebs units (KU), which are unique to Stormer viscometers.

Vibrating viscometers can also be used to measure viscosity. These models such as the *Dynatrol* use vibration rather than rotation to measure viscosity.

Extensional viscosity can be measured with various rheometers that apply extensional stress.

Volume viscosity can be measured with an acoustic rheometer.

Apparent viscosity is a calculation derived from tests performed on drilling fluid used in oil or gas well development. These calculations and tests help engineers develop and maintain the properties of the drilling fluid to the specifications required.

Units

Dynamic Viscosity μ

Both the physical unit of dynamic viscosity in SI units, the poiseuille (Pl), and cgs units, the poise (P), are named after Jean Léonard Marie Poiseuille. The poiseuille, which is rarely used, is equivalent to the pascal second (Pa·s), or (N·s)/m^2, or kg/(m·s). If a fluid is placed between two plates with distance one meter, and one plate is pushed sideways with a shear stress of one pascal, and it moves at x meters per second, then it has viscosity of $1/x$ pascal seconds. For example, water at 20 °C has a viscosity of 1.002 mPa·s, while a typical motor oil could have a viscosity of about 250 mPa·s. The units used in practice are either Pa·s and its submultiples or the cgs poise referred to below, and its submultiples.

The cgs physical unit for dynamic viscosity, the *poise* (P), is also named after Jean Poiseuille. It is more commonly expressed, particularly in ASTM standards, as *centipoise* (cP) since the latter is equal to the SI multiple millipascal seconds (mPa·s). For example, water at 20 °C has a viscosity of 1.002 mPa·s = 1.0020*nbsp;cP.

1 Pl = 1 Pa·s

1 P = 0.1 Pa·s= 0.1 $kg \cdot m^{-1} \cdot s^{-1}$

1 cP = 1 mPa·s = 0.001 Pa·s = 0.001 $N \cdot s \cdot m^{-2}$ = 0.001 $kg \cdot m^{-1} \cdot s^{-1}$.

Kinematic Viscosity *v*

The SI unit of kinematic viscosity is m^2/s.

The cgs physical unit for kinematic viscosity is the *stokes* (St), named after George Gabriel Stokes. It is sometimes expressed in terms of *centistokes* (cSt). In U.S. usage, *stoke* is sometimes used as the singular form.

1 St = 1 $cm^2 \cdot s^{-1} = 10^{-4}\ m^2 \cdot s^{-1}$.

1 cSt = 1 $mm^2 \cdot s^{-1} = 10^{-6}\ m^2 \cdot s^{-1}$.

Water at 20 °C has a kinematic viscosity of about $10^{-6}\ m^2 \cdot s^{-1}$ or 1 cSt.

The kinematic viscosity is sometimes referred to as *diffusivity of momentum*, because it is analogous to diffusivity of heat and diffusivity of mass. It is therefore used in dimensionless numbers which compare the ratio of the diffusivities.

Fluidity

The reciprocal of viscosity is *fluidity*, usually symbolized by $\varphi = 1/\mu$ or $F = 1/\mu$, depending on the convention used, measured in *reciprocal poise* (P^{-1}, or $cm \cdot s \cdot g^{-1}$), sometimes called the *rhe*. *Fluidity* is seldom used in engineering practice.

The concept of fluidity can be used to determine the viscosity of an ideal solution. For two components A and B, the fluidity when A and B are mixed is

$$F \approx \chi_A F_A + \chi_B F_B,$$

which is only slightly simpler than the equivalent equation in terms of viscosity:

$$\mu \approx \frac{1}{\frac{\chi_A}{\mu_A} + \frac{\chi_B}{\mu_B}},$$

where χ_A and χ_B are the mole fractions of components A and B respectively, and μ_A and μ_B are the components' pure viscosities.

Non-standard Units

The Reyn is a British unit of dynamic viscosity.

Viscosity index is a measure for the change of kinematic viscosity with temperature. It is used to characterise lubricating oil in the automotive industry.

At one time the petroleum industry relied on measuring kinematic viscosity by means of the Saybolt viscometer, and expressing kinematic viscosity in units of *Saybolt Universal Seconds* (SUS). Other abbreviations such as SSU (*Saybolt Seconds Universal*) or SUV (*Saybolt Universal Viscosity*) are sometimes used. Kinematic viscosity in centistoke can be converted from SUS according to the arithmetic and the reference table provided in ASTM D 2161.

Molecular Origins

Pitch has a viscosity approximately 230 billion (2.3×10^{11}) times that of water.

The viscosity of a system is determined by how molecules constituting the system interact. There are no simple but correct expressions for the viscosity of a fluid. The simplest exact expressions are the Green–Kubo relations for the linear shear viscosity or the Transient Time Correlation Function expressions derived by Evans and Morriss in 1985. Although these expressions are each exact, in order to calculate the viscosity of a dense fluid using these relations currently requires the use of molecular dynamics computer simulations.

Gases

Viscosity in gases arises principally from the molecular diffusion that transports momentum between layers of flow. The kinetic theory of gases allows accurate prediction of the behavior of gaseous viscosity.

Within the regime where the theory is applicable:

- Viscosity is independent of pressure and
- Viscosity increases as temperature increases.

James Clerk Maxwell published a famous paper in 1866 using the kinetic theory of gases to study gaseous viscosity. To understand why the viscosity is independent of pressure, consider two adjacent boundary layers (A and B) moving with respect to each other. The internal friction (the viscosity) of the gas is determined by the probability a particle of layer A enters layer B with a corresponding transfer of momentum. Maxwell's calculations show that the viscosity coefficient is proportional to the density, the mean free path, and the mean velocity of the atoms. On the other hand, the *mean free path* is inversely proportional to the density. So an increase in density due to an increase in pressure doesn't result in any change in viscosity.

Relation to Mean Free Path of Diffusing Particles

In relation to diffusion, the kinematic viscosity provides a better understanding of the behavior of mass transport of a dilute species. Viscosity is related to shear stress and the rate of shear in a fluid, which illustrates its dependence on the mean free path, λ, of the diffusing particles.

From fluid mechanics, for a Newtonian fluid, the shear stress, τ, on a unit area moving parallel to itself, is found to be proportional to the rate of change of velocity with distance perpendicular to the unit area:

$$\tau = \mu \frac{\mathrm{d}u_x}{\mathrm{d}y}$$

for a unit area parallel to the *xz*-plane, moving along the *x* axis. We will derive this formula and show how μ is related to λ.

Interpreting shear stress as the time rate of change of momentum, *p*, per unit area *A* (rate of momentum flux) of an arbitrary control surface gives

$$\tau = \frac{\dot{p}}{A} = \frac{\dot{m}\langle u_x \rangle}{A}.$$

where $\langle u_x \rangle$ is the average velocity, along the *x*-axis, of fluid molecules hitting the unit area, with respect to the unit area and $\dot{m}$ is the rate of fluid mass hitting the surface.

By making simplified assumption that the velocity of the molecules depends linearly on the distance they are coming from, the mean velocity depends linearly on the mean distance:

$$\langle u_x \rangle = \lambda \frac{\mathrm{d}u_x}{\mathrm{d}y}.$$

Further manipulation will show,

$$\dot{m} = \rho \bar{u} A$$

$$\tau = \underbrace{\rho \bar{u} \lambda}_{\mu} \cdot \frac{du_x}{dy} \Rightarrow \nu = \frac{\mu}{\rho} = \bar{u}\lambda,$$

where

ρ is the density of the fluid,

$\bar{u}$ is the root mean square molecular speed: $\bar{u} = \sqrt{\langle u^2 \rangle}$,

μ is the dynamic viscosity.

Note, that the mean free path itself typically depends (inversely) on the density.

Effect of Temperature on the Viscosity of a Gas

Sutherland's formula can be used to derive the dynamic viscosity of an ideal gas as a function of the temperature:

$$\mu = \mu_0 \frac{T_0 + C}{T + C} \left(\frac{T}{T_0} \right)^{\frac{3}{2}}.$$

This in turn is equal to

$$\mu = \lambda \frac{T^{\frac{3}{2}}}{T + C},$$

where

$$\lambda = \frac{\mu_0 (T_0 + C)}{T_0^{\frac{3}{2}}}$$

is a constant for the gas.

in Sutherland's formula:

- μ = dynamic viscosity (Pa·s or μPa·s) at input temperature T,
- μ_o = reference viscosity (in the same units as μ) at reference temperature T_o,
- T = input temperature (K),
- T_o = reference temperature (K),

- C = Sutherland's constant for the gaseous material in question.

Valid for temperatures between $0 < T < 555$ K with an error due to pressure less than 10% below 3.45 MPa.

According to Sutherland's formula, if the absolute temperature is less than C, the relative change in viscosity for a small change in temperature is greater than the relative change in the absolute temperature, but it is smaller when T is above C. The kinematic viscosity though always increases faster than the temperature (that is, d log(ν)/d log(T) is greater than 1).

Sutherland's constant, reference values and λ values for some gases:

Gas	C (K)	T_0 (K)	μ_0 (μPa·s)	λ (μPa·s·K$^{-1/2}$)
air	120	291.15	18.27	1.512041288
nitrogen	111	300.55	17.81	1.406732195
oxygen	127	292.25	20.18	1.693411300
carbon dioxide	240	293.15	14.8	1.572085931
carbon monoxide	118	288.15	17.2	1.428193225
hydrogen	72	293.85	8.76	0.636236562
ammonia	370	293.15	9.82	1.297443379
sulfur dioxide	416	293.65	12.54	1.768466086
helium	79.4	273	19	1.484381490

Viscosity of a Dilute Gas

The Chapman–Enskog equation may be used to estimate viscosity for a dilute gas. This equation is based on a semi-theoretical assumption by Chapman and Enskog. The equation requires three empirically determined parameters: the collision diameter (σ), the maximum energy of attraction divided by the Boltzmann constant (ε/κ) and the collision integral ($\omega(T^*)$).

$$\mu_0 \times 10^6 = 2.6693 \frac{\sqrt{MT}}{\sigma^2 \omega(T^*)},$$

with

- $T^* = \kappa T/\varepsilon$ is reduced temperature (dimensionless),
- μ_0 is viscosity for dilute gas (μPa·s),
- M is molecular mass (g/mol),
- T is temperature (K),

- σ is the collision diameter (Å),
- ε/κ is the maximum energy of attraction divided by the Boltzmann constant (K),
- ω_{μ} is the collision integral.

Liquids

Experiment showing the behaviour of a viscous fluid with blue dye for visibility. Performed by Prof. Oliver Zajkov at the Physics Institute at the Ss. Cyril and Methodius University of Skopje, Macedonia.

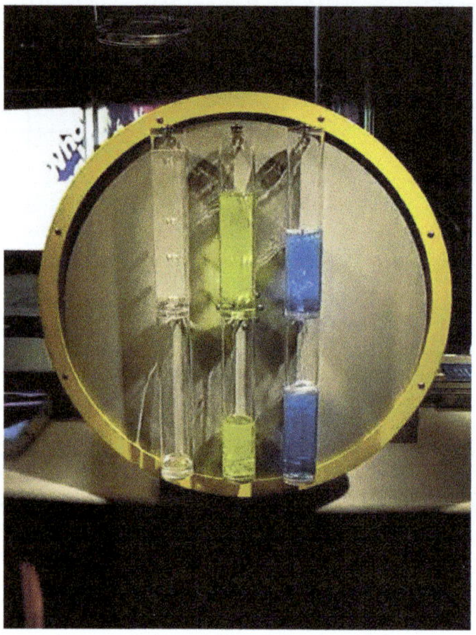

Video showing three liquids with different viscosities

In liquids, the additional forces between molecules become important. This leads to an additional contribution to the shear stress though the exact mechanics of this are still controversial. Thus, in liquids:

- Viscosity is independent of pressure (except at very high pressure); and
- Viscosity tends to fall as temperature increases (for example, water viscosity goes from 1.79 cP to 0.28 cP in the temperature range from 0 °C to 100 °C);

The dynamic viscosities of liquids are typically several orders of magnitude higher than dynamic viscosities of gases.

Viscosity of Blends of Liquids

The viscosity of the blend of two or more liquids can be estimated using the Refutas equation. The calculation is carried out in three steps.

The first step is to calculate the viscosity blending number (VBN) (also called the viscosity blending index) of each component of the blend:

$$\text{VBN} = 14.534 \times \ln(\ln(\nu + 0.8)) + 10.975 \qquad (1)$$

where ν is the kinematic viscosity in centistokes (cSt). It is important that the kinematic viscosity of each component of the blend be obtained at the same temperature.

The next step is to calculate the VBN of the blend, using this equation:

$$\text{VBN}_{\text{Blend}} = (x_{\text{A}} \times \text{VBN}_{\text{A}}) + (x_{\text{B}} \times \text{VBN}_{\text{B}}) + \cdots + (x_{\text{N}} \times \text{VBN}_{\text{N}}) \qquad (2)$$

where x_{X} is the mass fraction of each component of the blend.

Once the viscosity blending number of a blend has been calculated using equation (2), the final step is to determine the kinematic viscosity of the blend by solving equation (1) for ν:

$$\nu = \exp\left(\exp\left(\frac{\text{VBN}_{\text{Blend}} - 10.975}{14.534}\right)\right) - 0.8, \qquad (3)$$

where $\text{VBN}_{\text{Blend}}$ is the viscosity blending number of the blend.

Viscosity of Selected Substances

Air

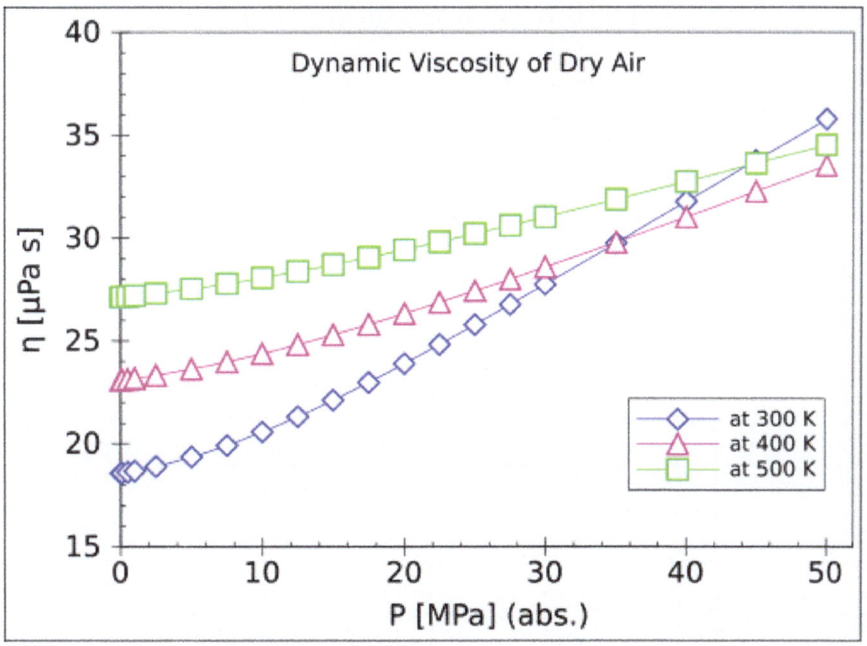

Pressure dependence of the dynamic viscosity of dry air at the temperatures of 300, 400 and 500 kelvins

The viscosity of air depends mostly on the temperature. At 15 °C, the viscosity of air is 1.81×10^{-5} kg/(m·s), 18.1 μPa·s or 1.81×10^{-5} Pa·s. The kinematic viscosity at 15 °C is 1.48×10^{-5} m²/s or 14.8 cSt. At 25 °C, the viscosity is 18.6 μPa·s and the kinematic viscosity 15.7 cSt.

Water

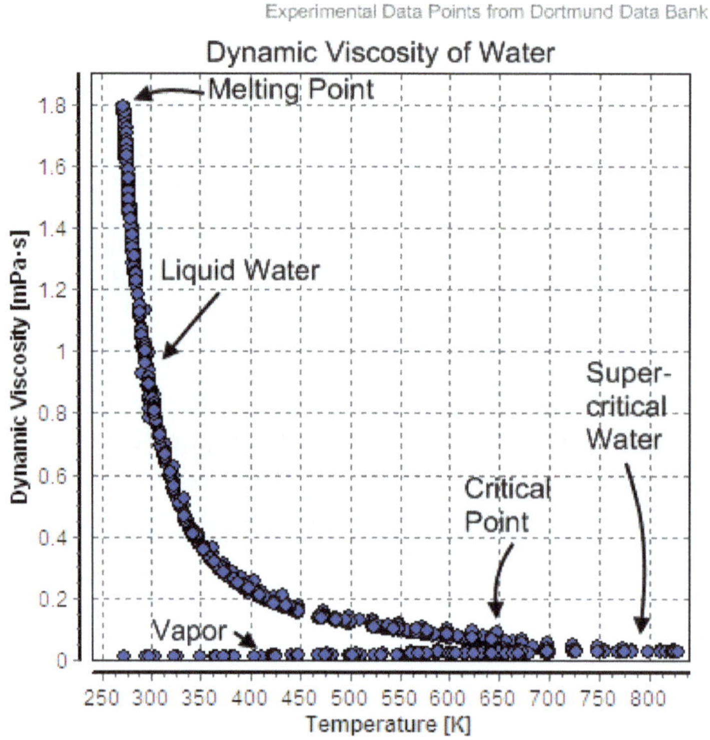

Dynamic viscosity of water

The dynamic viscosity of water is 8.90×10^{-4} Pa·s or 8.90×10^{-3} dyn·s/cm² or 0.890 cP at about 25 °C.

As a function of temperature T (in kelvins): $\mu = A \times 10^{B/(T-C)}$, where $A = 2.414\times10^{-5}$ Pa·s, B = 247.8 K, and C = 140 K.

Viscosity of liquid water at different temperatures up to the normal boiling point is listed below.

Viscosity of water at various temperatures	
Temperature (°C)	**Viscosity (mPa·s)**
10	1.308
20	1.002
30	0.7978
40	0.6531
50	0.5471
60	0.4658
70	0.4044
80	0.3550
90	0.3150
100	0.2822

Other Substances

Example of the viscosity of milk and water. Liquids with higher viscosities make smaller splashes when poured at the same velocity.

Peanut butter is a semi-solid and can therefore hold peaks.

Honey being drizzled.

Some dynamic viscosities of Newtonian fluids are listed below:

Viscosity of selected gases at 100 kPa, (μPa·s)		
Gas	**at 0 °C (273 K)**	**at 27 °C (300 K)**
air	17.4	18.6
hydrogen	8.4	9.0

helium		20.0
argon		22.9
xenon	21.2	23.2
carbon dioxide		15.0
methane		11.2
ethane		9.5

Viscosity of fluids with variable compositions		
Fluid	**Viscosity (Pa·s)**	**Viscosity (cP)**
blood (37 °C)	$3\times10^{-3} - 4\times10^{-3}$	3–4
honey	2–10	2000–10000
molasses	5–10	5000–10000
molten glass	10–1000	10000–1000000
chocolate syrup	10–25	10000–25000
molten chocolate[a]	45–130	45000–130000
ketchup[a]	50–100	50000–100000
lard	≈ 100	≈ 100000
peanut butter[a]	≈ 250	≈ 250000
shortening[a]	≈ 250	≈ 250000

Viscosity of liquids (at 25 °C unless otherwise specified)		
Liquid	**Viscosity (Pa·s)**	**Viscosity (cP)**
acetone	3.06×10^{-4}	0.306
benzene	6.04×10^{-4}	0.604
castor oil	0.985	985
corn syrup	1.3806	1380.6
ethanol	1.074×10^{-3}	1.074
ethylene glycol	1.61×10^{-2}	16.1
glycerol (at 20 °C)	1.2	1200
HFO-380	2.022	2022
mercury	1.526×10^{-3}	1.526
methanol	5.44×10^{-4}	0.544
motor oil SAE 10 (20 °C)	0.065	65
motor oil SAE 40 (20 °C)	0.319	319
nitrobenzene	1.863×10^{-3}	1.863
liquid nitrogen (−196 °C)	1.58×10^{-4}	0.158
propanol	1.945×10^{-3}	1.945
olive oil	0.081	81
pitch	2.3×10^{8}	2.30×10^{11}
sulfuric acid	2.42×10^{-2}	24.2
water	8.94×10^{-4}	0.894

Viscosity of solids		
Solid	**Viscosity (Pa·s)**	**Temperature (°C)**
asthenosphere	7.0×10^{19}	900
upper mantle	$7\times10^{20} - 1\times10^{21}$	1300–3000
lower mantle	$1\times10^{21} - 2\times10^{21}$	3000–4000

- These materials are highly non-Newtonian.

Viscosity of Slurry

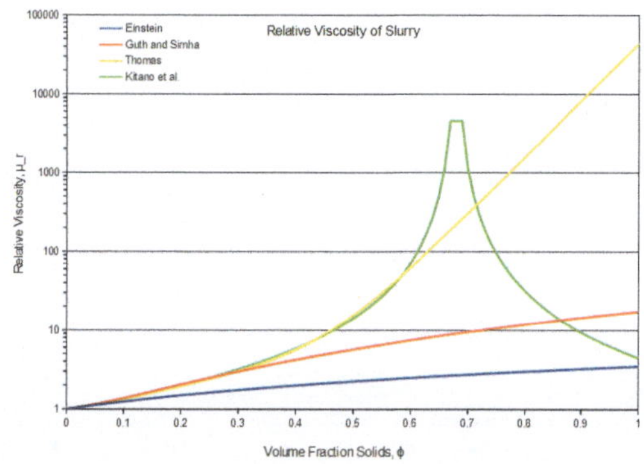

Plot of slurry relative viscosity μ_r as calculated by empirical correlations from Einstein, Guth and Simha, Thomas, and Kitano *et al.*.

The term slurry describes mixtures of a liquid and solid particles that retain some fluidity. The viscosity of slurry can be described as relative to the viscosity of the liquid phase:

$$\mu_s = \mu_r \mu_l,$$

where μ_s and μ_l are respectively the dynamic viscosity of the slurry and liquid (Pa·s), and μ_r is the relative viscosity (dimensionless).

Depending on the size and concentration of the solid particles, several models exist that describe the relative viscosity as a function of volume fraction φ of solid particles.

In the case of extremely low concentrations of fine particles, Einstein's equation may be used:

$$\mu_r = 1 + 2.5\varphi$$

In the case of higher concentrations, a modified equation was proposed by Guth and Simha, which takes into account interaction between the solid particles:

$$\mu_r = 1 + 2.5\varphi + 14.1\varphi^2$$

Further modification of this equation was proposed by Thomas from the fitting of empirical data:

$$\mu_r = 1 + 2.5\varphi + 10.05\varphi^2 + Ae^{B\varphi},$$

where $A = 0.00273$ and $B = 16.6$.

In the case of high shear stress (above 1 kPa), another empirical equation was proposed by Kitano *et al.* for polymer melts:

$$\mu_r = \left(1 - \frac{\varphi}{A}\right)^{-2},$$

where A = 0.68 for smooth spherical particles.

Viscosity of Amorphous Materials

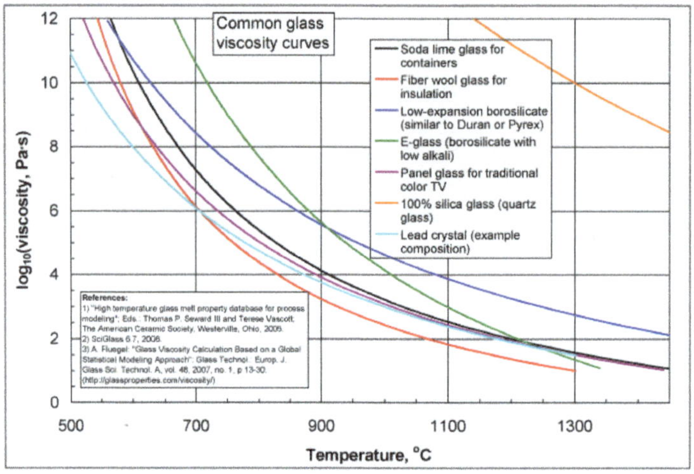

Common glass viscosity curves.

Viscous flow in amorphous materials (e.g. in glasses and melts) is a thermally activated process:

$$\mu = Ae^{\frac{Q}{RT}},$$

where Q is activation energy, T is temperature, R is the molar gas constant and A is approximately a constant.

The viscous flow in amorphous materials is characterized by a deviation from the Arrhenius-type behavior: Q changes from a high value Q_H at low temperatures (in the glassy state) to a low value Q_L at high temperatures (in the liquid state). Depending on this change, amorphous materials are classified as either

- strong when: $Q_H - Q_L < Q_L$ or
- fragile when: $Q_H - Q_L \geq Q_L$.

The fragility of amorphous materials is numerically characterized by Doremus' fragility ratio:

$$R_D = \frac{Q_H}{Q_L}$$

and strong materials have $R_D < 2$ whereas fragile materials have $R_D \geq 2$.

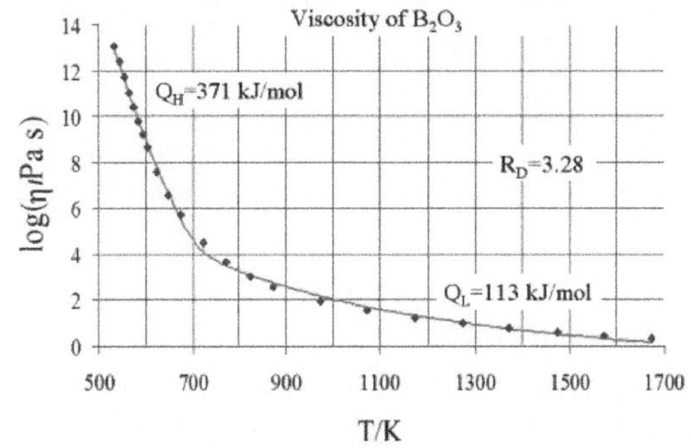

Common logarithm of viscosity against temperature for B_2O_3, showing two regimes

The viscosity of amorphous materials is quite exactly described by a two-exponential equation:

$$\mu = A_1 T\left(1 + A_2 e^{\frac{B}{RT}}\right)\left(1 + C e^{\frac{D}{RT}}\right),$$

with constants A_1, A_2, B, C and D related to thermodynamic parameters of joining bonds of an amorphous material.

Not very far from the glass transition temperature, T_g, this equation can be approximated by a Vogel–Fulcher–Tammann (VFT) equation.

If the temperature is significantly lower than the glass transition temperature, $T \ll T_g$, then the two-exponential equation simplifies to an Arrhenius-type equation:

$$A\ Te^{\overline{RT}}$$

with

$$Q_H = H_d + H_m,$$

where H_d is the enthalpy of formation of broken bonds (termed configurons) and H_m is the enthalpy of their motion.

When the temperature is less than the glass transition temperature, $T < T_g$, the activation energy of viscosity is high because the amorphous materials are in the glassy state and most of their joining bonds are intact.

If the temperature is much higher than the glass transition temperature, $T \gg T_g$, the two-exponential equation also simplifies to an Arrhenius-type equation:

$$\mu = A_H T e^{\frac{Q_L}{RT}},$$

with

$$Q_L = H_m.$$

When the temperature is higher than the glass transition temperature, $T > T_g$, the activation energy of viscosity is low because amorphous materials are melted and have most of their joining bonds broken, which facilitates flow.

Eddy Viscosity

In the study of turbulence in fluids, a common practical strategy for calculation is to ignore the small-scale vortices (or eddies) in the motion and to calculate a large-scale motion with an eddy viscosity that characterizes the transport and dissipation of energy in the smaller-scale flow. Values of eddy viscosity used in modeling ocean circulation may be from 5×10^4 to 1×10^6 Pa·s depending upon the resolution of the numerical grid.

Low Emissivity

Low emissivity (low *e* or low thermal emissivity) refers to a surface condition that emits low levels of radiant thermal (heat) energy. All materials absorb, reflect and emit radiant energy according to Planck's Law but here, the primary concern is a special wavelength interval of radiant energy, namely thermal radiation of materials. In common use, especially building application, the temperature range of approximately -40 to +80 degrees Celsius is the focus, but in aerospace and industrial process engineering, much broader ranges are of practical concern.

Definition

Emissivity is the value given to materials based on the ratio of heat emitted compared to a perfect blackbody, on a scale from zero to one. A blackbody would have an emissivity of 1 and a perfect reflector would have a value of 0.

Kirchhoff's law of thermal radiation states that absorption equals emissivity opaque for every specific wavelength/frequency (materials often have quite different emissivities at different wavelengths). Therefore, if asphalt has an emissivity value of 0.90 at a specific wavelength (say wavelength of 10 micrometer, or room temperature thermal radiation), its thermal absorptance value would also be 0.90. This means that it absorbs and emits 90 percent of radiant thermal energy. As it is an opaque material, the remaining 10 percent must be reflected. Conversely, a low-*e* material such as aluminum foil has a thermal emissivity/absorptance value of 0.03 and as an opaque material the thermal reflectance value must be 1.0 - 0.03 =0.97, meaning it reflects 97 percent of radiant thermal energy. Low-emissivity building materials include window glass manufactured with metal-oxide coatings as well as housewrap materials, reflective thermal insulations and other forms of radiant thermal barriers.

The thermal emissivity of various surfaces is listed in the following table.

Materials surface	Thermal emissivity
Silver, polished	0.02
Aluminum foil	0.03
Marble, smooth	0.56

Paper, roofing or white	0.88 to 0.86
Asphalt	0.88
Plaster, rough	0.89
Brick	0.90
Marble, polished or white	0.89 to 0.92
Concrete, rough	0.91
Glass, smooth (uncoated)	0.91
Limestone	0.92

Low-emissivity Windows

Window glass is by nature highly thermally emissive, as indicated in the table above. To improve thermal control (insulation and solar optical properties) thin film coatings are applied to the raw soda-lime glass. There are two primary methods in use: pyrolytic CVD and magnetron sputtering. The first involves deposition of fluorinated tin oxide at high temperatures. Pyrolytic coatings are usually applied at the float glass plant when the glass is manufactured. The second involves depositing thin silver layers with antireflection layers. Magnetron sputtering uses large vacuum chambers with multiple deposition chambers depositing 5 to 10 or more layers in succession. Silver-based films are environmentally unstable and must be enclosed in insulated glazing or an Insulated Glass Unit (IGU) to maintain their properties over time. Specially designed coatings may be applied to one or more surfaces of insulated glass. One type of coating (low-e coatings) reduces the emission of radiant infrared energy, thus tending to keep heat on the side of the glass where it originated, while letting visible light pass. This results in glazing with better control of energy - heat originating from indoors in winter remains inside (the warm side), while heat during summer does not emit from the exterior, keeping it cooler inside.

Glass can be made with differing thermal emissivities, but this is not used for windows. Certain properties such as the iron content may be controlled, changing the thermal emissivity properties of glass. This “naturally” low thermal emissivity is found in some formulations of borosilicate or Pyrex. Naturally low-e glass does not have the property of reflecting near infrared (NIR)/thermal radiation; instead, this type of glass has higher NIR transmission, leading to undesirable heat loss (or gain) in a building with that type of window.

Criticism of Low-E Windows

Since energy-efficient windows reflect much more sunlight than simple glass windows, when these windows are somewhat concave they can focus sunlight and cause damage. Damage to the sidings of homes and to automobiles has been reported in news stories.

Low-E windows may also block radio frequency signals. Buildings without distributed antenna systems may then suffer degraded cell phone reception.

Reflective Thermal Insulation

Reflective thermal insulation is typically fabricated from aluminum foil with a variety of core materials such as low-density polyethylene foam, polyethylene bubbles, fiberglass, or similar materials. Each core material presents its own set of benefits and drawbacks based on its ability to provide a thermal break, deaden sound, absorb moisture, and resist combustion during a fire. When aluminum foil is used as the facing material, reflective thermal insulation can stop 97% of radiant heat transfer. Recently, some reflective thermal insulation manufacturers have switched to a metalized polyethylene facing. The long-term efficiency and durability of such facings are still undetermined.

Reflective thermal insulation can be installed in a variety of applications and locations including residential, agricultural, commercial, and industrial structures. Some common installations include house wraps, duct wraps, pipe wraps, under radiant floors, inside wall cavities, roof systems, attic systems and crawl spaces. Reflective thermal insulation can be used as a stand-alone product in many applications but can also be used in combination systems with mass insulation where higher R-values are required.

Heatable Glass

Heatable glass unit at work

Electrically heatable glass and windows are relatively new products, which help solve problems in the design of buildings and vehicles. The idea of heating glass is based on the use of energy-efficient low-emissive glass, which is generally simple silicate glass with a special metallic oxides coating. Low-emissive coating decreases heat loss by approximately 30%. Heatable glass can be used in all kinds of standard glazing systems, whether wood, plastic, aluminum or steel.

Heatable glass based on low-emissive coatings was first produced in high volume in the early 1980s. Today, heating glass is used in the construction of many kinds of buildings and in mass production of vehicles, ships and trains. Heatable glass removes discomfort and other disadvantages induced by the low heat-insulating features of silicate glass. The effect of "cold glass" disappears when the surface of the glass is heated. Condensation is eliminated, along with ice and snow covering, the window's heat losses are compensated and room comfort is improved.

Heatable glass can be used as the principal system of heating and can be combined with floor and ceiling heating. Such combination helps reduce the total rate of heat loss of the building, thereby lowering heating expenses. Also, the active area of the room can be used more efficiently, as massive window-sill radiators are not needed. Initially, heating glass was produced by sputtering ordinary glass, and stable quality could not be guaranteed. A technological breakthrough took place in 1989 when the mass production of low-emissive glass began. The glass was coated during the manufacturing process.

Standard Windows

- Window construction

Windows play a significant role in room comfort. As a result, the area of glazing of buildings is constantly being increased. Window technologies always in progress and it is common today to use low-emissive glass. In spite of progress the low temperature of glass surface is still the problem of constructive glazing. Heatable glass helps to solve problems concerning low surface temperature and increase the level of comfort in the room significantly. Heatable glass can be used in practically all kinds of glazing systems made of wood, plastic or aluminum. Heatable glass and multiple glass panes can be used both in blind and openable constructions. Multiple glass panes made of heating glass can have one or two chambers. The advantages of multiple glass panes are their hermiticity and ability to decrease heat transfer significantly.

- Optical transmission and heat losses of windows

If the temperature in the building is higher than the temperature outside, the heat leaks through the elements of construction. Windows are usually the most vulnerable elements of the building in terms of heat losses. The heat loss though window constructions is about 20-25% of total heat loss. Heat insulation of translucent constructions can be improved by increasing the number of glasses and chambers of multiple glass panes, but will result in increased construction cost and decreased optical transmission. The reasonable alternative is the usage of low-emissivity glass, which is practically the same as ordinary glass in terms of optical transmission, but it also reflects the heat radiation back into the room. The major indicator which characterizes the ability of glass to reflect heat radiation is its emissivity (E) or the "emission factor". The emission factor of ordinary glass is 0.83; the factor for low-emissive glass can reach 0.03, so that more than 90% of accumulated heat will be reflected back into the room. The lower the emission factor is, the more effective is the material to reflect the heat, and the more heat it will accumulate. To compare, the emission factor of a multiple glass pane with two chambers, which is made of ordinary glass, is the same as the emission factor of a multiple glass pane with one chamber, which was produced with usage of low-emissive glass. Besides energy-efficient functions in cold seasons of the year, low-emissive glass possesses the ability to reflect the excess of outside heat energy in summer seasons; the optical transmission coefficient is affected insignificantly in this way. The additional factor of reduce of heat transfer of multiple glass panes is the usage of low thermal conductivity gases – Ar or Kr – to fill the chambers. In present-day multiple glass panes Ar is more often used, which helps to reduce heat losses by 10-20%, though the cost of multiple glass panes is insignificantly increased.

- Influence of window surface temperature on comfort

There are two reasons why people feel discomfort when they are close to a cold window surface. First, a cold window is the reason for outflow of heat, which is produced by the cutaneous covering of the individual. Second, a cold window provokes the circulation of air, which is felt like a draft. In order to reduce these factors the heating radiators are always placed under window sills. As far as people can feel cold and heat, the actual temperature of environment is not the only factor which defines the total level of comfort. In reality, the heat radiation of surrounding surfaces has a greater influence than air temperature. If the window surface is cold then to maintain the comfort atmosphere it is necessary to increase the heating temperature, but it will also increase energy consumption. The problem of cold window can be solved effectively with the help of heatable glass. These windows allow to maintain the optimal comfort level and temperature of the room. The air temperature can be decreased at least by 1 degree if the temperature of surrounding surfaces has the same significance. You also do not have to install heating radiators and free the additional space for that. Besides, when turned off the multiple glass panes made of heatable glass act like ordinary low-emissive glass.

Heatable Glass and Multiple Glass Panes Made of Heating Glass

- Structure of heatable glass

The idea of heatable glass is based on usage of energy-efficient low-emissive glass, where the coating plays the role of heating element. It can be used both in production of multiple glass panes and as a part of triplex, which has also the function of protective glazing. The technological process of production of multiple glass panes made of heatable glass is practically the same as the process of ordinary multiple glass panes production. The main difference is the presence of power supply and, if necessary, temperature sensor. The temperature sensor allows to track the temperature of heating glass and eliminates the possibility of overheating of the product. In order to prevent shocks, the conductive coating is always placed inside the multiple glass pane or laminated unit. Only safe tempered glass, the strength of which is a lot higher than the strength of ordinary glass, is used in production of heatable glass. When the hardened glass is destroyed there are safe splittings. Also the current-carrying coating loses its integrity and the automatic fuse, which turns off the power supply of the glass, is activated. The electrodes are placed inside the lamination and no one can reach them without destruction of the product.

- Usage of heatable glass

Alarm Glass

Transport Glass

Heating glass is mostly used for heating of windows. It is especially useful for rooms where people spend much time by the windows, at home or at work. The most common usage of heating glass—windows of cottages, office buildings and also big areas—leaded panes, translucent roofing, garret windows, canopies and so on.

Heating glass is used for defogging and prevention of frosting of windows of pools, saunas and other buildings of such kind. Insofar as heatable glass has a current-carrying coating, it can be used as the sensor of alarm systems. When the glass is destroyed the system of protection is activated and it results in activation of alarm system. This kind of product is widely used on objects of tightened standards in questions of protection: nuclear power plants, stations of air navigation control, museums, special storehouses, etc. Heatable glass is also used in production of windows for different kinds of vehicles: electric and diesel locomotives, vessels and boats, various kinds of aircraft and automobiles. One of well-known examples of application of heating glass is armored windows, because the protective glazing is very thick and is disposed to frosting. The usage of heating glass is especially urgent in terms of being the part of armored multiple glass of Smart Glass of switchable transparency, because the heating significantly decreases the period of reaction of liquid crystals structure. The power consumed by products depends on the type of use. Power of about 50-100 watts per square meter of the window is generally enough for maintenance of comfort temperature in the room and for maintenance of glass surface temperature at the rate from +20 degrees to +30 degrees. When the heatable glass is used as the only source of heat it is necessary to maintain the glass surface temperature at the rate from +30 degrees to -45 degrees and provide the power of 100 to 300 watt for 1 square meter of the window. The power needed for vehicle windows reaches 1.5 kilowatts per square meter or more, which is why there are such tight standards in terms of sputtering of current-carrying components. Heat power of about 500-700 watts per square meter of glazing is necessary for snow unloading and taking ice-covering off the outside protective translucent constructions in low temperatures and windy environments.

Technology of Production

Heatable glass is produced by lamination of two or more sheets of silicate glass. The most widespread technologies are the following technologies of panel production according to the type of materials used:

- EVA - ethylene-vinyl-acetate film with good adhesion to glass. Major advantages: low cost of both film and equipment. One needs only a primitive furnace with vacuum bags for production. Disadvantages: high rate of opalescence, especially after multcoat lamination, with the lapse of time yellowness appears. EVA has low shear strength, especially in low temperatures; it results in delamination (layering).
- PVB - polyvinyl butyral film with high rate of adhesion to glass. Major advantages: low cost of mass production of laminated glass, insignificant rate of opalescence, high quality of product. Disadvantages: high initial cost of equipment, it is necessary to have autoclave, press for preliminary hot pressing, "clean" room, and qualified personnel. Besides that, triplex made with the help of PVB technology can not be use in wet environment.
- TPU - thermoplastic polyurethane film with very high rate of adhesion to glass. Major advantages: insignificant rate of opalescence, insensible to humidity, mechanical effects

and severe atmosphere; very high quality of the product. Disadvantages: high cost of film and equipment, it is necessary to have autoclave, “clean” room, and qualified personnel.

- Photocurable polymers (resins) – so-called “filling technology”. Major advantages: low cost of both resin and equipment. Only an ultraviolet furnace and a minimum of additional equipment is needed to produce laminated glass. Disadvantages: it is necessary to have qualified personnel for the work. Heating laminated glass that is produced with the help of this technology is insensible to humidity and temperature influence, has high shear strength.

Corrosion

Mechanical failure modes
• Buckling • Corrosion • Corrosion fatigue • Creep • Fatigue • Fouling
• Fracture • Hydrogen embrittlement • Impact • Mechanical overload • Stress corrosion cracking • Thermal shock • Wear • Yielding

Rust, the most familiar example of corrosion

Volcanic gases have accelerated the extensive corrosion of this abandoned mining machinery, rendering it almost unrecognizable

Corrosion on exposed metal, including a bolt and nut

Corrosion is a natural process, which converts a refined metal to a more chemically-stable form, such as its oxide, hydroxide, or sulfide. It is the gradual destruction of materials (usually metals) by chemical and/or electrochemical reaction with their environment. Corrosion engineering is the field dedicated to controlling and stopping corrosion.

In the most common use of the word, this means electrochemical oxidation of metal in reaction with an oxidant such as oxygen or sulfur. Rusting, the formation of iron oxides, is a well-known example of electrochemical corrosion. This type of damage typically produces oxide(s) or salt(s) of the original metal, and results in a distinctive orange colouration. Corrosion can also occur in materials other than metals, such as ceramics or polymers, although in this context, the term "degradation" is more common. Corrosion degrades the useful properties of materials and structures including strength, appearance and permeability to liquids and gases.

Many structural alloys corrode merely from exposure to moisture in air, but the process can be strongly affected by exposure to certain substances. Corrosion can be concentrated locally to form a pit or crack, or it can extend across a wide area more or less uniformly corroding the surface. Because corrosion is a diffusion-controlled process, it occurs on exposed surfaces. As a result, methods to reduce the activity of the exposed surface, such as passivation and chromate conversion, can increase a material's corrosion resistance. However, some corrosion mechanisms are less visible and less predictable.

Galvanic Corrosion

Galvanic corrosion of aluminium. A 5-mm-thick aluminium alloy plate is physically (and hence, electrically) connected to a 10-mm-thick mild steel structural support. Galvanic corrosion occurred on the aluminium plate along the joint with the steel. Perforation of aluminium plate occurred within 2 years.

Galvanic corrosion occurs when two different metals have physical or electrical contact with each other and are immersed in a common electrolyte, or when the same metal is exposed to electrolyte with different concentrations. In a galvanic couple, the more active metal (the anode) corrodes at an accelerated rate and the more noble metal (the cathode) corrodes at a slower rate. When immersed separately, each metal corrodes at its own rate. What type of metal(s) to use is readily determined by following the galvanic series. For example, zinc is often used as a sacrificial anode for steel structures. Galvanic corrosion is of major interest to the marine industry and also anywhere water (containing salts) contacts pipes or metal structures.

Factors such as relative size of anode, types of metal, and operating conditions (temperature, humidity, salinity, etc.) affect galvanic corrosion. The surface area ratio of the anode and cathode directly affects the corrosion rates of the materials. Galvanic corrosion is often prevented by the use of sacrificial anodes.

Galvanic Series

In any given environment (one standard medium is aerated, room-temperature seawater), one metal will be either more *noble* or more *active* than others, based on how strongly its ions are bound to the surface. Two metals in electrical contact share the same electrons, so that the "tug-of-war" at each surface is analogous to competition for free electrons between the two materials. Using the electrolyte as a host for the flow of ions in the same direction, the noble metal will take electrons from the active one. The resulting mass flow or electric current can be measured to establish a hierarchy of materials in the medium of interest. This hierarchy is called a *galvanic series* and is useful in predicting and understanding corrosion.

Corrosion Removal

Often it is possible to chemically remove the products of corrosion. For example, phosphoric acid in the form of naval jelly is often applied to ferrous tools or surfaces to remove rust. Corrosion removal should not be confused with electropolishing, which removes some layers of the underlying metal to make a smooth surface. For example, phosphoric acid may also be used to electropolish copper but it does this by removing copper, not the products of copper corrosion.

Resistance to Corrosion

Some metals are more intrinsically resistant to corrosion than others. There are various ways of protecting metals from corrosion (oxidation) including painting, hot dip galvanizing, and combinations of these.

Intrinsic Chemistry

Gold nuggets do not naturally corrode, even on a geological time scale.

The materials most resistant to corrosion are those for which corrosion is thermodynamically unfavorable. Any corrosion products of gold or platinum tend to decompose spontaneously into pure metal, which is why these elements can be found in metallic form on Earth and have long been valued. More common "base" metals can only be protected by more temporary means.

Some metals have naturally slow reaction kinetics, even though their corrosion is thermodynamically favorable. These include such metals as zinc, magnesium, and cadmium. While corrosion of these metals is continuous and ongoing, it happens at an acceptably slow rate. An extreme example is graphite, which releases large amounts of energy upon oxidation, but has such slow kinetics that it is effectively immune to electrochemical corrosion under normal conditions.

Passivation

Passivation refers to the spontaneous formation of an ultrathin film of corrosion products, known as a passive film, on the metal's surface that act as a barrier to further oxidation. The chemical composition and microstructure of a passive film are different from the underlying metal. Typical passive film thickness on aluminium, stainless steels, and alloys is within 10 nanometers. The passive film is different from oxide layers that are formed upon heating and are in the micrometer thickness range – the passive film recovers if removed or damaged whereas the oxide layer does not. Passivation in natural environments such as air, water and soil at moderate pH is seen in such materials as aluminium, stainless steel, titanium, and silicon.

Passivation is primarily determined by metallurgical and environmental factors. The effect of pH is summarized using Pourbaix diagrams, but many other factors are influential. Some conditions that inhibit passivation include high pH for aluminium and zinc, low pH or the presence of chloride ions for stainless steel, high temperature for titanium (in which case the oxide dissolves into the metal, rather than the electrolyte) and fluoride ions for silicon. On the other hand, unusual conditions may result in passivation of materials that are normally unprotected, as the alkaline environment of concrete does for steel rebar. Exposure to a liquid metal such as mercury or hot solder can often circumvent passivation mechanisms.

Corrosion in Passivated Materials

Passivation is extremely useful in mitigating corrosion damage, however even a high-quality alloy will corrode if its ability to form a passivating film is hindered. Proper selection of the right grade of material for the specific environment is important for the long-lasting performance of this group of materials. If breakdown occurs in the passive film due to chemical or mechanical factors, the resulting major modes of corrosion may include pitting corrosion, crevice corrosion, and stress corrosion cracking.

Pitting Corrosion

Certain conditions, such as low concentrations of oxygen or high concentrations of species such as chloride which complete as anions, can interfere with a given alloy's ability to re-form a passivating film. In the worst case, almost all of the surface will remain protected, but tiny local fluctuations will degrade the oxide film in a few critical points. Corrosion at these points will be greatly amplified, and can cause *corrosion pits* of several types, depending upon conditions. While the corrosion pits only nucleate under fairly extreme circumstances, they can continue to grow even when conditions return to normal, since the interior of a pit is naturally deprived of oxygen and locally the pH decreases to very low values and the corrosion rate increases due to an autocatalytic process. In extreme cases, the sharp tips of extremely long and narrow corrosion pits can cause stress concentration to the point that otherwise tough alloys can shatter; a thin film pierced by an invisibly small hole can hide a thumb sized pit from view. These problems are especially dangerous because they are difficult to detect before a part or structure fails. Pitting remains among the most common and damaging forms of corrosion in passivated alloys, but it can be prevented by control of the alloy's environment.

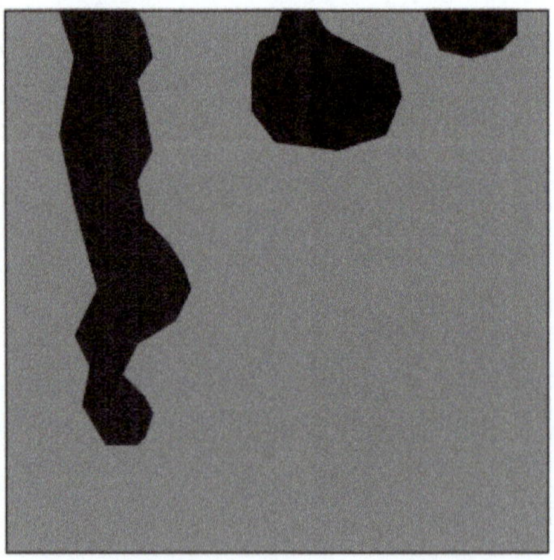

Diagram showing cross-section of pitting corrosion

Pitting results when a small hole, or cavity, forms in the metal, usually as a result of depassivation of a small area. This area becomes anodic, while part of the remaining metal becomes cathodic, producing a localized galvanic reaction. The deterioration of this small area penetrates the metal and can lead to failure. This form of corrosion is often difficult to detect due to the fact that it is usually relatively small and may be covered and hidden by corrosion-produced compounds.

Weld Decay and Knifeline Attack

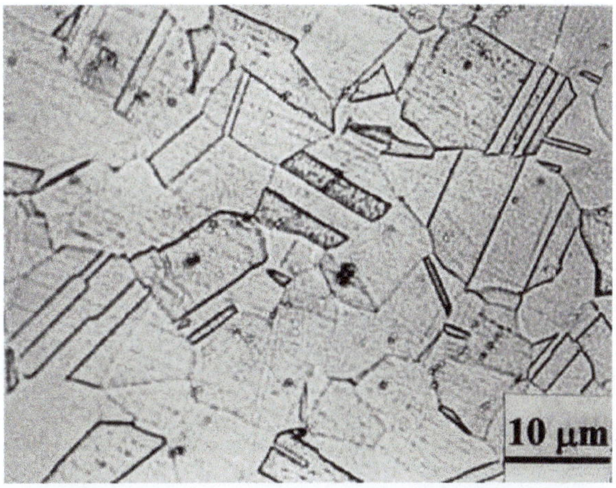

Normal microstructure of Type 304 stainless steel surface

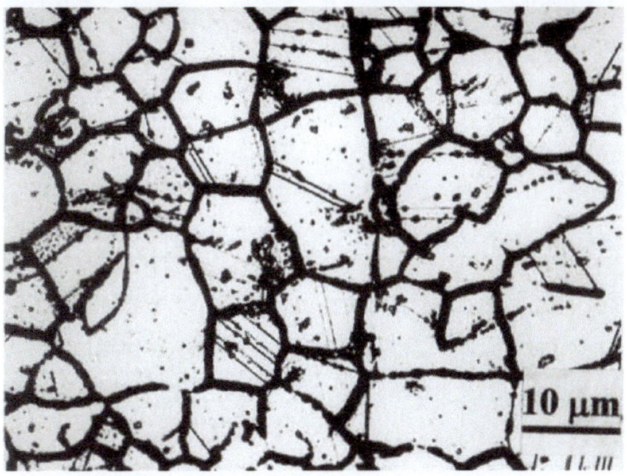

Sensitized metallic microstructure, showing wider intergranular boundaries

Stainless steel can pose special corrosion challenges, since its passivating behavior relies on the presence of a major alloying component (chromium, at least 11.5%). Because of the elevated temperatures of welding and heat treatment, chromium carbides can form in the grain boundaries of stainless alloys. This chemical reaction robs the material of chromium in the zone near the grain boundary, making those areas much less resistant to corrosion. This creates a galvanic couple with the well-protected alloy nearby, which leads to "weld decay" (corrosion of the grain boundaries in the heat affected zones) in highly corrosive environments. This process can seriously reduce the mechanical strength of welded joints over time.

A stainless steel is said to be "sensitized" if chromium carbides are formed in the microstructure. A typical microstructure of a normalized type 304 stainless steel shows no signs of sensitization, while a heavily-sensitized steel shows the presence of grain boundary precipitates. The dark lines in the sensitized microstructure are networks of chromium carbides formed along the grain boundaries.

Special alloys, either with low carbon content or with added carbon "getters" such as titanium and niobium (in types 321 and 347, respectively), can prevent this effect, but the latter require special heat treatment after welding to prevent the similar phenomenon of "knifeline attack". As its name implies, corrosion is limited to a very narrow zone adjacent to the weld, often only a few micrometers across, making it even less noticeable.

Crevice Corrosion

Corrosion in the crevice between the tube and tube sheet
(both made of type 316 stainless steel) of a heat exchanger in a seawater desalination plant

Crevice corrosion is a localized form of corrosion occurring in confined spaces (crevices), to which the access of the working fluid from the environment is limited. Formation of a differential aeration cell leads to corrosion inside the crevices. Examples of crevices are gaps and contact areas between parts, under gaskets or seals, inside cracks and seams, spaces filled with deposits and under sludge piles.

Crevice corrosion is influenced by the crevice type (metal-metal, metal-nonmetal), crevice geometry (size, surface finish), and metallurgical and environmental factors. The susceptibility to crevice corrosion can be evaluated with ASTM standard procedures. A critical crevice corrosion temperature is commonly used to rank a material's resistance to crevice corrosion.

Microbial Corrosion

Microbial corrosion, or commonly known as microbiologically influenced corrosion (MIC), is a corrosion caused or promoted by microorganisms, usually chemoautotrophs. It can apply to both metallic and non-metallic materials, in the presence or absence of oxygen. Sulfate-reducing bacteria are active in the absence of oxygen (anaerobic); they produce hydrogen sulfide, causing sulfide stress cracking. In the presence of oxygen (aerobic), some bacteria may directly oxidize iron to iron oxides and hydroxides, other bacteria oxidize sulfur and produce sulfuric acid causing biogenic sulfide corrosion. Concentration cells can form in the deposits of corrosion products, leading to localized corrosion.

Accelerated low-water corrosion (ALWC) is a particularly aggressive form of MIC that affects steel piles in seawater near the low water tide mark. It is characterized by an orange sludge, which smells of hydrogen sulfide when treated with acid. Corrosion rates can be very high and design corrosion allowances can soon be exceeded leading to premature failure of the steel pile. Piles that have been coated and have cathodic protection installed at the time of construction are not susceptible to ALWC. For unprotected piles, sacrificial anodes can be installed local to the affected areas to inhibit the corrosion or a complete retrofitted sacrificial anode system can be installed. Affected areas can also be treated electrochemically by using an electrode to first produce chlorine to kill the bacteria, and then to produce a calcareous deposit, which will help shield the metal from further attack.

High-temperature Corrosion

High-temperature corrosion is chemical deterioration of a material (typically a metal) as a result of heating. This non-galvanic form of corrosion can occur when a metal is subjected to a hot atmosphere containing oxygen, sulfur, or other compounds capable of oxidizing (or assisting the oxidation of) the material concerned. For example, materials used in aerospace, power generation and even in car engines have to resist sustained periods at high temperature in which they may be exposed to an atmosphere containing potentially highly corrosive products of combustion.

The products of high-temperature corrosion can potentially be turned to the advantage of the engineer. The formation of oxides on stainless steels, for example, can provide a protective layer preventing further atmospheric attack, allowing for a material to be used for sustained periods at both room and high temperatures in hostile conditions. Such high-temperature corrosion products, in the form of compacted oxide layer glazes, prevent or reduce wear during high-temperature sliding contact of metallic (or metallic and ceramic) surfaces.

Metal Dusting

Metal dusting is a catastrophic form of corrosion that occurs when susceptible materials are exposed to environments with high carbon activities, such as synthesis gas and other high-CO environments. The corrosion manifests itself as a break-up of bulk metal to metal powder. The suspected mechanism is firstly the deposition of a graphite layer on the surface of the metal, usually from carbon monoxide (CO) in the vapor phase. This graphite layer is then thought to form metastable M_3C species (where M is the metal), which migrate away from the metal surface. However, in some regimes no M_3C species is observed indicating a direct transfer of metal atoms into the graphite layer.

Protection from Corrosion

The US Army shrink wraps equipment such as helicopters to protect them from corrosion and thus save millions of dollars

Various treatments are used to slow corrosion damage to metallic objects which are exposed to the weather, salt water, acids, or other hostile environments. Some unprotected metallic alloys are extremely vulnerable to corrosion, such as those used in neodymium magnets, which can spall or crumble into powder even in dry, temperature-stable indoor environments unless properly treated to discourage corrosion.

Surface Treatments

When surface treatments are used to retard corrosion, great care must be taken to ensure complete coverage, without gaps, cracks, or pinhole defects. Small defects can act as an "Achilles' heel", allowing corrosion to penetrate the interior and causing extensive damage even while the outer protective layer remains apparently intact for a period of time.

Applied Coatings

Plating, painting, and the application of enamel are the most common anti-corrosion treatments. They work by providing a barrier of corrosion-resistant material between the damaging environment and the structural material. Aside from cosmetic and manufacturing issues, there may be tradeoffs in mechanical flexibility versus resistance to abrasion and high temperature. Platings usually fail only in small sections, but if the plating is more noble than the substrate (for example, chromium on steel), a galvanic couple will cause any exposed area to corrode much more rapidly than an unplated surface would. For this reason, it is often wise to plate with active metal such as zinc or cadmium.

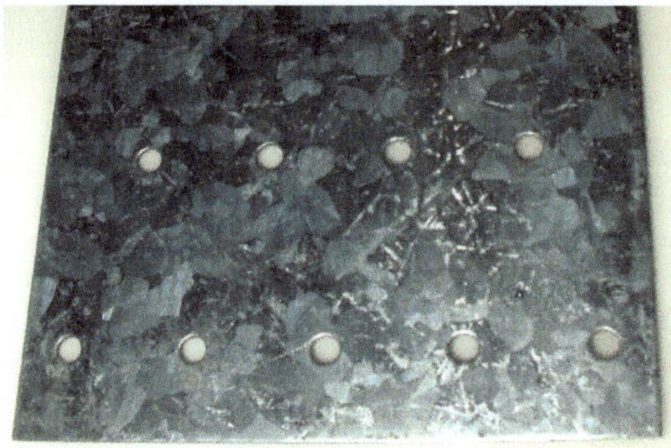

Galvanized surface

Painting either by roller or brush is more desirable for tight spaces; spray would be better for larger coating areas such as steel decks and waterfront applications. Flexible polyurethane coatings, like Durabak-M26 for example, can provide an anti-corrosive seal with a highly durable slip resistant membrane. Painted coatings are relatively easy to apply and have fast drying times although temperature and humidity may cause dry times to vary.

Reactive Coatings

If the environment is controlled (especially in recirculating systems), corrosion inhibitors can often be added to it. These chemicals form an electrically insulating or chemically impermeable coating on exposed metal surfaces, to suppress electrochemical reactions. Such methods make the system less sensitive to scratches or defects in the coating, since extra inhibitors can be made available wherever metal becomes exposed. Chemicals that inhibit corrosion include some of the salts in hard water (Roman water systems are famous for their mineral deposits), chromates, phosphates, polyaniline, other conducting polymers and a wide range of specially-designed chemicals that resemble surfactants (i.e. long-chain organic molecules with ionic end groups).

Anodization

This climbing descender is anodized with a yellow finish.

Aluminium alloys often undergo a surface treatment. Electrochemical conditions in the bath are carefully adjusted so that uniform pores, several nanometers wide, appear in the metal's oxide film. These pores allow the oxide to grow much thicker than passivating conditions would allow. At the end of the treatment, the pores are allowed to seal, forming a harder-than-usual surface layer. If this coating is scratched, normal passivation processes take over to protect the damaged area.

Anodizing is very resilient to weathering and corrosion, so it is commonly used for building facades and other areas where the surface will come into regular contact with the elements. While being resilient, it must be cleaned frequently. If left without cleaning, panel edge staining will naturally occur. Anodization is the process of converting an anode into cathode by bringing a more active anode in contact with it.

Biofilm Coatings

A new form of protection has been developed by applying certain species of bacterial films to the surface of metals in highly corrosive environments. This process increases the corrosion resistance substantially. Alternatively, antimicrobial-producing biofilms can be used to inhibit mild steel corrosion from sulfate-reducing bacteria.

Controlled Permeability Formwork

Controlled permeability formwork (CPF) is a method of preventing the corrosion of reinforcement by naturally enhancing the durability of the cover during concrete placement. CPF has been used in environments to combat the effects of carbonation, chlorides, frost and abrasion.

Cathodic Protection

Cathodic protection (CP) is a technique to control the corrosion of a metal surface by making that surface the cathode of an electrochemical cell. Cathodic protection systems are most commonly used to protect steel, and pipelines and tanks; steel pier piles, ships, and offshore oil platforms.

Sacrificial Anode Protection

Sacrificial anode attached to the hull of a ship

For effective CP, the potential of the steel surface is polarized (pushed) more negative until the metal surface has a uniform potential. With a uniform potential, the driving force for the corrosion reaction is halted. For galvanic CP systems, the anode material corrodes under the influence of the steel, and eventually it must be replaced. The polarization is caused by the current flow from the anode to the cathode, driven by the difference in electrode potential between the anode and the cathode.

Impressed Current Cathodic Protection

For larger structures, galvanic anodes cannot economically deliver enough current to provide complete protection. Impressed current cathodic protection (ICCP) systems use anodes connected to a DC power source (such as a cathodic protection rectifier). Anodes for ICCP systems are tubular and solid rod shapes of various specialized materials. These include high silicon cast iron, graphite, mixed metal oxide or platinum coated titanium or niobium coated rod and wires.

Anodic Protection

Anodic protection impresses anodic current on the structure to be protected (opposite to the cathodic protection). It is appropriate for metals that exhibit passivity (e.g. stainless steel) and suitably small passive current over a wide range of potentials. It is used in aggressive environments, such as solutions of sulfuric acid.

Rate of Corrosion

These neodymium magnets corroded extremely rapidly after only 5 months of outside exposure

A simple test for measuring corrosion is the weight loss method. The method involves exposing a clean weighed piece of the metal or alloy to the corrosive environment for a specified time followed by cleaning to remove corrosion products and weighing the piece to determine the loss of weight. The rate of corrosion (R) is calculated as

$$R = \frac{kW}{\rho A t}$$

where k is a constant, W is the weight loss of the metal in time t, A is the surface area of the metal exposed, and ρ is the density of the metal (in g/cm^3).

Other common expressions for the corrosion rate is penetration depth and change of mechanical properties.

Economic Impact

The collapsed Silver Bridge, as seen from the Ohio side

In 2002, the US Federal Highway Administration released a study titled "Corrosion Costs and Preventive Strategies in the United States" on the direct costs associated with metallic corrosion in the US industry. In 1998, the total annual direct cost of corrosion in the U.S. was ca. $276 billion (ca. 3.2% of the US gross domestic product). Broken down into five specific industries, the economic losses are $22.6 billion in infrastructure; $17.6 billion in production and manufacturing; $29.7 billion in transportation; $20.1 billion in government; and $47.9 billion in utilities.

Rust is one of the most common causes of bridge accidents. As rust has a much higher volume than the originating mass of iron, its build-up can also cause failure by forcing apart adjacent parts. It was the cause of the collapse of the Mianus river bridge in 1983, when the bearings rusted internally and pushed one corner of the road slab off its support. Three drivers on the roadway at the time died as the slab fell into the river below. The following NTSB investigation showed that a drain in the road had been blocked for road re-surfacing, and had not been unblocked; as a result, runoff water penetrated the support hangers. Rust was also an important factor in the Silver Bridge disaster of 1967 in West Virginia, when a steel suspension bridge collapsed within a minute, killing 46 drivers and passengers on the bridge at the time.

Similarly, corrosion of concrete-covered steel and iron can cause the concrete to spall, creating severe structural problems. It is one of the most common failure modes of reinforced concrete bridges. Measuring instruments based on the half-cell potential can detect the potential corrosion spots before total failure of the concrete structure is reached.

Until 20–30 years ago, galvanized steel pipe was used extensively in the potable water systems for single and multi-family residents as well as commercial and public construction. Today, these systems have long ago consumed the protective zinc and are corroding internally resulting in poor water quality and pipe failures. The economic impact on homeowners, condo dwellers, and the public infrastructure is estimated at 22 billion dollars as the insurance industry braces for a wave of claims due to pipe failures.

Corrosion in Nonmetals

Most ceramic materials are almost entirely immune to corrosion. The strong chemical bonds that hold them together leave very little free chemical energy in the structure; they can be thought of as already corroded. When corrosion does occur, it is almost always a simple dissolution of the material or chemical reaction, rather than an electrochemical process. A common example of corrosion protection in ceramics is the lime added to soda-lime glass to reduce its solubility in water; though it is not nearly as soluble as pure sodium silicate, normal glass does form sub-microscopic flaws when exposed to moisture. Due to its brittleness, such flaws cause a dramatic reduction in the strength of a glass object during its first few hours at room temperature.

Corrosion of Polymers

Polymer degradation involves several complex and often poorly understood physiochemical processes. These are strikingly different from the other processes discussed here, and so the term “corrosion” is only applied to them in a loose sense of the word. Because of their large molecular weight, very little entropy can be gained by mixing a given mass of polymer with another substance, making them generally quite difficult to dissolve. While dissolution is a problem in some polymer applications, it is relatively simple to design against.

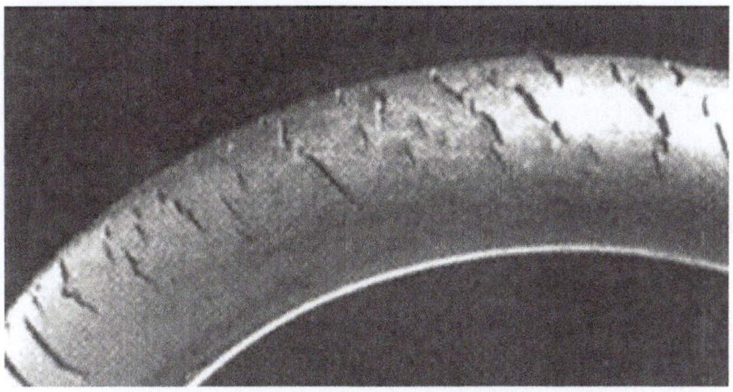

Ozone cracking in natural rubber tubing

A more common and related problem is "swelling", where small molecules infiltrate the structure, reducing strength and stiffness and causing a volume change. Conversely, many polymers (notably flexible vinyl) are intentionally swelled with plasticizers, which can be leached out of the structure, causing brittleness or other undesirable changes.

The most common form of degradation, however, is a decrease in polymer chain length. Mechanisms which break polymer chains are familiar to biologists because of their effect on DNA: ionizing radiation (most commonly ultraviolet light), free radicals, and oxidizers such as oxygen, ozone, and chlorine. Ozone cracking is a well-known problem affecting natural rubber for example. Plastic additives can slow these process very effectively, and can be as simple as a UV-absorbing pigment (e.g. titanium dioxide or carbon black). Plastic shopping bags often do not include these additives so that they break down more easily as ultrafine particles of litter.

Corrosion of Glasses

Glass corrosion

Glass is characterized by a high degree of corrosion-resistance. Because of its high water-resistance it is often used as primary packaging material in the pharma industry since most medicines are preserved in a watery solution. Besides its water-resistance, glass is also very robust when being exposed to chemically aggressive liquids or gases. While other materials like metal or plastics quickly reach their limits, special glass-types can easily hold up.

Glass disease is the corrosion of silicate glasses in aqueous solutions. It is governed by two mechanisms: diffusion-controlled leaching (ion exchange) and hydrolytic dissolution of the glass network. Both mechanisms strongly depend on the pH of contacting solution: the rate of ion exchange decreases with pH as $10^{-0.5pH}$ whereas the rate of hydrolytic dissolution increases with pH as $10^{0.5pH}$.

Mathematically, corrosion rates of glasses are characterized by normalized corrosion rates of elements NR_i (g/cm²·d) which are determined as the ratio of total amount of released species into the water M_i (g) to the water-contacting surface area S (cm²), time of contact t (days) and weight fraction content of the element in the glass f_i:

$$\mathrm{NR}_i = \frac{M_i}{Sf_i t}..$$

The overall corrosion rate is a sum of contributions from both mechanisms (leaching + dissolution) $NR_i = NRx_i + NRh$. Diffusion-controlled leaching (ion exchange) is characteristic of the initial phase of corrosion and involves replacement of alkali ions in the glass by a hydronium (H_3O^+) ion from the solution. It causes an ion-selective depletion of near surface layers of glasses and gives an inverse square root dependence of corrosion rate with exposure time. The diffusion-controlled normalized leaching rate of cations from glasses (g/cm²·d) is given by:

$$\mathrm{NR}x_i = 2\rho\sqrt{\frac{D_i}{\pi t}},,$$

where t is time, D_i is the i-th cation effective diffusion coefficient (cm²/d), which depends on pH of contacting water as $D_i = D_{io} \cdot 10^{-pH}$, and ρ is the density of the glass (g/cm³).

Glass network dissolution is characteristic of the later phases of corrosion and causes a congruent release of ions into the water solution at a time-independent rate in dilute solutions (g/cm²·d):

$$\mathrm{NR}h = \rho r_h,,$$

where r_h is the stationary hydrolysis (dissolution) rate of the glass (cm/d). In closed systems the consumption of protons from the aqueous phase increases the pH and causes a fast transition to hydrolysis. However, a further saturation of solution with silica impedes hydrolysis and causes the glass to return to an ion-exchange, e.g. diffusion-controlled regime of corrosion.

In typical natural conditions normalized corrosion rates of silicate glasses are very low and are of the order of 10^{-7}–10^{-5} g/(cm²·d). The very high durability of silicate glasses in water makes them suitable for hazardous and nuclear waste immobilisation.

Glass Corrosion Tests

There exist numerous standardized procedures for measuring the corrosion (also called chemical durability) of glasses in neutral, basic, and acidic environments, under simulated environmental conditions, in simulated body fluid, at high temperature and pressure, and under other conditions.

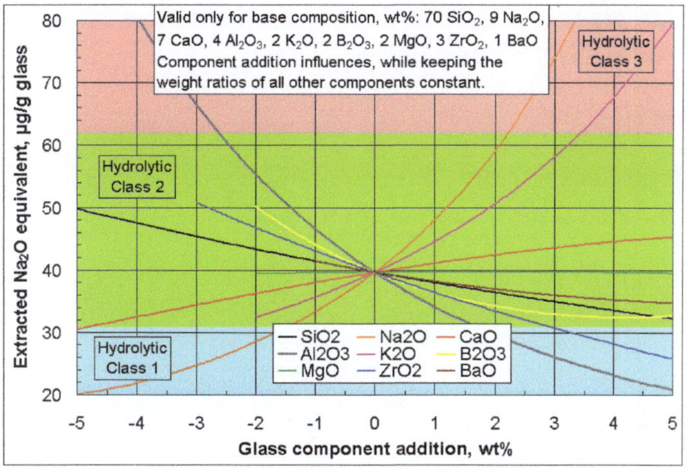

Effect of addition of a certain glass component on the chemical durability against water corrosion of a specific base glass (corrosion test ISO 719).

The standard procedure ISO 719 describes a test of the extraction of water-soluble basic compounds under neutral conditions: 2 g of glass, particle size 300–500 µm, is kept for 60 min in 50 ml deionized water of grade 2 at 98 °C; 25 ml of the obtained solution is titrated against 0.01 mol/l HCl solution. The volume of HCl required for neutralization is classified according to the table below.

Amount of 0.01M HCl needed to neutralize extracted basic oxides, ml	**Extracted Na_2O equivalent, µg**	**Hydrolytic class**
< 0.1	< 31	1
0.1-0.2	31-62	2
0.2-0.85	62-264	3
0.85-2.0	264-620	4
2.0-3.5	620-1085	5
> 3.5	> 1085	> 5

The standardized test ISO 719 is not suitable for glasses with poor or not extractable alkaline components, but which are still attacked by water, e.g. quartz glass, B_2O_3 glass or P_2O_5 glass.

Usual glasses are differentiated into the following classes:

Hydrolytic class 1 (Type I):

This class, which is also called neutral glass, includes borosilicate glasses (e.g. Duran, Pyrex, Fiolax).

Glass of this class contains essential quantities of boron oxides, aluminium oxides and alkaline earth oxides. Through its composition neutral glass has a high resistance against temperature shocks and the highest hydrolytic resistance. Against acid and neutral solutions it shows high chemical resistance, because of its poor alkali content against alkaline solutions.

Hydrolytic class 2 (Type II):

This class usually contains sodium silicate glasses with a high hydrolytic resistance through surface finishing. Sodium silicate glass is a silicate glass, which contains alkali- and alkaline earth oxide and primarily sodium oxide and Calcium oxide.

Hydrolytic class 3 (Type III):

Glass of the 3rd hydrolytic class usually contains sodium silicate glasses and has a mean hydrolytic resistance, which is two times poorer than of type 1 glasses.

Acid class DIN 12116 and alkali class DIN 52322 (ISO 695) are to be distinguished from the hydrolytic class DIN 12111 (ISO 719).

Glass Transition

The glass–liquid transition or glass transition for short is the reversible transition in amorphous materials (or in amorphous regions within semicrystalline materials) from a hard and relatively brittle "glassy" state into a viscous or rubbery state as the temperature is increased. An amorphous solid that exhibits a glass transition is called a glass. The reverse transition, achieved by supercooling a viscous liquid into the glass state, is called vitrification.

The glass-transition temperature T_g of a material characterizes the range of temperatures over which this glass transition occurs. It is always lower than the melting temperature, T_m, of the crystalline state of the material, if one exists.

Hard plastics like polystyrene and poly(methyl methacrylate) are used well below their glass transition temperatures, that is in their glassy state. Their T_g values are well above room temperature, both at around 100 °C (212 °F). Rubber elastomers like polyisoprene and polyisobutylene are used above their T_g, that is, in the rubbery state, where they are soft and flexible.

Despite the change in the physical properties of a material through its glass transition, the transition is not considered a phase transition; rather it is a phenomenon extending over a range of temperature and defined by one of several conventions. Such conventions include a constant cooling rate (20 kelvins per minute (36 °F/min)) and a viscosity threshold of 10^{12} Pa·s, among others. Upon cooling or heating through this glass-transition range, the material also exhibits a smooth step in the thermal-expansion coefficient and in the specific heat, with the location of these effects again being dependent on the history of the material. The question of whether some phase transition underlies the glass transition is a matter of continuing research.

Introduction

The glass transition of a liquid to a solid-like state may occur with either cooling or compression. The transition comprises a smooth increase in the viscosity of a material by as much as 17 orders of magnitude without any pronounced change in material structure. The consequence of this dramatic increase is a glass exhibiting solid-like mechanical properties on the timescale of practical observation. This transition is in contrast to the freezing or crystallization transition, which is a first-order phase transition in the Ehrenfest classification and involves discontinuities in thermodynamic and dynamic properties such as volume, energy, and viscosity. In many materials that normally undergo a freezing transition, rapid cooling will avoid this phase transition and instead result in a glass transition at some lower temperature. Other materials, such as many polymers, lack a well defined crystalline state and easily form glasses, even upon very slow cooling or compression. The tendency for a material to form a glass while quenched is called glass forming ability. This ability depends on the composition of the material and can be predicted by the rigidity theory.

Below the transition temperature range, the glassy structure does not relax in accordance with the cooling rate used. The expansion coefficient for the glassy state is roughly equivalent to that of the crystalline solid. If slower cooling rates are used, the increased time for structural relaxation (or intermolecular rearrangement) to occur may result in a higher density glass product. Similarly, by annealing (and thus allowing for slow structural relaxation) the glass structure in time approaches an equilibrium density corresponding to the supercooled liquid at this same temperature. T_g is located at the intersection between the cooling curve (volume versus temperature) for the glassy state and the supercooled liquid.

The configuration of the glass in this temperature range changes slowly with time towards the equilibrium structure. The principle of the minimization of the Gibbs free energy provides the thermodynamic driving force necessary for the eventual change. It should be noted here that at somewhat higher temperatures than T_g, the structure corresponding to equilibrium at any temperature is achieved quite rapidly. In contrast, at considerably lower temperatures, the configuration of the glass remains sensibly stable over increasingly extended periods of time.

Thus, the liquid-glass transition is not a transition between states of thermodynamic equilibrium. It is widely believed that the true equilibrium state is always crystalline. Glass is believed to exist in a kinetically locked state, and its entropy, density, and so on, depend on the thermal history. Therefore, the glass transition is primarily a dynamic phenomenon. Time and temperature are interchangeable quantities (to some extent) when dealing with glasses, a fact often expressed in the time–temperature superposition principle. On cooling a liquid, *internal degrees of freedom successively fall out of equilibrium.* However, there is a longstanding debate whether there is an underlying second-order phase transition in the hypothetical limit of infinitely long relaxation times.

Transition Temperature T_g

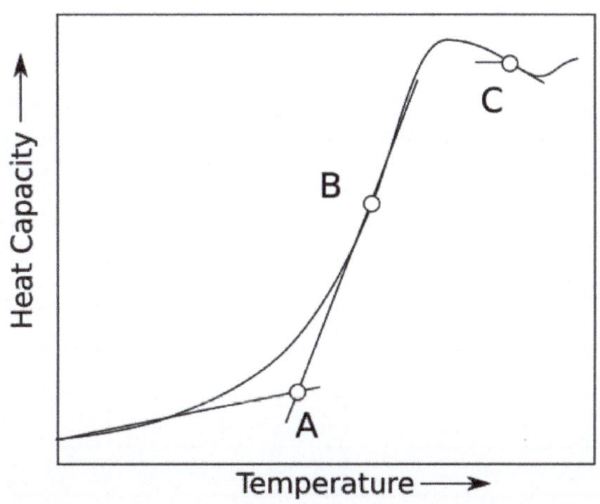

Measurement of T_g (the temperature at the point A) by differential scanning calorimetry

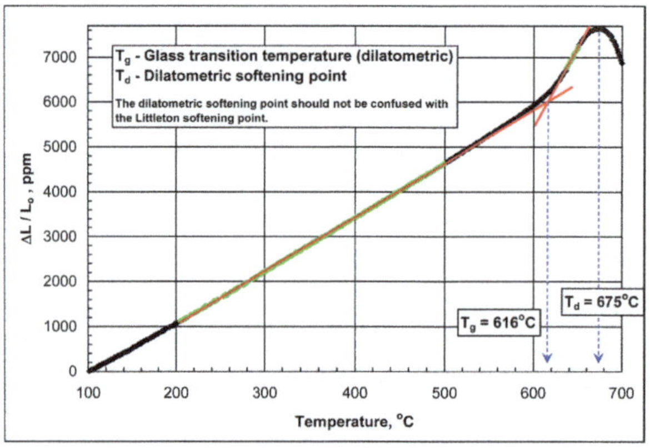

Determination of T_g by dilatometry.

Refer to the figure on the right plotting the heat capacity as a function of temperature. In this context, T_g is the temperature corresponding to point A on the curve. The linear sections below and above T_g are colored green. T_g is the temperature at the intersection of the red regression lines.

Different operational definitions of the glass transition temperature T_g are in use, and several of them are endorsed as accepted scientific standards. Nevertheless, all definitions are arbitrary, and all yield different numeric results: at best, values of T_g for a given substance agree within a few kelvins. One definition refers to the viscosity, fixing T_g at a value of 10^{13} poise (or 10^{12} Pa·s). As evidenced experimentally, this value is close to the annealing point of many glasses.

In contrast to viscosity, the thermal expansion, heat capacity, shear modulus, and many other properties of inorganic glasses show a relatively sudden change at the glass transition temperature. Any such step or kink can be used to define T_g. To make this definition reproducible, the cooling or heating rate must be specified.

The most frequently used definition of T_g uses the energy release on heating in differential scanning calorimetry. Typically, the sample is first cooled with 10 K/min and then heated with that same speed.

Yet another definition of T_g uses the kink in dilatometry (a.k.a. thermal expansion). Here, heating rates of 3–5 K/min (5.4–9.0 °F/min) are common. Summarized below are T_g values characteristic of certain classes of materials.

Polymers

Material	T_g (°C)	T_g (°F)	Commercial name
Tire rubber	−70	−94	
Polyvinylidene fluoride (PVDF)	−35	−31	
Polypropylene (PP atactic)	−20	−4	
Polyvinyl fluoride (PVF)	−20	−4	
Polypropylene (PP isotactic)	0	32	
Poly-3-hydroxybutyrate (PHB)	15	59	

Poly(vinyl acetate) (PVAc)	30	86	
Polychlorotrifluoroethylene (PCTFE)	45	113	
Polyamide (PA)	47–60	117–140	Nylon-6,x
Polylactic acid (PLA)	60–65	140–149	
Polyethylene terephthalate (PET)	70	158	
Poly(vinyl chloride) (PVC)	80	176	
Poly(vinyl alcohol) (PVA)	85	185	
Polystyrene (PS)	95	203	
Poly(methyl methacrylate) (PMMA atactic)	105	221	Plexiglas
Acrylonitrile butadiene styrene (ABS)	105	221	
Polytetrafluoroethylene (PTFE)	115	239	Teflon
Poly(carbonate) (PC)	145	293	Lexan
Polysulfone	185	365	
Polynorbornene	215	419	

Dry nylon-6 has a glass transition temperature of 47 °C (117 °F). Nylon-6,6 in the dry state has a glass transition temperature of about 70 °C (158 °F). Whereas polyethene has a glass transition range of −130 – −80 °C (−202 – −112 °F) The above are only mean values, as the glass transition temperature depends on the cooling rate and molecular weight distribution and could be influenced by additives. For a semi-crystalline material, such as polyethene that is 60–80% crystalline at room temperature, the quoted glass transition refers to what happens to the amorphous part of the material upon cooling.

Silicates and other Covalent Network Glasses

Material	T_g (°C)	T_g (°F)
Chalcogenide GeSbTe	150	302
Chalcogenide AsGeSeTe	245	473
ZBLAN fluoride glass	235	455
Tellurium dioxide	280	536
Fluoroaluminate	400	752
Soda-lime glass	520–600	968–1,112
Fused quartz (approximate)	1,200	2,200

Kauzmann's Paradox

As a liquid is supercooled, the difference in entropy between the liquid and solid phase decreases. By extrapolating the heat capacity of the supercooled liquid below its glass transition temperature, it is possible to calculate the temperature at which the difference in entropies becomes zero. This temperature has been named the Kauzmann temperature.

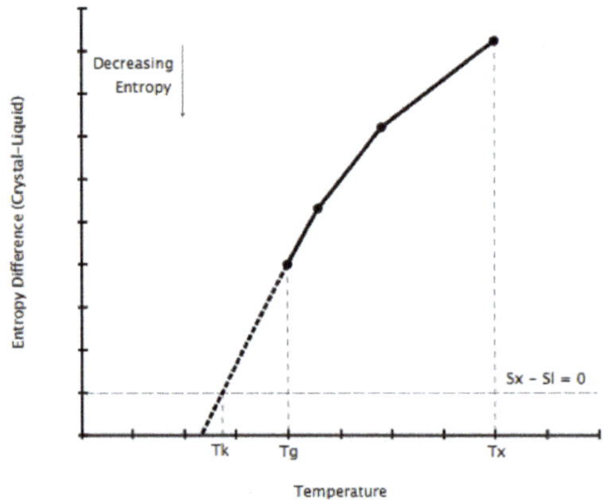

Entropy difference between crystal and undercooled melt

If a liquid could be supercooled below its Kauzmann temperature, and it did indeed display a lower entropy than the crystal phase, the consequences would be paradoxical. This Kauzmann paradox has been the subject of much debate and many publications since it was first put forward by Walter Kauzmann in 1948.

One resolution of the Kauzmann paradox is to say that there must be a phase transition before the entropy of the liquid decreases. In this scenario, the transition temperature is known as the *calorimetric ideal glass transition temperature* T_{oc}. In this view, the glass transition is not merely a kinetic effect, i.e. merely the result of fast cooling of a melt, but there is an underlying thermodynamic basis for glass formation. The glass transition temperature:

$$T_g \to T_{0c} \text{ as } \frac{dT}{dt} \to 0.$$

There are at least three other possible resolutions to the Kauzmann paradox. It could be that the heat capacity of the supercooled liquid near the Kauzmann temperature smoothly decreases to a smaller value. It could also be that a first order phase transition to another liquid state occurs before the Kauzmann temperature with the heat capacity of this new state being less than that obtained by extrapolation from higher temperature. Finally, Kauzmann himself resolved the entropy paradox by postulating that all supercooled liquids must crystallize before the Kauzmann temperature is reached.

The Glass Transition in Specific Materials

Silica, SiO_2

Silica (the chemical compound SiO_2) has a number of distinct crystalline forms in addition to the quartz structure. Nearly all of the crystalline forms involve tetrahedral SiO_4 units linked together by *shared vertices* in different arrangements. Si-O bond lengths vary between the different crystal forms. For example, in α-quartz the bond length is 161 picometres (6.3×10^{-9} in), whereas in α-tridymite it ranges from 154–171 pm (6.1×10^{-9}–6.7×10^{-9} in). The Si-O-Si bond angle also varies from 140° in α-tridymite to 144° in α-quartz to 180° in β-tridymite. Any deviations from these standard parameters constitute microstructural differences or variations that represent an approach to an amorphous, vitreous or glassy solid. The transition temperature T_g in silicates is related to the energy required to break and re-form covalent bonds in an amorphous (or random network) lattice of covalent bonds. The T_g is clearly influenced by the chemistry of the glass. For example, addition of elements such as B, Na, K or Ca to a silica glass, which have a valency less than 4, helps in breaking up the network structure, thus reducing the T_g. Alternatively, P, which has a valency of 5, helps to reinforce an ordered lattice, and thus increases the T_g.

T_g is directly proportional to bond strength, e.g. it depends on quasi-equilibrium thermodynamic parameters of the bonds e.g. on the enthalpy H_d and entropy S_d of configurons – broken bonds: $T_g = H_d / [S_d + R\ln[(1-f_c)/ f_c]$ where R is the gas constant and f_c is the percolation threshold. For strong melts such as SiO_2 the percolation threshold in the above equation is the universal Scher-Zallen critical density in the 3-D space e.g. $f_c = 0.15$, however for fragile materials the percolation thresholds are material-dependent and $f_c << 1$. The enthalpy H_d and the entropy S_d of configurons – broken bonds can be found from available experimental data on viscosity.

In ironing, a fabric is heated through the glass-rubber transition.

Polymers

In polymers the glass transition temperature, T_g, is often expressed as the temperature at which the Gibbs free energy is such that the activation energy for the cooperative movement of 50 or so elements of the polymer is exceeded. This allows molecular chains to slide past each other when a force is applied. From this definition, we can see that the introduction of relatively stiff chemical groups (such as benzene rings) will interfere with the flowing process and hence increase T_g. The stiffness of thermoplastics decreases due to this effect. When the glass temperature has been reached, the stiffness stays the same for a while, i.e., at or near E_2, until the temperature exceeds T_m, and the material melts. This region is called the rubber plateau.

In ironing, a fabric is heated through this transition so that the polymer chains become mobile. The weight of the iron then imposes a preferred orientation. T_g can be significantly decreased by addition of plasticizers into the polymer matrix. Smaller molecules of plasticizer embed themselves between the polymer chains, increasing the spacing and free volume, and allowing them to move past one another even at lower temperatures. The "new-car smell" is due to the initial outgassing of volatile small-molecule plasticizers (most commonly known as phthalates) used to modify interior plastics (e.g., dashboards) to keep them from cracking in the cold of winter weather. The addition of nonreactive side groups to a polymer can also make the chains stand off from one another, reducing T_g. If a plastic with some desirable properties has a T_g that is too high, it can sometimes be combined with another in a copolymer or composite material with a T_g below the temperature of intended use. Note that some plastics are used at high temperatures, e.g., in automobile engines, and others at low temperatures.

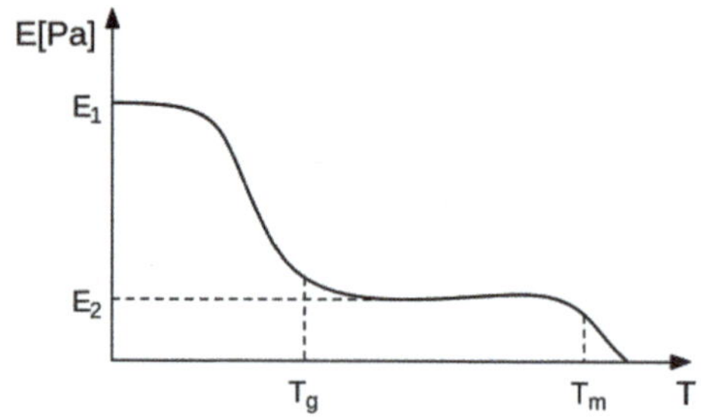

Stiffness versus temperature

In viscoelastic materials, the presence of liquid-like behavior depends on the properties of and so varies with rate of applied load, i.e., how quickly a force is applied. The silicone toy Silly Putty behaves quite differently depending on the time rate of applying a force: pull slowly and it flows, acting as a heavily viscous liquid; hit it with a hammer and it shatters, acting as a glass.

On cooling, rubber undergoes a *liquid-glass transition*, which has also been called a *rubber-glass transition*.

Mechanics of Vitrification

Molecular motion in condensed matter can be represented by a Fourier series whose physical interpretation consists of a superposition of longitudinal and transverse waves of atomic displacement with varying directions and wavelengths. In monatomic systems, these waves are called *density fluctuations*. (In polyatomic systems, they may also include compositional fluctuations.)

Thus, thermal motion in liquids can be decomposed into elementary longitudinal vibrations (or acoustic phonons) while transverse vibrations (or shear waves) were originally described only in elastic solids exhibiting the highly ordered crystalline state of matter. In other words, simple liquids cannot support an applied force in the form of a shearing stress, and will yield mechanically via macroscopic plastic deformation (or viscous flow). Furthermore, the fact that a solid deforms locally while retaining its rigidity – while a liquid yields to macroscopic viscous flow in response to the application of an applied shearing force – is accepted by many as the mechanical distinction between the two.

The inadequacies of this conclusion, however, were pointed out by Frenkel in his revision of the kinetic theory of solids and the theory of elasticity in liquids. This revision follows directly from the continuous characteristic of the structural transition from the liquid state into the solid one when this transition is not accompanied by crystallization—ergo the supercooled viscous liquid. Thus we see the intimate correlation between transverse acoustic phonons (or shear waves) and the onset of rigidity upon vitrification, as described by Bartenev in his mechanical description of the vitrification process.

The velocities of longitudinal acoustic phonons in condensed matter are directly responsible for the thermal conductivity that levels out temperature differentials between compressed and expanded volume elements. Kittel proposed that the behavior of glasses is interpreted in terms of an approximately constant "mean free path" for lattice phonons, and that the value of the mean free path is of the order of magnitude of the scale of disorder in the molecular structure of a liquid or solid. The thermal phonon mean free paths or relaxation lengths of a number of glass formers have been plotted versus the glass transition temperature, indicating a linear relationship between the two. This has suggested a new criterion for glass formation based on the value of the phonon mean free path.

It has often been suggested that heat transport in dielectric solids occurs through elastic vibrations of the lattice, and that this transport is limited by elastic scattering of acoustic phonons by lattice defects (e.g. randomly spaced vacancies). These predictions were confirmed by experiments on commercial glasses and glass ceramics, where mean free paths were apparently limited by "internal boundary scattering" to length scales of 10–100 micrometres (0.00039–0.00394 in). The relationship between these transverse waves and the mechanism of vitrification has been described by several authors who proposed that the onset of correlations between such phonons results in an orientational ordering or "freezing" of local shear stresses in glass-forming liquids, thus yielding the glass transition.

Electronic Structure

The influence of thermal phonons and their interaction with electronic structure is a topic that was appropriately introduced in a discussion of the resistance of liquid metals. Lindemann's theory of melting is referenced, and it is suggested that the drop in conductivity in going from the crystalline to the liquid state is due to the increased scattering of conduction electrons as a result of the increased amplitude of atomic vibration. Such theories of localization have been applied to transport in metallic glasses, where the mean free path of the electrons is very small (on the order of the interatomic spacing).

The formation of a non-crystalline form of a gold-silicon alloy by the method of splat quenching from the melt led to further considerations of the influence of electronic structure on glass forming ability, based on the properties of the metallic bond.

Other work indicates that the mobility of localized electrons is enhanced by the presence of dynamic phonon modes. One claim against such a model is that if chemical bonds are important, the nearly free electron models should not be applicable. However, if the model includes the buildup of a charge distribution between all pairs of atoms just like a chemical bond (e.g., silicon, when a band is just filled with electrons) then it should apply to solids.

Thus, if the electrical conductivity is low, the mean free path of the electrons is very short. The electrons will only be sensitive to the short-range order in the glass since they do not get a chance to scatter from atoms spaced at large distances. Since the short-range order is similar in glasses and crystals, the electronic energies should be similar in these two states. For alloys with lower resistivity and longer electronic mean free paths, the electrons could begin to sense that there is disorder in the glass, and this would raise their energies and destabilize the glass with respect to crystallization. Thus, the glass formation tendencies of certain alloys may therefore be due in part to the fact that the electron mean free paths are very short, so that only the short-range order is ever important for the energy of the electrons.

It has also been argued that glass formation in metallic systems is related to the "softness" of the interaction potential between unlike atoms. Some authors, emphasizing the strong similarities between the local structure of the glass and the corresponding crystal, suggest that chemical bonding helps to stabilize the amorphous structure.

Other authors have suggested that the electronic structure yields its influence on glass formation through the directional properties of bonds. Non-crystallinity is thus favored in elements with a large number of polymorphic forms and a high degree of bonding anisotropy. Crystallization becomes more unlikely as bonding anisotropy is increased from isotropic metallic to anisotropic metallic to covalent bonding, thus suggesting a relationship between the group number in the periodic table and the glass forming ability in elemental solids.

Thermal Expansion

Expansion joint in a road bridge used to avoid damage from thermal expansion.

Thermal expansion is the tendency of matter to change in shape, area, and volume in response to a change in temperature, through heat transfer.

Temperature is a monotonic function of the average molecular kinetic energy of a substance. When a substance is heated, the kinetic energy of its molecules increases. Thus, the molecules begin moving more and usually maintain a greater average separation. Materials which contract with increasing temperature are unusual; this effect is limited in size, and only occurs within limited temperature ranges. The degree of expansion divided by the change in temperature is called the material's coefficient of thermal expansion and generally varies with temperature.

Overview

Predicting Expansion

If an equation of state is available, it can be used to predict the values of the thermal expansion at all the required temperatures and pressures, along with many other state functions.

Contraction Effects (Negative Thermal Expansion)

A number of materials contract on heating within certain temperature ranges; this is usually called negative thermal expansion, rather than "thermal contraction". For example, the coefficient of thermal expansion of water drops to zero as it is cooled to 3.983 °C and then becomes negative below this temperature; this means that water has a maximum density at this temperature, and this leads to bodies of water maintaining this temperature at their lower depths during extended periods of sub-zero weather. Also, fairly pure silicon has a negative coefficient of thermal expansion for temperatures between about 18 and 120 Kelvin.

Factors Affecting Thermal Expansion

Unlike gases or liquids, solid materials tend to keep their shape when undergoing thermal expansion.

Thermal expansion generally decreases with increasing bond energy, which also has an effect on the melting point of solids, so, high melting point materials are more likely to have lower thermal expansion. In general, liquids expand slightly more than solids. The thermal expansion of glasses is higher compared to that of crystals. At the glass transition temperature, rearrangements that occur in an amorphous material lead to characteristic discontinuities of coefficient of thermal expansion or specific heat. These discontinuities allow detection of the glass transition temperature where a supercooled liquid transforms to a glass.

Absorption or desorption of water (or other solvents) can change the size of many common materials; many organic materials change size much more due to this effect than they do to thermal expansion. Common plastics exposed to water can, in the long term, expand by many percent.

Coefficient of Thermal Expansion

The coefficient of thermal expansion describes how the size of an object changes with a change in temperature. Specifically, it measures the fractional change in size per degree change in temperature at a constant pressure. Several types of coefficients have been developed: volumetric, area, and linear. Which is used depends on the particular application and which dimensions are considered important. For solids, one might only be concerned with the change along a length, or over some area.

The volumetric thermal expansion coefficient is the most basic thermal expansion coefficient, and the most relevant for fluids. In general, substances expand or contract when their temperature changes, with expansion or contraction occurring in all directions. Substances that expand at the same rate in every direction are called isotropic. For isotropic materials, the area and volumetric thermal expansion coefficient are, respectively, approximately twice and three times larger than the linear thermal expansion coefficient.

Mathematical definitions of these coefficients are defined below for solids, liquids, and gases.

General Volumetric Thermal Expansion Coefficient

In the general case of a gas, liquid, or solid, the volumetric coefficient of thermal expansion is given by

$$\alpha_V = \frac{1}{V}\left(\frac{\partial V}{\partial T}\right)_p$$

The subscript p indicates that the pressure is held constant during the expansion, and the subscript V stresses that it is the volumetric (not linear) expansion that enters this general definition. In the case of a gas, the fact that the pressure is held constant is important, because the volume of a gas will vary appreciably with pressure as well as temperature. For a gas of low density this can be seen from the ideal gas law.

Expansion in Solids

When calculating thermal expansion it is necessary to consider whether the body is free to expand or is constrained. If the body is free to expand, the expansion or strain resulting from an increase in temperature can be simply calculated by using the applicable coefficient of thermal expansion.

If the body is constrained so that it cannot expand, then internal stress will be caused (or changed) by a change in temperature. This stress can be calculated by considering the strain that would occur if the body were free to expand and the stress required to reduce that strain to zero, through the stress/strain relationship characterised by the elastic or Young's modulus. In the special case of solid materials, external ambient pressure does not usually appreciably affect the size of an object and so it is not usually necessary to consider the effect of pressure changes.

Common engineering solids usually have coefficients of thermal expansion that do not vary significantly over the range of temperatures where they are designed to be used, so where extremely high accuracy is not required, practical calculations can be based on a constant, average, value of the coefficient of expansion.

Linear Expansion

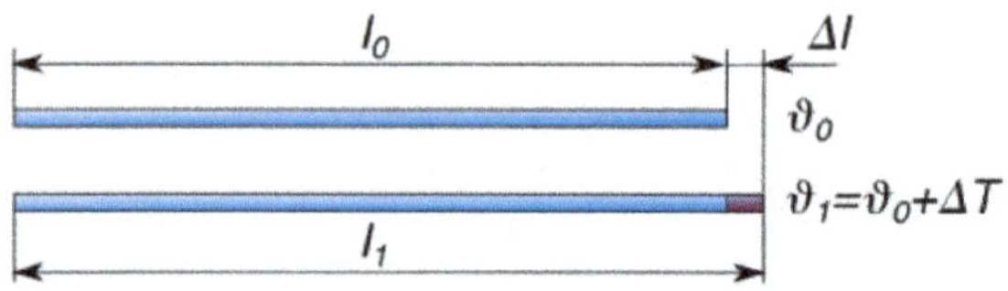

Change in length of a rod due to thermal expansion.

Linear expansion means change in one dimension (length) as opposed to change in volume (volumetric expansion). To a first approximation, the change in length measurements of an object due to thermal expansion is related to temperature change by a "linear expansion coefficient". It is the fractional change in length per degree of temperature change. Assuming negligible effect of pressure, we may write:

$$\alpha_L = \frac{1}{L}\frac{dL}{dT}$$

where L is a particular length measurement and $dL >$ is the rate of change of that linear dimension per unit change in temperature.

The change in the linear dimension can be estimated to be:

$$\frac{\Delta L}{L} = \alpha_L \Delta T$$

This equation works well as long as the linear-expansion coefficient does not change much over the change in temperature $\ddot{A}T$, and the fractional change in length is small $\Delta L / L \ll 1$.. If either of these conditions does not hold, the equation must be integrated.

Effects on Strain

For solid materials with a significant length, like rods or cables, an estimate of the amount of thermal expansion can be described by the material strain, given by $\grave{o}_{\text{thermal}}$ and defined as:

$$\epsilon_{\text{thermal}} = \frac{(L_{\text{final}} - L_{\text{initial}})}{L_{\text{initial}}}$$

where L_{initial} is the length before the change of temperature and L_{final} is the length after the change of temperature.

For most solids, thermal expansion is proportional to the change in temperature:

$$\epsilon_{\text{thermal}} \propto \Delta T$$

Thus, the change in either the strain or temperature can be estimated by:

$$\epsilon_{\text{thermal}} = \alpha_L \Delta T$$

where

$$\Delta \mathrm{T} = (\mathrm{T}_{final} - \mathrm{T}_{initial})$$

is the difference of the temperature between the two recorded strains, measured in degrees Celsius or Kelvin, and α_L is the linear coefficient of thermal expansion in "per degree Celsius" or "per Kelvin", denoted by $°C^{-1}$ or K^{-1}, respectively. In the field of continuum mechanics, the thermal expansion and its effects are treated as eigenstrain and eigenstress.

Area Expansion

The area thermal expansion coefficient relates the change in a material's area dimensions to a change in temperature. It is the fractional change in area per degree of temperature change. Ignoring pressure, we may write:

$$\alpha_A = \frac{1}{A}\frac{dA}{dT}$$

where A is some area of interest on the object, and $dA >$ is the rate of change of that area per unit change in temperature.

The change in the area can be estimated as:

$$\frac{\Delta A}{A} = \alpha_A \Delta T$$

This equation works well as long as the area expansion coefficient does not change much over the change in temperature $\ddot{A}T$, and the fractional change in area is small $\ddot{A}A / A \ll 1$. If either of these conditions does not hold, the equation must be integrated.

Volume Expansion

For a solid, we can ignore the effects of pressure on the material, and the volumetric thermal expansion coefficient can be written:

$$\alpha_V = \frac{1}{V}\frac{dV}{dT}$$

where V is the volume of the material, and $dV >$ is the rate of change of that volume with temperature.

This means that the volume of a material changes by some fixed fractional amount. For example, a steel block with a volume of 1 cubic meter might expand to 1.002 cubic meters when the temperature is raised by 50 K. This is an expansion of 0.2%. If we had a block of steel with a volume of 2 cubic meters, then under the same conditions, it would expand to 2.004 cubic meters, again an expansion of 0.2%. The volumetric expansion coefficient would be 0.2% for 50 K, or 0.004% K^{-1}.

If we already know the expansion coefficient, then we can calculate the change in volume

$$\frac{\Delta V}{V} = \alpha_V \Delta T$$

where $\Delta V / V$ is the fractional change in volume (e.g., 0.002) and ΔT is the change in temperature (50 °C).

The above example assumes that the expansion coefficient did not change as the temperature changed and the increase in volume is small compared to the original volume. This is not always true, but for small changes in temperature, it is a good approximation. If the volumetric expansion coefficient does change appreciably with temperature, or the increase in volume is significant, then the above equation will have to be integrated:

$$\ln\left(\frac{V + \Delta V}{V}\right) = \int_{T_i}^{T_f} \alpha_V(T)dT$$

$$\frac{\Delta V}{V} = \exp\left(\int_{T_i}^{T_f} \alpha_V(T)dT\right) - 1$$

where $\alpha_V(T)$ is the volumetric expansion coefficient as a function of temperature T, and T_i, T_f are the initial and final temperatures respectively.

Isotropic Materials

For isotropic materials the volumetric thermal expansion coefficient is three times the linear coefficient:

$$\alpha_V = 3\alpha_L$$

This ratio arises because volume is composed of three mutually orthogonal directions. Thus, in an isotropic material, for small differential changes, one-third of the volumetric expansion is in a single axis. As an example, take a cube of steel that has sides of length L. The original volume will be $V = L^3$ and the new volume, after a temperature increase, will be

$$V + \Delta V = (L + \Delta L)^3 = L^3 + 3L^2\Delta L + 3L\Delta L^2 + \Delta L^3 \approx L^3 + 3L^2\Delta L = V + 3V\frac{\Delta L}{L}$$

We can make the substitutions $\Delta V = \alpha_V L^3 \Delta T$ and, for isotropic materials, $\Delta L = \alpha_L L \Delta T$. We now have:

$$V + \Delta V = (L + L\alpha_L \Delta T)^3 = L^3 + 3L^3\alpha_L\Delta T + 3L^3\alpha_L^2\Delta T^2 + L^3\alpha_L^3\Delta T^3 \approx L^3 + 3L^3\alpha_L\Delta T$$

Since the volumetric and linear coefficients are defined only for extremely small temperature and dimensional changes (that is, when ΔT and ÄL are small), the last two terms can be ignored and we get the above relationship between the two coefficients. If we are trying to go back and forth between volumetric and linear coefficients using larger values of ΔT then we will need to take into account the third term, and sometimes even the fourth term.

Similarly, the area thermal expansion coefficient is two times the linear coefficient:

$$\alpha_A = 2\alpha_L$$

This ratio can be found in a way similar to that in the linear example above, noting that the area of a face on the cube is just L^2. Also, the same considerations must be made when dealing with large values of ΔT.

Anisotropic Materials

Materials with anisotropic structures, such as crystals (with less than cubic symmetry) and many composites, will generally have different linear expansion coefficients $-\alpha_L$ in different directions. As a result, the total volumetric expansion is distributed unequally among the three axes. If the crystal symmetry is monoclinic or triclinic, even the angles between these axes are subject to thermal changes. In such cases it is necessary to treat the coefficient of thermal expansion as a tensor with up to six independent elements. A good way to determine the elements of the tensor is to study the expansion by powder diffraction.

Isobaric Expansion in Gases

For an ideal gas, the volumetric thermal expansion (i.e., relative change in volume due to temperature change) depends on the type of process in which temperature is changed. Two simple cases are isobaric change, where pressure is held constant, and adiabatic change, where no heat is exchanged with the environment.

The ideal gas law can be written as:

$$pv = T$$

where p is the pressure, v is the specific volume, and t is temperature measured in *energy units*. By taking the logarithm of this equation:

$$\ln(v) + \ln(p) = \ln(T)$$

Then by definition of isobaric thermal expansion coefficient, with the above equation of state:

$$\gamma_p \equiv \frac{1}{v}\left(\frac{\partial v}{\partial T}\right)_p = \left(\frac{d(\ln v)}{dT}\right)_p = \frac{d(\ln T)}{dT} = \frac{1}{T}.$$

The index p denotes an isobaric process.

Expansion in Liquids

Theoretically, the coefficient of linear expansion can be found from the coefficient of volumetric expansion ($\alpha_V \approx 3\alpha$). However, for liquids, α is calculated through the experimental determination of α_V.

Liquids in general, expand on heating . but water is an exception to this general behaviour . it contracts from 0°C-4°C on heating. the thermal expansions of liquid is more compared to solids because of weak intermolecular forces present in liquids. Expansion of liquid is different from the expansions of solids in the following ways.

→ Liquids, unlike solids have no definite shape and they take the shape of the container. consequently, liquids have no definite length and area, so linear and areal expansions of liquids have no significance.

→ Expansion of solid of almost uniform in any temperature change whereas liquids expand at different rates at different temperature.

→ Solids can be heated directly but liquids cannot be heated directly. when a liquid expand in a vessel, the vessel expands along with the liquid. hence the observed increase in volume of the liquid is not actual increase in its volume. the expansion of the liquid relative to the container is called as *Apparent Expansion*. the actual expansion of the liquid is called *Real (or) Absolute expansion.*

For small and same rise in temperature . increase in volume (real expansion) of liquid is equal to the sum of apparent increase in volume (apparent expansion) of liquid and the increase in volume of the containing vessel. Thus a liquid has to coefficients of expansion

Coefficient of Apparent Expansion of a Liquid

Expansion in Mixtures and Alloys

The expansivity of the components of the mixture can cancel each other like in invar.

The thermal expansivity of a mixture from the expansivities of the pure components and their excess expansivities follow from:

$$\frac{\partial V}{\partial T} = \sum_i \frac{\partial V_i}{\partial T} + \sum_i \frac{\partial V_i^E}{\partial T}$$

$$\alpha = \sum_i \alpha_i V_i + \sum_i \alpha_i^E V_i^E$$

$$\frac{\partial \overline{V^E}_i}{\partial T} = R\frac{\partial(ln(\gamma_i))}{\partial P} + RT\frac{\partial^2}{\partial T \partial P} ln(\gamma_i)$$

Apparent and Absolute Expansion

When measuring the expansion of a liquid, the measurement must account for the expansion of the container as well. For example, a flask that has been constructed with a long narrow stem filled with enough liquid that the stem itself is partially filled, when placed in a heat bath will initially show the column of liquid in the stem to drop followed by the immediate increase of that column until the flask-liquid-heat bath system has thermalized. The initial observation of the column of liquid dropping is not due to an initial contraction of the liquid but rather the expansion of the flask as it contacts the heat bath first. Soon after, the liquid in the flask is heated by the flask itself and begins to expand. Since liquids typically have a greater expansion over solids, the liquid in the flask eventually exceeds that of the flask, causing the column of liquid in the flask to rise. A direct measurement of the height of the liquid column is a measurement of the apparent expansion of the liquid. The absolute expansion of the liquid is the apparent expansion corrected for the expansion of the containing vessel.

Examples and Applications

The expansion and contraction of materials must be considered when designing large structures, when using tape or chain to measure distances for land surveys, when designing molds for casting hot material, and in other engineering applications when large changes in dimension due to temperature are expected.

Thermal expansion of long continuous sections of rail tracks is the driving force for rail buckling. This phenomenon resulted in 190 train derailments during 1998–2002 in the US alone.

Thermal expansion is also used in mechanical applications to fit parts over one another, e.g. a bushing can be fitted over a shaft by making its inner diameter slightly smaller than the diameter of the shaft, then heating it until it fits over the shaft, and allowing it to cool after it has been pushed over the shaft, thus achieving a 'shrink fit'. Induction shrink fitting is a common industrial method to pre-heat metal components between 150 °C and 300 °C thereby causing them to expand and allow for the insertion or removal of another component.

There exist some alloys with a very small linear expansion coefficient, used in applications that demand very small changes in physical dimension over a range of temperatures. One of these is Invar 36, with α approximately equal to 0.6×10^{-6} K^{-1}. These alloys are useful in aerospace applications where wide temperature swings may occur.

Pullinger's apparatus is used to determine the linear expansion of a metallic rod in the laboratory. The apparatus consists of a metal cylinder closed at both ends (called a steam jacket). It is provided with an inlet and outlet for the steam. The steam for heating the rod is supplied by a boiler which is connected by a rubber tube to the inlet. The center of the cylinder contains a hole to insert a thermometer. The rod under investigation is enclosed in a steam jacket. One of its ends is free, but the other end is pressed against a fixed screw. The position of the rod is determined by a micrometer screw gauge or spherometer.

Drinking glass with fracture due to uneven thermal expansion after pouring of hot liquid into the otherwise cool glass

The control of thermal expansion in brittle materials is a key concern for a wide range of reasons. For example, both glass and ceramics are brittle and uneven temperature causes uneven expansion which again causes thermal stress and this might lead to fracture. Ceramics need to be joined or work in consort with a wide range of materials and therefore their expansion must be matched to the application. Because glazes need to be firmly attached to the underlying porcelain (or other body type) their thermal expansion must be tuned to 'fit' the body so that crazing or shivering do not occur. Good example of products whose thermal expansion is the key to their success are CorningWare and the spark plug. The thermal expansion of ceramic bodies can be controlled by firing to create crystalline species that will influence the overall expansion of the material in the desired direction. In addition or instead the formulation of the body can employ materials delivering particles of the desired expansion to the matrix. The thermal expansion of glazes is controlled by their chemical composition and the firing schedule to which they were subjected. In most cases there are complex issues involved in controlling body and glaze expansion, adjusting for thermal expansion must be done with an eye to other properties that will be affected, generally trade-offs are required.

Thermal expansion can have a noticeable effect in gasoline stored in above ground storage tanks which can cause gasoline pumps to dispense gasoline which may be more compressed than gasoline held in underground storage tanks in the winter time or less compressed than gasoline held in underground storage tanks in the summer time.

Heat-induced expansion has to be taken into account in most areas of engineering. A few examples are:

- Metal framed windows need rubber spacers
- Rubber tires
- Metal hot water heating pipes should not be used in long straight lengths
- Large structures such as railways and bridges need expansion joints in the structures to avoid sun kink
- One of the reasons for the poor performance of cold car engines is that parts have inefficiently large spacings until the normal operating temperature is achieved.
- A gridiron pendulum uses an arrangement of different metals to maintain a more temperature stable pendulum length.
- A power line on a hot day is droopy, but on a cold day it is tight. This is because the metals expand under heat.
- Expansion joints that absorb the thermal expansion in a piping system.
- Precision engineering nearly always requires the engineer to pay attention to the thermal expansion of the product. For example, when using a scanning electron microscope even small changes in temperature such as 1 degree can cause a sample to change its position relative to the focus point.

Thermometers are another application of thermal expansion – most contain a liquid (usually mercury or alcohol) which is constrained to flow in only one direction (along the tube) due to changes in volume brought about by changes in temperature. A bi-metal mechanical thermometer uses a bimetallic strip and bends due to the differing thermal expansion of the two metals.

Metal pipes made of different materials are heated by passing steam through them. While each pipe is being tested, one end is securely fixed and the other rests on a rotating shaft, the motion of which is indicated with a pointer. The linear expansion of the different metals is compared qualitatively and the coefficient of linear thermal expansion is calculated.

Thermal Expansion Coefficients for Various Materials

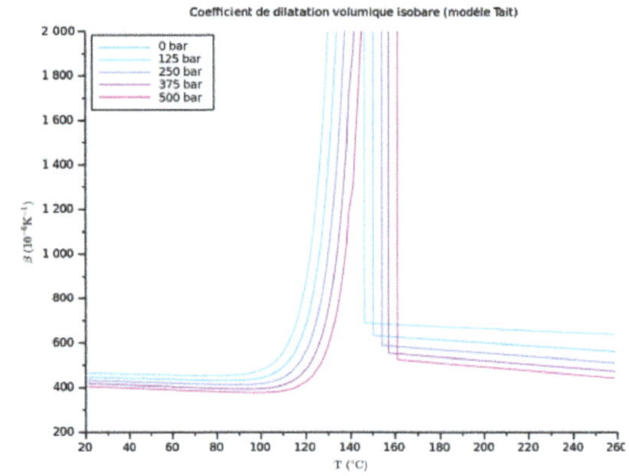

Volumetric thermal expansion coefficient for a semicrystalline polypropylene.

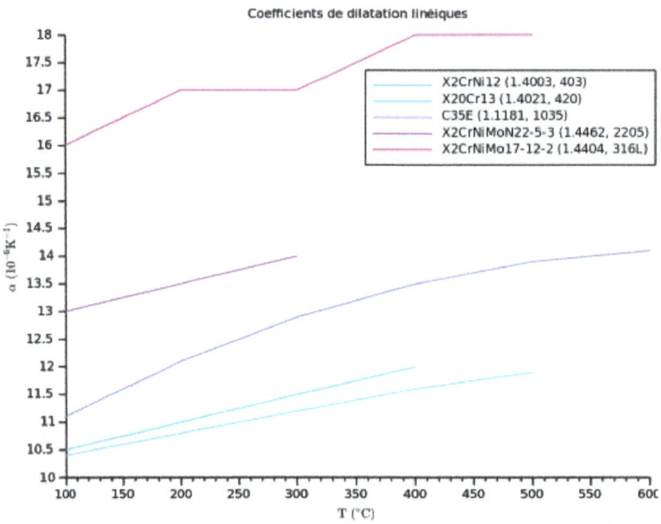

Linear thermal expansion coefficient for some steel grades.

For isotropic materials the coefficients linear thermal expansion α and volumetric thermal expansion α_V are related by $\alpha_V = 3\alpha$. For liquids usually the coefficient of volumetric expansion is listed and linear expansion is calculated here for comparison.

For common materials like many metals and compounds, the thermal expansion coefficient is inversely proportional to the melting point. In particular for metals the relation is:

$$\alpha \approx \frac{0.020}{M_P}$$

for halides and oxides

$$\alpha \approx \frac{0.038}{M_P} - 7.0 \cdot 10^{-6} \mathrm{K}^{-1}$$

In the table below, the range for α is from 10^{-7} K^{-1} for hard solids to 10^{-3} K^{-1} for organic liquids. The coefficient α varies with the temperature and some materials have a very high variation ; see for example the variation vs. temperature of the volumetric coefficient for a semicrystalline polypropylene (PP) at different pressure, and the variation of the linear coefficient vs. temperature for some steel grades (from bottom to top: ferritic stainless steel, martensitic stainless steel, carbon steel, duplex stainless steel, austenitic steel).

(The formula $\alpha_V \approx 3\alpha$ is usually used for solids.)

Material	Linear coefficient α at 20 °C (10^{-6} K^{-1})	Volumetric coefficient α_V at 20 °C (10^{-6} K^{-1})	Notes
Aluminium	23.1	69	
Aluminium nitride	5.3	4.2	
Benzocyclobutene	42	126	
Brass	19	57	
Carbon steel	10.8	32.4	
CFRP	– 0.8	Anisotropic	Fiber direction
Concrete	12	36	
Copper	17	51	
Diamond	1	3	
Ethanol	250	750	
Gallium(III) arsenide	5.8	17.4	
Gasoline	317	950	
Glass	8.5	25.5	
Glass, borosilicate	3.3	9.9	matched sealing partner for tungsten, molybdenum and kovar.
Glass (Pyrex)	3.2		
Glycerine		485	
Gold	14	42	
Helium		36.65	
Indium phosphide	4.6	13.8	
Invar	1.2	3.6	
Iron	11.8	33.3	
Kapton	20	60	DuPont Kapton 200EN
Lead	29	87	
Macor	9.3		
Magnesium	26	78	
Mercury	61	182	
Molybdenum	4.8	14.4	

Nickel	13	39	
Oak	54		Perpendicular to the grain
Douglas-fir	27	75	radial
Douglas-fir	45	75	tangential
Douglas-fir	3.5	75	parallel to grain
Platinum	9	27	
PP	150	450	
PVC	52	156	
Quartz (fused)	0.59	1.77	
Quartz	0.33	1	
Rubber	*disputed*	*disputed*	
Sapphire	5.3		Parallel to C axis, or
Silicon Carbide	2.77	8.31	
Silicon	2.56	9	
Silver	18	54	
Sitall	0±0.15	0±0.45	average for −60 °C to 60 °C
Stainless steel	10.1 ~ 17.3	30.3 ~ 51.9	
Steel	11.0 ~ 13.0	33.0 ~ 39.0	Depends on composition
Titanium	8.6	26	
Tungsten	4.5	13.5	
Turpentine		90	
Water	69	207	
YbGaGe	≐0	≐0	Refuted
Zerodur	≈0.02		at 0...50 °C

Density

The density, or more precisely, the volumetric mass density, of a substance is its mass per unit volume. The symbol most often used for density is ρ (the lower case Greek letter rho), although the Latin letter D can also be used. Mathematically, density is defined as mass divided by volume:

Density	
Common symbols	ρ D
SI unit	kg/m^3

where ρ is the density, m is the mass, and V is the volume. In some cases (for instance, in the United States oil and gas industry), density is loosely defined as its weight per unit volume, although this is scientifically inaccurate – this quantity is more specifically called specific weight.

For a pure substance the density has the same numerical value as its mass concentration. Different materials usually have different densities, and density may be relevant to buoyancy, purity and packaging. Osmium and iridium are the densest known elements at standard conditions for temperature and pressure but certain chemical compounds may be denser.

A graduated cylinder containing various coloured liquids with different densities.

$$\rho = \frac{m}{V},$$

To simplify comparisons of density across different systems of units, it is sometimes replaced by the dimensionless quantity "relative density" or "specific gravity", i.e. the ratio of the density of the material to that of a standard material, usually water. Thus a relative density less than one means that the substance floats in water.

The density of a material varies with temperature and pressure. This variation is typically small for solids and liquids but much greater for gases. Increasing the pressure on an object decreases the volume of the object and thus increases its density. Increasing the temperature of a substance (with a few exceptions) decreases its density by increasing its volume. In most materials, heating the bottom of a fluid results in convection of the heat from the bottom to the top, due to the decrease in the density of the heated fluid. This causes it to rise relative to more dense unheated material.

The reciprocal of the density of a substance is occasionally called its specific volume, a term sometimes used in thermodynamics. Density is an intensive property in that increasing the amount of a substance does not increase its density; rather it increases its mass.

History

In a well-known but probably apocryphal tale, Archimedes was given the task of determining whether King Hiero's goldsmith was embezzling gold during the manufacture of a golden wreath dedicated to the gods and replacing it with another, cheaper alloy. Archimedes knew that the irregularly shaped wreath could be crushed into a cube whose volume could be calculated easily and compared with the mass; but the king did not approve of this. Baffled, Archimedes is said to have taken an immersion bath and observed from the rise of the water upon entering that he could calculate the volume of the gold wreath through the displacement of the water. Upon this discovery, he leapt from his bath and ran naked through the streets shouting, "Eureka! Eureka!" (Εὕρηκα! Greek "I have found it"). As a result, the term "eureka" entered common parlance and is used today to indicate a moment of enlightenment.

The story first appeared in written form in Vitruvius' *books of architecture*, two centuries after it supposedly took place. Some scholars have doubted the accuracy of this tale, saying among other things that the method would have required precise measurements that would have been difficult to make at the time.

From the equation for density ($\rho = m / V$), mass density has units of mass divided by volume. As there are many units of mass and volume covering many different magnitudes there are a large number of units for mass density in use. The SI unit of kilogram per cubic metre (kg/m^3) and the cgs unit of gram per cubic centimetre (g/cm^3) are probably the most commonly used units for density.1,000 kg/m^3 equals 1 g/cm^3. (The cubic centimeter can be alternately called a *millilitre* or a *cc*.) In industry, other larger or smaller units of mass and or volume are often more practical and US customary units may be used.

Measurement of Density

Homogeneous Materials

The density at all points of a homogeneous object equals its total mass divided by its total volume. The mass is normally measured with a scale or balance; the volume may be measured directly (from the geometry of the object) or by the displacement of a fluid. To determine the density of a liquid or a gas, a hydrometer, a dasymeter or a Coriolis flow meter may be used, respectively. Similarly, hydrostatic weighing uses the displacement of water due to a submerged object to determine the density of the object.

Heterogeneous Materials

If the body is not homogeneous, then its density varies between different regions of the object. In that case the density around any given location is determined by calculating the density of a small volume around that location. In the limit of an infinitesimal volume the density of an inhomogeneous object at a point becomes: ρ() = dm/dV, where dV is an elementary volume at position r. The mass of the body then can be expressed as

$$m = \int_V \rho(\vec{r})dV.$$

Non-compact Materials

In practice, bulk materials such as sugar, sand, or snow contain voids. Many materials exist in nature as flakes, pellets, or granules.

Voids are regions which contain something other than the considered material. Commonly the void is air, but it could also be vacuum, liquid, solid, or a different gas or gaseous mixture.

The bulk volume of a material—inclusive of the void fraction—is often obtained by a simple measurement (e.g. with a calibrated measuring cup) or geometrically from known dimensions.

Mass divided by *bulk* volume determines bulk density. This is not the same thing as volumetric mass density.

To determine volumetric mass density, one must first discount the volume of the void fraction. Sometimes this can be determined by geometrical reasoning. For the close-packing of equal spheres the non-void fraction can be at most about 74%. It can also be determined empirically. Some bulk materials, however, such as sand, have a *variable* void fraction which depends on how the material is agitated or poured. It might be loose or compact, with more or less air space depending on handling.

In practice, the void fraction is not necessarily air, or even gaseous. In the case of sand, it could be water, which can be advantageous for measurement as the void fraction for sand saturated in water—once any air bubbles are thoroughly driven out—is potentially more consistent than dry sand measured with an air void.

In the case of non-compact materials, one must also take care in determining the mass of the material sample. If the material is under pressure (commonly ambient air pressure at the earth's surface) the determination of mass from a measured sample weight might need to account for buoyancy effects due to the density of the void constituent, depending on how the measurement was conducted. In the case of dry sand, sand is so much denser than air that the buoyancy effect is commonly neglected (less than one part in one thousand).

Mass change upon displacing one void material with another while maintaining constant volume can be used to estimate the void fraction, if the difference in density of the two voids materials is reliably known.

Changes of Density

In general, density can be changed by changing either the pressure or the temperature. Increasing the pressure always increases the density of a material. Increasing the temperature generally decreases the density, but there are notable exceptions to this generalization. For example, the density of water increases between its melting point at 0 °C and 4 °C; similar behavior is observed in silicon at low temperatures.

The effect of pressure and temperature on the densities of liquids and solids is small. The compressibility for a typical liquid or solid is 10^{-6} bar^{-1} (1 bar = 0.1 MPa) and a typical thermal expansivity is 10^{-5} K^{-1}. This roughly translates into needing around ten thousand times atmospheric pressure to reduce the volume of a substance by one percent. (Although the pressures needed may be around a thousand times smaller for sandy soil and some clays.) A one percent expansion of volume typically requires a temperature increase on the order of thousands of degrees Celsius.

In contrast, the density of gases is strongly affected by pressure. The density of an ideal gas is

$$\rho = \frac{MP}{RT},$$

where M is the molar mass, P is the pressure, R is the universal gas constant, and T is the absolute temperature. This means that the density of an ideal gas can be doubled by doubling the pressure, or by halving the absolute temperature.

In the case of volumic thermal expansion at constant pressure and small intervals of temperature the temperature dependence of density is :

$$\rho = \frac{\rho_{T_0}}{1+\alpha \cdot \Delta T}$$

where ρ_{T_0} is the density at a reference temperature, α is the thermal expansion coefficient of the material at temperatures close to T_0.

Density of Solutions

The density of a solution is the sum of mass (massic) concentrations of the components of that solution.

Mass (massic) concentration of each given component ρ_i in a solution sums to density of the solution.

$$\rho = \sum_i \varrho_i$$

Expressed as a function of the densities of pure components of the mixture and their volume participation, it allows the determination of excess molar volumes:

$$\rho = \sum_i \rho_i \frac{V_i}{V} = \sum_i \rho_i \varphi_i = \sum_i \rho_i \frac{V_i}{\sum_i V_i + \sum_i {V^E}_i}$$

provided that there is no interaction between the components.

Knowing the relation between excess volumes and activity coefficients of the components, one can determine the activity coefficients.

$$\overline{V^E}_i = RT \frac{\partial(ln(\gamma_i))}{\partial P}$$

Densities

Water

Density of liquid water at 1 atm pressure	
Temp. (°C)	**Density (kg/m³)**
−30	983.854
−20	993.547
−10	998.117
0	999.8395
4	999.9720
10	999.7026
15	999.1026
20	998.2071

22	997.7735
25	997.0479
30	995.6502
40	992.2
60	983.2
80	971.8
100	958.4
Notes: 1. Values below 0 °C refer to supercooled water.	

Air

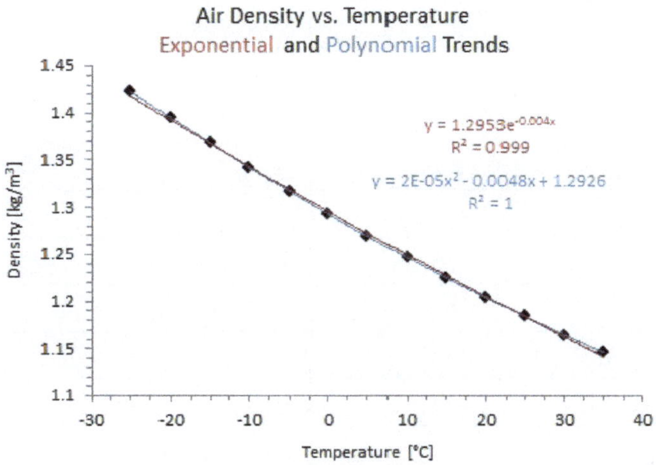

Air density *vs.* temperature

Density of air at 1 atm pressure	
T **(°C)**	***ρ*** **(kg/m³)**
−25	1.423
−20	1.395
−15	1.368
−10	1.342
−5	1.316
0	1.293
5	1.269
10	1.247
15	1.225
20	1.204
25	1.184
30	1.164
35	1.146

Various Materials

Densities of various materials covering a range of values		
Material	**ρ (kg/m³)**	**Notes**
Helium	0.179	
Aerographite	0.2	
Metallic microlattice	0.9	
Aerogel	1.0	
Air	1.2	At sea level
Tungsten hexafluoride	12.4	One of the heaviest known gases at standard conditions
Liquid hydrogen	70	At approx. −255 °C
Styrofoam	75	Approx.
Cork	240	Approx.
Pine	373	
Lithium	535	
Wood	700	Seasoned, typical
Oak	710	
Potassium	860	
Sodium	970	
Ice	916.7	At temperature < 0 °C
Water (fresh)	1,000	At 4 °C, the temperature of its maximum density
Water (salt)	1,030	
Nylon	1,150	
Plastics	1,175	Approx.; for polypropylene and PETE/PVC
Tetrachloroethene	1,622	
Magnesium	1,740	
Beryllium	1,850	
Glycerol	1,261	
Concrete	2,000	
Silicon	2,330	
Aluminium	2,700	
Diiodomethane	3,325	Liquid at room temperature
Diamond	3,500	
Titanium	4,540	
Selenium	4,800	

Vanadium	6,100	
Antimony	6,690	
Zinc	7,000	
Chromium	7,200	
Tin	7,310	
Manganese	7,325	Approx.
Iron	7,870	
Niobium	8,570	
Brass	8,600	
Cadmium	8,650	
Cobalt	8,900	
Nickel	8,900	
Copper	8,940	
Bismuth	9,750	
Molybdenum	10,220	
Silver	10,500	
Lead	11,340	
Thorium	11,700	
Rhodium	12,410	
Mercury	13,546	
Tantalum	16,600	
Uranium	18,800	
Tungsten	19,300	
Gold	19,320	
Plutonium	19,840	
Platinum	21,450	
Iridium	22,420	
Osmium	22,570	

Notes:

1. Unless otherwise noted, all densities given are at standard conditions for temperature and pressure, that is, 273.15 K (0.00 °C) and 100 kPa (0.987 atm).
2. Air contained in material excluded when calculating density

Others

Entity	ρ (kg/m³)	Notes

Interstellar medium	1×10^{-19}	Assuming 90% H, 10% He; variable T
The Earth	5,515	Mean density.
The inner core of the Earth	13,000	Approx., as listed in Earth.
The core of the Sun	33,000–160,000	Approx.
Super-massive black hole	9×10^{5}	Density of a 4.5-million-solar-mass black hole Event horizon radius is 13.5 million km.
White dwarf star	2.1×10^{9}	Approx.
Atomic nuclei	2.3×10^{17}	Does not depend strongly on size of nucleus
Neutron star	1×10^{18}	
Stellar-mass black hole	1×10^{18}	Density of a 4-solar-mass black hole Event horizon radius is 12 km.

Common Units

The SI unit for density is:

- kilograms per cubic meter (kg/m^3)

Litres and metric tons are not part of the SI, but are acceptable for use with it, leading to the following units:

- kilograms per liter (kg/L)
- grams per milliliter (g/mL)
- metric tons per cubic meter (t/m^3)

Densities using the following metric units all have exactly the same numerical value, one thousandth of the value in (kg/m^3). Liquid water has a density of about 1 kg/dm^3, making any of these SI units numerically convenient to use as most solids and liquids have densities between 0.1 and 20 kg/dm^3.

- kilograms per cubic decimetre (kg/dm^3)
- grams per cubic centimetre (g/cm^3)
    - 1 $gram/cm^3$ = 1000 kg/m^3
- megagrams (metric tons) per cubic metre (Mg/m^3)

In US customary units density can be stated in:

- Avoirdupois ounces per cubic inch (1 g/cc ≈ 0.578036672 oz/cu in)
- Avoirdupois ounces per fluid ounce (1 g/cc ≈ 1.04317556 oz/fl. oz = 1.04317556 lbs/pint)
- Avoirdupois pounds per cubic inch (1 g/cc ≈ 0.036127292 lb/cu in)
- pounds per cubic foot (1 g/cc ≈ 62.427961 lb/cu ft)

- pounds per cubic yard (1 g/cc ≈ 1685.5549 lb/cu yd)
- pounds per US liquid gallon (1 g/cc ≈ 8.34540445 lb/gal)
- pounds per US bushel (1 g/cc ≈ 77.6888513 lb/bu)
- slugs per cubic foot

Imperial units differing from the above (as the Imperial gallon and bushel differ from the US units) in practice are rarely used, though found in older documents. The Imperial gallon was based on the concept that an Imperial fluid ounce of water would have a mass of one Avoirdupois ounce, and indeed 1 g/cc ≈ 1.00224129 ounces per Imperial fluid ounce = 10.0224129 lbs per Imperial gallon. The density of precious metals could conceivably be based on Troy ounces and pounds, a possible cause of confusion.

Refractive Index

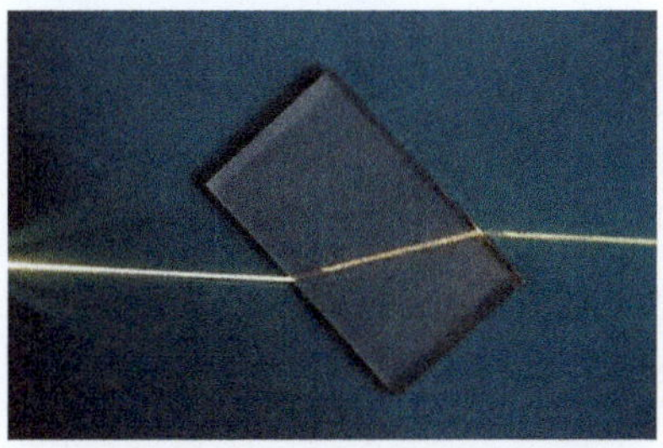

A ray of light being refracted in a plastic block

In optics, the refractive index or index of refraction n of a material is a dimensionless number that describes how light propagates through that medium. It is defined as

$$n = \frac{c}{v},$$

where c is the speed of light in vacuum and v is the phase velocity of light in the medium. For example, the refractive index of water is 1.333, meaning that light travels 1.333 times faster in a vacuum than it does in water.

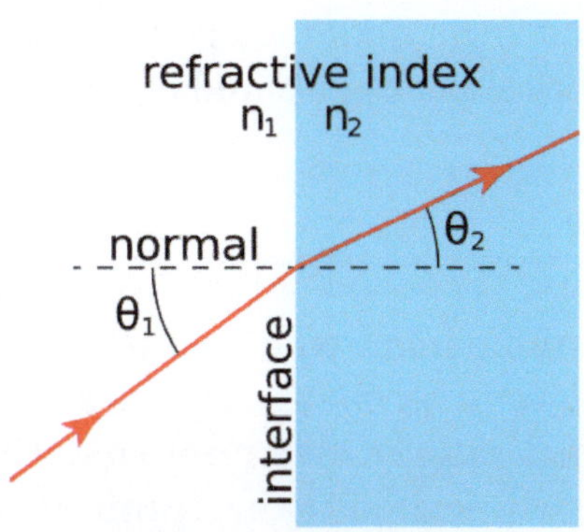

Refraction of a light ray

The refractive index determines how much light is bent, or refracted, when entering a material. This is the first documented use of refractive indices and is described by Snell's law of refraction, $n_1 \sin\theta_1 = n_2 \sin\theta_2$, where θ_1 and θ_2 are the angles of incidence and refraction, respectively, of a ray crossing the interface between two media with refractive indices n_1 and n_2. The refractive indices also determine the amount of light that is reflected when reaching the interface, as well as the critical angle for total internal reflection and Brewster's angle.

The refractive index can be seen as the factor by which the speed and the wavelength of the radiation are reduced with respect to their vacuum values: the speed of light in a medium is $v = c/n$, and similarly the wavelength in that medium is $\lambda = \lambda_0/n$, where λ_0 is the wavelength of that light in vacuum. This implies that vacuum has a refractive index of 1, and that the frequency ($f = v/\lambda$) of the wave is not affected by the refractive index.

The refractive index varies with the wavelength of light. This is called dispersion and causes the splitting of white light into its constituent colors in prisms and rainbows, and chromatic aberration in lenses. Light propagation in absorbing materials can be described using a complex-valued refractive index. The imaginary part then handles the attenuation, while the real part accounts for refraction.

The concept of refractive index is widely used within the full electromagnetic spectrum, from X-rays to radio waves. It can also be used with wave phenomena such as sound. In this case the speed of sound is used instead of that of light and a reference medium other than vacuum must be chosen.

Definition

The refractive index n of an optical medium is defined as the ratio of the speed of light in vacuum, $c = 299792458$ m/s, and the phase velocity v of light in the medium,

$$n = \frac{c}{v}.$$

The phase velocity is the speed at which the crests or the phase of the wave moves, which may be different from the group velocity, the speed at which the pulse of light or the envelope of the wave moves.

The definition above is sometimes referred to as the absolute refractive index or the absolute index of refraction to distinguish it from definitions where the speed of light in other reference media than vacuum is used. Historically air at a standardized pressure and temperature has been common as a reference medium.

History

Thomas Young was presumably the person who first used, and invented, the name "index of refraction", in 1807. At the same time he changed this value of refractive power into a single number, instead of the traditional ratio of two numbers. The ratio had the disadvantage of different appearances. Newton, who called it the "proportion of the sines of incidence and refraction", wrote it as a ratio of two numbers, like "529 to 396" (or "nearly 4 to 3"; for water).

Hauksbee, who called it the "ratio of refraction", wrote it as a ratio with a fixed numerator, like "10000 to 7451.9" (for urine). Hutton wrote it as a ratio with a fixed denominator, like 1.3358 to 1 (water).

Thomas Young coined the term *index of refraction.*

Young did not use a symbol for the index of refraction, in 1807. In the next years, others started using different symbols: n, m, and μ. The symbol n gradually prevailed.

Typical Values

Diamonds have a very high refractive index of 2.42.

Selected refractive indices at λ=589 nm.	
Material	***n***
Vacuum	1
Gases at 0 °C and 1 atm	
Air	1.000293

Helium	1.000036
Hydrogen	1.000132
Carbon dioxide	1.00045
Liquids at 20 °C	
Water	1.333
Ethanol	1.36
Olive oil	1.47
Solids	
Ice	1.31
Soda-lime glass	1.46
PMMA (Plexiglas)	1.49
Crown glass (typical)	1.52
Flint glass (typical)	1.62
Cubic zirconia	2.15
Diamond	2.42
Moissanite	2.65
Zinc selenide	2.67

For visible light most transparent media have refractive indices between 1 and 2. A few examples are given in the adjacent table. These values are measured at the yellow doublet D-line of sodium, with a wavelength of 589 nanometers, as is conventionally done. Gases at atmospheric pressure have refractive indices close to 1 because of their low density. Almost all solids and liquids have refractive indices above 1.3, with aerogel as the clear exception. Aerogel is a very low density solid that can be produced with refractive index in the range from 1.002 to 1.265. Moissanite lies at the other end of the range with a refractive index as high as 2.65. Most plastics have refractive indices in the range from 1.3 to 1.7, but some high-refractive-index polymers can have values as high as 1.76.

For infrared light refractive indices can be considerably higher. Germanium is transparent in the wavelength region from 2 to 14 μm and has a refractive index of about 4, making it an important material for infrared optics.

Refractive Index Below Unity

According to the theory of relativity, no information can travel faster than the speed of light in vacuum, but this does not mean that the refractive index cannot be lower than 1. The refractive index measures the phase velocity of light, which does not carry information. The phase velocity is the speed at which the crests of the wave move and can be faster than the speed of light in vacuum, and thereby give a refractive index below 1. This can occur close to resonance frequencies, for absorbing media, in plasmas, and for X-rays. In the X-ray regime the refractive indices are lower than but very close to 1 (exceptions close to some resonance frequencies). As an example, water has a refractive index of $0.99999974 = 1 - 2.6 \times 10^{-7}$ for X-ray radiation at a photon energy of 30 keV (0.04 nm wavelength).

Negative Refractive Index

A split-ring resonator array arranged to produce a negative index of refraction for microwaves

Recent research has also demonstrated the existence of materials with a negative refractive index, which can occur if permittivity and permeability have simultaneous negative values. This can be achieved with periodically constructed metamaterials. The resulting negative refraction (i.e., a reversal of Snell's law) offers the possibility of the superlens and other exotic phenomena.

Microscopic Explanation

At the microscale, an electromagnetic wave's phase velocity is slowed in a material because the electric field creates a disturbance in the charges of each atom (primarily the electrons) proportional to the electric susceptibility of the medium. (Similarly, the magnetic field creates a disturbance proportional to the magnetic susceptibility.) As the electromagnetic fields oscillate in the wave, the charges in the material will be "shaken" back and forth at the same frequency. The charges thus radiate their own electromagnetic wave that is at the same frequency, but usually with a phase delay, as the charges may move out of phase with the force driving them. The light wave traveling in the medium is the macroscopic superposition (sum) of all such contributions in the material: the original wave plus the waves radiated by all the moving charges. This wave is typically a wave with the same frequency but shorter wavelength than the original, leading to a slowing of the wave's phase velocity. Most of the radiation from oscillating material charges will modify the incoming wave, changing its velocity. However, some net energy will be radiated in other directions or even at other frequencies.

Depending on the relative phase of the original driving wave and the waves radiated by the charge motion, there are several possibilities:

- If the electrons emit a light wave which is 90° out of phase with the light wave shaking them, it will cause the total light wave to travel slower. This is the normal refraction of transparent materials like glass or water, and corresponds to a refractive index which is real and greater than 1.
- If the electrons emit a light wave which is 270° out of phase with the light wave shaking them, it will cause the wave to travel faster. This is called "anomalous refraction", and is observed

close to absorption lines (typically in infrared spectra), with X-rays in ordinary materials, and with radio waves in Earth's ionosphere. It corresponds to a permittivity less than 1, which causes the refractive index to be also less than unity and the phase velocity of light greater than the speed of light in vacuum *c* (note that the signal velocity is still less than *c*, as discussed above). If the response is sufficiently strong and out-of-phase, the result is a negative value of permittivity and imaginary index of refraction, as observed in metals or plasma.

- If the electrons emit a light wave which is 180° out of phase with the light wave shaking them, it will destructively interfere with the original light to reduce the total light intensity. This is light absorption in opaque materials and corresponds to an imaginary refractive index.
- If the electrons emit a light wave which is in phase with the light wave shaking them, it will amplify the light wave. This is rare, but occurs in lasers due to stimulated emission. It corresponds to an imaginary index of refraction, with the opposite sign to that of absorption.

For most materials at visible-light frequencies, the phase is somewhere between 90° and 180°, corresponding to a combination of both refraction and absorption.

Dispersion

The refractive index of materials varies with the wavelength (and frequency) of light. This is called dispersion and causes prisms and rainbows to divide white light into its constituent spectral colors. As the refractive index varies with wavelength, so will the refraction angle as light goes from one material to another. Dispersion also causes the focal length of lenses to be wavelength dependent. This is a type of chromatic aberration, which often needs to be corrected for in imaging systems. In regions of the spectrum where the material does not absorb light, the refractive index tends to decrease with increasing wavelength, and thus increase with frequency. This is called "normal dispersion", in contrast to "anomalous dispersion", where the refractive index increases with wavelength. For visible light normal dispersion means that the refractive index is higher for blue light than for red.

Light of different colors has slightly different refractive indices in water and therefore shows up at different positions in the rainbow.

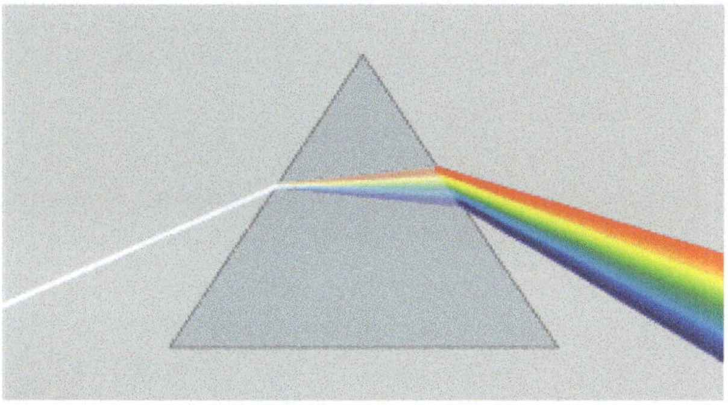

In a prism, dispersion causes different colors to refract at different angles, splitting white light into a rainbow of colors.

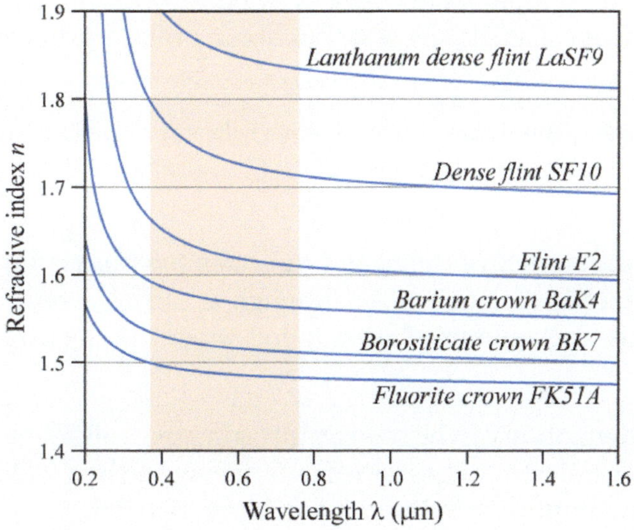

The variation of refractive index with wavelength for various glasses

For optics in the visual range, the amount of dispersion of a lens material is often quantified by the Abbe number:

$$V = \frac{n_{\text{yellow}} - 1}{n_{\text{blue}} - n_{\text{red}}}.$$

For a more accurate description of the wavelength dependence of the refractive index, the Sellmeier equation can be used. It is an empirical formula that works well in describing dispersion. *Sellmeier coefficients* are often quoted instead of the refractive index in tables.

Because of dispersion, it is usually important to specify the vacuum wavelength of light for which a refractive index is measured. Typically, measurements are done at various well-defined spectral emission lines; for example, n_D usually denotes the refractive index at the Fraunhofer "D" line, the centre of the yellow sodium double emission at 589.29 nm wavelength.

Complex Refractive Index

When light passes through a medium, some part of it will always be attenuated. This can be conveniently taken into account by defining a complex refractive index,

A graduated neutral density filter showing light absorption in the upper half

$$\underline{n} = n + i\kappa.$$

Here, the real part n is the refractive index and indicates the phase velocity, while the imaginary part κ is called the extinction coefficient — although κ can also refer to the mass attenuation coefficient— and indicates the amount of attenuation when the electromagnetic wave propagates through the material.

That κ corresponds to attenuation can be seen by inserting this refractive index into the expression for electric field of a plane electromagnetic wave traveling in the z-direction. We can do this by relating the complex wave number k to the complex refractive index n through $k = 2\pi n/\lambda_0$, with λ_0 being the vacuum wavelength; this can be inserted into the plane wave expression as

$$\mathbf{E}(z,t) = \mathrm{Re}\left[\mathbf{E}_0 e^{i(\underline{k}z-\omega t)}\right] = \mathrm{Re}\left[\mathbf{E}_0 e^{i(2\pi(n+i\kappa)z/\lambda_0-\omega t)}\right] = e^{-2\pi\kappa z/\lambda_0}\,\mathrm{Re}\left[\mathbf{E}_0 e^{i(kz-\omega t)}\right].$$

Here we see that κ gives an exponential decay, as expected from the Beer–Lambert law. Since intensity is proportional to the square of the electric field, it will depend on the depth into the material as $\exp(-4\pi\kappa z/\lambda_0)$, and the attenuation coefficient becomes $\alpha = 4\pi\kappa/\lambda_0$. This also relates it to the penetration depth, the distance after which the intensity is reduced by $1/e$, $\delta_p = 1/\alpha = \lambda_0/(4\pi\kappa)$.

Both n and κ are dependent on the frequency. In most circumstances $\kappa > 0$ (light is absorbed) or $\kappa = 0$ (light travels forever without loss). In special situations, especially in the gain medium of lasers, it is also possible that $\kappa < 0$, corresponding to an amplification of the light.

An alternative convention uses $n = n - i\kappa$ instead of $n = n + i\kappa$, but where $\kappa > 0$ still corresponds to loss. Therefore these two conventions are inconsistent and should not be confused. The difference is related to defining sinusoidal time dependence as $\mathrm{Re}[\exp(-i\omega t)]$ versus $\mathrm{Re}[\exp(+i\omega t)]$.

Dielectric loss and non-zero DC conductivity in materials cause absorption. Good dielectric materials such as glass have extremely low DC conductivity, and at low frequencies the dielectric loss is also negligible, resulting in almost no absorption. However, at higher frequencies (such as visible light), dielectric loss may increase absorption significantly, reducing the material's transparency to these frequencies.

The real, n, and imaginary, κ, parts of the complex refractive index are related through the Kramers–Kronig relations. In 1986 A.R. Forouhi and I. Bloomer deduced an equation describing κ as a function of photon energy, E, applicable to amorphous materials. Forouhi and Bloomer then applied the Kramers–Kronig relation to derive the corresponding equation for n as a function of E. The same formalism was applied to crystalline materials by Forouhi and Bloomer in 1988.

The refractive index and extinction coefficient, n and κ, cannot be measured directly. They must be determined indirectly from measurable quantities that depend on them, such as reflectance, R, or transmittance, T, or ellipsometric parameters, ψ and δ. The determination of n and κ from such measured quantities will involve developing a theoretical expression for R or T, or ψ and δ in terms of a valid physical model for n and κ. By fitting the theoretical model to the measured R or T, or ψ and δ using regression analysis, n and κ can be deduced.

For X-ray and extreme ultraviolet radiation the complex refractive index deviates only slightly from unity and usually has a real part smaller than 1. It is therefore normally written as $n = 1 - \delta + i\beta$ (or $n = 1 - \delta - i\beta$ with the alternative convention mentioned above).

Relations to Other Quantities

Optical Path Length

The colors of a soap bubble are determined by the optical path length through the thin soap film in a phenomenon called thin-film interference.

Optical path length (OPL) is the product of the geometric length d of the path light follows through a system, and the index of refraction of the medium through which it propagates,

$$\mathrm{OPL} = nd.$$

This is an important concept in optics because it determines the phase of the light and governs interference and diffraction of light as it propagates. According to Fermat's principle, light rays can be characterized as those curves that optimize the optical path length.

Refraction

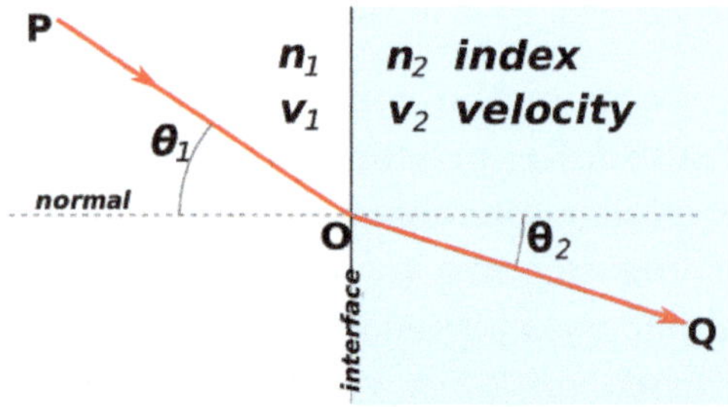

Refraction of light at the interface between two media of different refractive indices, with $n_2 > n_1$. Since the phase velocity is lower in the second medium ($v_2 < v_1$), the angle of refraction θ_2 is less than the angle of incidence θ_1; that is, the ray in the higher-index medium is closer to the normal.

When light moves from one medium to another, it changes direction, i.e. it is refracted. If it moves from a medium with refractive index n_1 to one with refractive index n_2, with an incidence angle to the surface normal of θ_1, the refraction angle θ_2 can be calculated from Snell's law:

$$n_1 \sin\theta_1 = n_2 \sin\theta_2.$$

When light enters a material with higher refractive index, the angle of refraction will be smaller than the angle of incidence and the light will be refracted towards the normal of the surface. The higher the refractive index, the closer to the normal direction the light will travel. When passing into a medium with lower refractive index, the light will instead be refracted away from the normal, towards the surface.

Total Internal Reflection

Total internal reflection can be seen at the air-water boundary.

If there is no angle θ_2 fulfilling Snell's law, i.e.,

$$\frac{n_1}{n_2}\sin\theta_1 > 1,$$

the light cannot be transmitted and will instead undergo total internal reflection. This occurs only when going to a less optically dense material, i.e., one with lower refractive index. To get total internal reflection the angles of incidence θ_1 must be larger than the critical angle

$$\theta_c = \arcsin\left(\frac{n_2}{n_1}\right).$$

Reflectivity

Apart from the transmitted light there is also a reflected part. The reflection angle is equal to the incidence angle, and the amount of light that is reflected is determined by the reflectivity of the surface. The reflectivity can be calculated from the refractive index and the incidence angle with the Fresnel equations, which for normal incidence reduces to

$$R_0 = \left|\frac{n_1 - n_2}{n_1 + n_2}\right|^2.$$

For common glass in air, $n_1 = 1$ and $n_2 = 1.5$, and thus about 4% of the incident power is reflected. At other incidence angles the reflectivity will also depend on the polarization of the incoming light. At a certain angle called Brewster's angle, p-polarized light (light with the electric field in the plane of incidence) will be totally transmitted. Brewster's angle can be calculated from the two refractive indices of the interface as

$$\theta_B = \arctan\left(\frac{n_2}{n_1}\right).$$

Lenses

The power of a magnifying glass is determined by the shape and refractive index of the lens.

The focal length of a lens is determined by its refractive index n and the radii of curvature R_1 and R_2 of its surfaces. The power of a thin lens in air is given by the Lensmaker's formula:

$$\frac{1}{f} = (n-1)\left(\frac{1}{R_1} - \frac{1}{R_2}\right),$$

where f is the focal length of the lens.

Microscope Resolution

The resolution of a good optical microscope is mainly determined by the numerical aperture (NA) of its objective lens. The numerical aperture in turn is determined by the refractive index n of the medium filling the space between the sample and the lens and the half collection angle of light θ according to

$$\mathrm{NA} = n \sin \theta.$$

For this reason oil immersion is commonly used to obtain high resolution in microscopy. In this technique the objective is dipped into a drop of high refractive index immersion oil on the sample under study.

Relative Permittivity and Permeability

The refractive index of electromagnetic radiation equals

$$n = \sqrt{\varepsilon_\mathrm{r} \mu_\mathrm{r}},$$

where ε_r is the material's relative permittivity, and μ_r is its relative permeability. The refractive index is used for optics in Fresnel equations and Snell's law; while the relative permittivity and permeability are used in Maxwell's equations and electronics. Most naturally occurring materials are non-magnetic at optical frequencies, that is μ_r is very close to 1, therefore n is approximately $\sqrt{\varepsilon_\mathrm{r}}$. In this particular case, the complex relative permittivity ε_r, with real and imaginary parts ε_r and $\tilde{\varepsilon}_\mathrm{r}$, and the complex refractive index n, with real and imaginary parts n and κ (the latter called the "extinction coefficient"), follow the relation

$$\underline{\varepsilon}_\mathrm{r} = \varepsilon_\mathrm{r} + i\tilde{\varepsilon}_\mathrm{r} = \underline{n}^2 = (n + i\kappa)^2,$$

and their components are related by:

$$\varepsilon_\mathrm{r} = n^2 - \kappa^2,$$

$$\tilde{\varepsilon}_\mathrm{r} = 2n\kappa,$$

and:

$$n = \sqrt{\frac{|\underline{\varepsilon}_\mathrm{r}| + \varepsilon_\mathrm{r}}{2}},$$

$$\kappa = \sqrt{\frac{|\underline{\varepsilon}_\mathrm{r}| - \varepsilon_\mathrm{r}}{2}}.$$

where $|\underline{\varepsilon}_\mathrm{r}| = \sqrt{\varepsilon_\mathrm{r}^2 + \tilde{\varepsilon}_\mathrm{r}^2}$ is the complex modulus.

Density

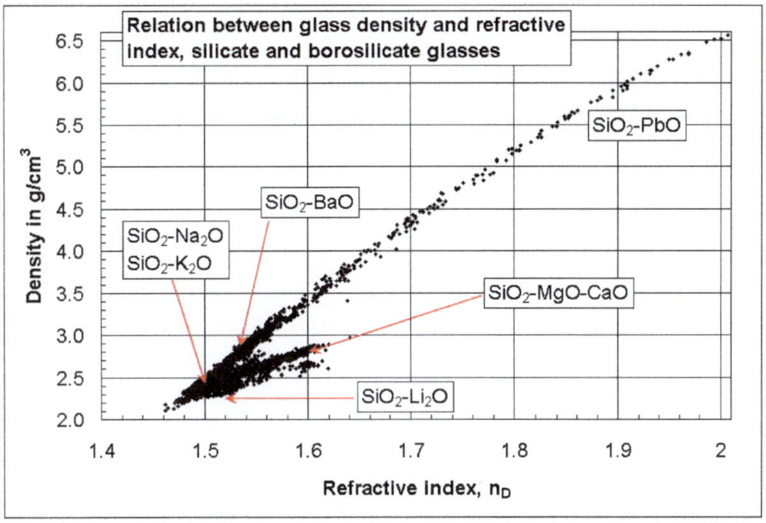

Relation between the refractive index and the density of silicate and borosilicate glasses

In general, the refractive index of a glass increases with its density. However, there does not exist an overall linear relation between the refractive index and the density for all silicate and borosilicate glasses. A relatively high refractive index and low density can be obtained with glasses containing light metal oxides such as Li_2O and MgO, while the opposite trend is observed with glasses containing PbO and BaO as seen in the diagram at the right.

Many oils (such as olive oil) and ethyl alcohol are examples of liquids which are more refractive, but less dense, than water, contrary to the general correlation between density and refractive index.

For gases, $n - 1$ is proportional to the density of the gas as long as the chemical composition does not change. This means that it is also proportional to the pressure and inversely proportional to the temperature for ideal gases.

Group Index

Sometimes, a "group velocity refractive index", usually called the *group index* is defined:

$$n_g = \frac{c}{v_g},$$

where v_g is the group velocity. This value should not be confused with n, which is always defined with respect to the phase velocity. When the dispersion is small, the group velocity can be linked to the phase velocity by the relation

$$v_g = v - \lambda \frac{dv}{d\lambda},$$

where λ is the wavelength in the medium. In this case the group index can thus be written in terms of the wavelength dependence of the refractive index as

$$n_g = \frac{n}{1 + \frac{\lambda}{n}\frac{dn}{d\lambda}}.$$

When the refractive index of a medium is known as a function of the vacuum wavelength (instead of the wavelength in the medium), the corresponding expressions for the group velocity and index are (for all values of dispersion)

$$v_g = c\left(n - \lambda_0 \frac{dn}{d\lambda_0}\right)^{-1},$$

$$n_g = n - \lambda_0 \frac{dn}{d\lambda_0},$$

where λ_0 is the wavelength in vacuum.

Momentum (Abraham–Minkowski Controversy)

In 1908, Hermann Minkowski calculated the momentum p of a refracted ray as follows:

$$p = \frac{nE}{c},$$

where E is energy of the photon, c is the speed of light in vacuum and n is the refractive index of the medium. In 1909, Max Abraham proposed the following formula for this calculation:

$$p = \frac{E}{nc}.$$

A 2010 study suggested that *both* equations are correct, with the Abraham version being the kinetic momentum and the Minkowski version being the canonical momentum, and claims to explain the contradicting experimental results using this interpretation.

Other Relations

As shown in the Fizeau experiment, when light is transmitted through a moving medium, its speed relative to a stationary observer is:

$$V = \frac{c}{n} + v\left(1 - \frac{1}{n^2}\right)$$

The refractive index of a substance can be related to its polarizability with the Lorentz–Lorenz equation or to the molar refractivities of its constituents by the Gladstone–Dale relation.

Refractivity

In atmospheric applications, the *refractivity* is taken as $N = n - 1$. Atmospheric refractivity is often expressed as either $N = 10^6(n - 1)$ or $N = 10^8(n - 1)$ The multiplication factors are used because the refractive index for air, n deviates from unity by at most a few parts per ten thousand.

Molar refractivity, on the other hand, is a measure of the total polarizability of a mole of a substance and can be calculated from the refractive index as

$$A = \frac{M}{\rho}\frac{n^2 - 1}{n^2 + 2},$$

where ρ is the density, and M is the molar mass.

Nonscalar, Nonlinear, or Nonhomogeneous Refraction

So far, we have assumed that refraction is given by linear equations involving a spatially constant, scalar refractive index. These assumptions can break down in different ways, to be described in the following subsections.

Birefringence

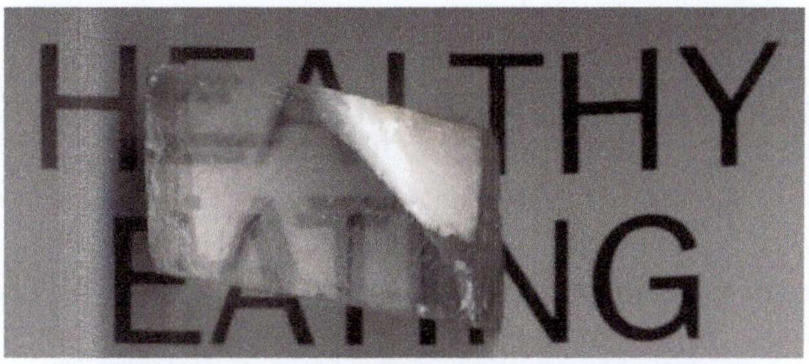

A calcite crystal laid upon a paper with some letters showing double refraction

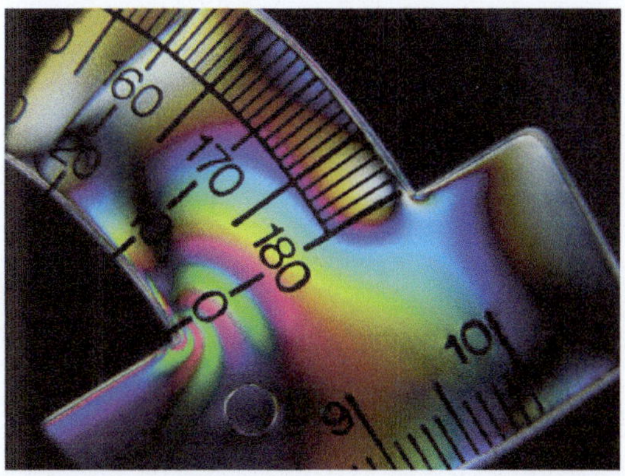

Birefringent materials can give rise to colors when placed between crossed polarizers. This is the basis for photoelasticity.

In some materials the refractive index depends on the polarization and propagation direction of the light. This is called birefringence or optical anisotropy.

In the simplest form, uniaxial birefringence, there is only one special direction in the material.

This axis is known as the optical axis of the material. Light with linear polarization perpendicular to this axis will experience an *ordinary* refractive index n_o while light polarized in parallel will experience an *extraordinary* refractive index n_e. The birefringence of the material is the difference between these indices of refraction, $\Delta n = n_e - n_o$. Light propagating in the direction of the optical axis will not be affected by the birefringence since the refractive index will be n_o independent of polarization. For other propagation directions the light will split into two linearly polarized beams. For light traveling perpendicularly to the optical axis the beams will have the same direction. This can be used to change the polarization direction of linearly polarized light or to convert between linear, circular and elliptical polarizations with waveplates.

Many crystals are naturally birefringent, but isotropic materials such as plastics and glass can also often be made birefringent by introducing a preferred direction through, e.g., an external force or electric field. This can be utilized in the determination of stresses in structures using photoelasticity. The birefringent material is then placed between crossed polarizers. A change in birefringence will alter the polarization and thereby the fraction of light that is transmitted through the second polarizer.

In the more general case of trirefringent materials described by the field of crystal optics, the *dielectric constant* is a rank-2 tensor (a 3 by 3 matrix). In this case the propagation of light cannot simply be described by refractive indices except for polarizations along principal axes.

Nonlinearity

The strong electric field of high intensity light (such as output of a laser) may cause a medium's refractive index to vary as the light passes through it, giving rise to nonlinear optics. If the index varies quadratically with the field (linearly with the intensity), it is called the optical Kerr effect and causes phenomena such as self-focusing and self-phase modulation. If the index varies linearly with the field (a nontrivial linear coefficient is only possible in materials that do not possess inversion symmetry), it is known as the Pockels effect.

Inhomogeneity

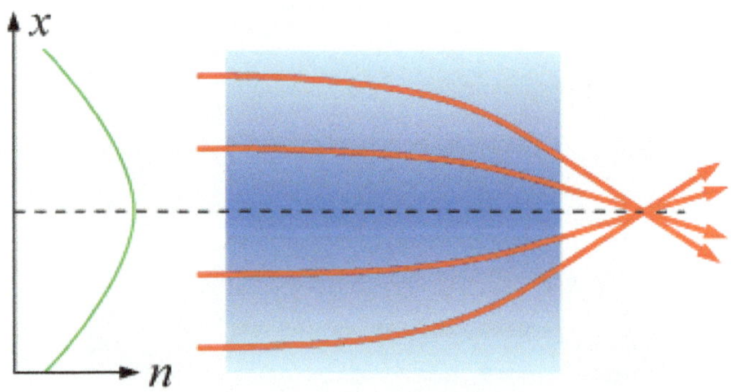

A gradient-index lens with a parabolic variation of refractive index (n) with radial distance (x). The lens focuses light in the same way as a conventional lens.

If the refractive index of a medium is not constant, but varies gradually with position, the material is known as a gradient-index or GRIN medium and is described by gradient index optics. Light traveling through such a medium can be bent or focused, and this effect can be exploited to produce lenses, some optical fibers and other devices. Introducing GRIN elements in the design of an optical system can greatly simplify the system, reducing the number of elements by as much as a third while maintaining overall performance. The crystalline lens of the human eye is an example of a GRIN lens with a refractive index varying from about 1.406 in the inner core to approximately 1.386 at the less dense cortex. Some common mirages are caused by a spatially varying refractive index of air.

Refractive Index Measurement

Homogeneous Media

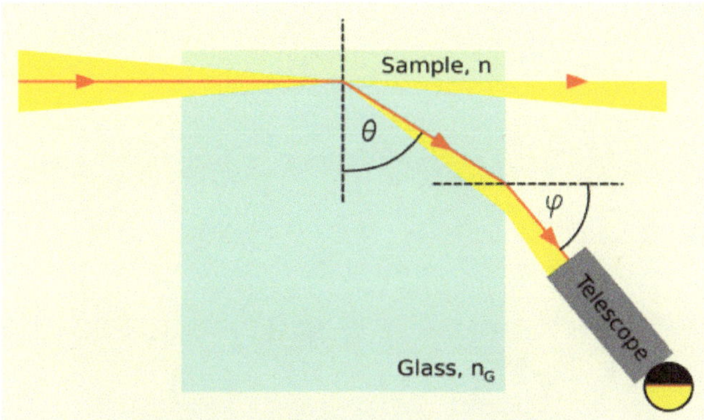

The principle of many refractometers

The refractive index of liquids or solids can be measured with refractometers. They typically measure some angle of refraction or the critical angle for total internal reflection. The first laboratory refractometers sold commercially were developed by Ernst Abbe in the late 19th century. The same principles are still used today. In this instrument a thin layer of the liquid to be measured is placed between two prisms. Light is shone through the liquid at incidence angles all the way up to 90°, i.e., light rays parallel to the surface. The second prism should have an index of refraction higher than that of the liquid, so that light only enters the prism at angles smaller than the critical angle for total reflection. This angle can then be measured either by looking through a telescope, or with a digital photodetector placed in the focal plane of a lens. The refractive index n of the liquid can then be calculated from the maximum transmission angle θ as $n = n_G \sin\theta$, where n_G is the refractive index of the prism.

A handheld refractometer used to measure sugar content of fruits

This type of devices are commonly used in chemical laboratories for identification of substances and for quality control. Handheld variants are used in agriculture by, e.g., wine makers to determine sugar content in grape juice, and inline process refractometers are used in, e.g., chemical and pharmaceutical industry for process control.

In gemology a different type of refractometer is used to measure index of refraction and birefringence of gemstones. The gem is placed on a high refractive index prism and illuminated from below. A high refractive index contact liquid is used to achieve optical contact between the gem and the prism. At small incidence angles most of the light will be transmitted into the gem, but at high angles total internal reflection will occur in the prism. The critical angle is normally measured by looking through a telescope.

Refractive Index Variations

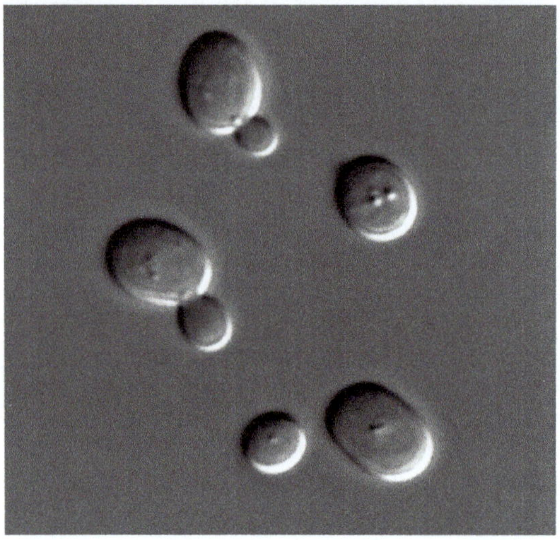

A differential interference contrast microscopy image of yeast cells

Unstained biological structures appear mostly transparent under Bright-field microscopy as most cellular structures do not attenuate appreciable quantities of light. Nevertheless, the variation in the materials that constitutes these structures also corresponds to a variation in the refractive index. The following techniques convert such variation into measurable amplitude differences:

To measure the spatial variation of refractive index in a sample phase-contrast imaging methods are used. These methods measure the variations in phase of the light wave exiting the sample. The phase is proportional to the optical path length the light ray has traversed, and thus gives a measure of the integral of the refractive index along the ray path. The phase cannot be measured directly at optical or higher frequencies, and therefore needs to be converted into intensity by interference with a reference beam. In the visual spectrum this is done using Zernike phase-contrast microscopy, differential interference contrast microscopy (DIC) or interferometry.

Zernike phase-contrast microscopy introduces a phase shift to the low spatial frequency components of the image with a phase-shifting annulus in the Fourier plane of the sample, so that high-spatial-frequency parts of the image can interfere with the low-frequency reference beam. In DIC the illumination is split up into two beams that are given different polarizations, are phase shifted differently, and are shifted transversely with slightly different amounts. After the specimen, the two parts are made to interfere, giving an image of the derivative of the optical path length in the direction of the difference in transverse shift. In interferometry the illumination is split up into two beams by a partially reflective mirror. One of the beams is let through the sample before they are combined to interfere and give a direct image of the phase shifts. If the optical path length variations are more than a wavelength the image will contain fringes.

There exist several phase-contrast X-ray imaging techniques to determine 2D or 3D spatial distribution of refractive index of samples in the X-ray regime.

Applications

The refractive index is a very important property of the components of any optical instrument that uses refraction. It determines the focusing power of lenses, the dispersive power of prisms, and generally the path of light through the system. It is the increase in refractive index in the core that guides the light in an optical fiber, and the variations in refractive index that reduces the reflectivity of a surface treated with an anti-reflective coating.

Since refractive index is a fundamental physical property of a substance, it is often used to identify a particular substance, confirm its purity, or measure its concentration. Refractive index is used to measure solids, liquids, and gases. Most commonly it is used to measure the concentration of a solute in an aqueous solution. It can also be used as a useful tool to differentiate between different types of gemstone, due to the unique chatoyance each individual stone displays. A refractometer is the instrument used to measure refractive index. For a solution of sugar, the refractive index can be used to determine the sugar content.

Dispersion (Optics)

In a dispersive prism, material dispersion (a wavelength-dependent refractive index) causes different colors to refract at different angles, splitting white light into a rainbow.

A compact fluorescent lamp seen through an Amici prism.

In optics, dispersion is the phenomenon in which the phase velocity of a wave depends on its frequency. Media having this common property may be termed *dispersive media*. Sometimes the term *chromatic* dispersion is used for specificity. Although the term is used in the field of optics to describe light and other electromagnetic waves, dispersion in the same sense can apply to any sort of wave motion such as acoustic dispersion in the case of sound and seismic waves, in gravity waves (ocean waves), and for telecommunication signals propagating along transmission lines (such as coaxial cable) or optical fiber.

In optics, one important and familiar consequence of dispersion is the change in the angle of refraction of different colors of light, as seen in the spectrum produced by a dispersive prism and in chromatic aberration of lenses. Design of compound achromatic lenses, in which chromatic aberration is largely cancelled, uses a quantification of a glass's dispersion given by its Abbe number V, where *lower* Abbe numbers correspond to *greater* dispersion over the visible spectrum. In some applications such as telecommunications, the absolute phase of a wave is often not important but only the propagation of wave packets or "pulses"; in that case one is interested only in variations of group velocity with frequency, so-called group-velocity dispersion (GVD).

Examples of Dispersion

The most familiar example of dispersion is probably a rainbow, in which dispersion causes the spatial separation of a white light into components of different wavelengths (different colors). However, dispersion also has an effect in many other circumstances: for example, group velocity dispersion (GVD) causes pulses to spread in optical fibers, degrading signals over long distances; also, a cancellation between group-velocity dispersion and nonlinear effects leads to soliton waves.

Material and Waveguide Dispersion

Most often, chromatic dispersion refers to bulk material dispersion, that is, the change in refractive index with optical frequency. However, in a waveguide there is also the phenomenon of *waveguide dispersion*, in which case a wave's phase velocity in a structure depends on its frequency simply due to the structure's geometry. More generally, "waveguide" dispersion can occur for waves propagating through any inhomogeneous structure (e.g., a photonic crystal), whether or not the waves are confined to some region. In a waveguide, *both* types of dispersion will generally be present, although they are not strictly additive. For example, in fiber optics the material and waveguide dispersion can effectively cancel each other out to produce a zero-dispersion wavelength, important for fast fiber-optic communication.

Material Dispersion in Optics

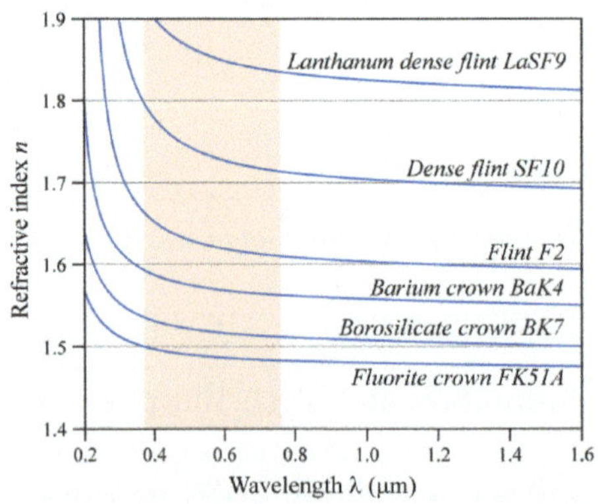

The variation of refractive index vs. vacuum wavelength for various glasses. The wavelengths of visible light are shaded in red.

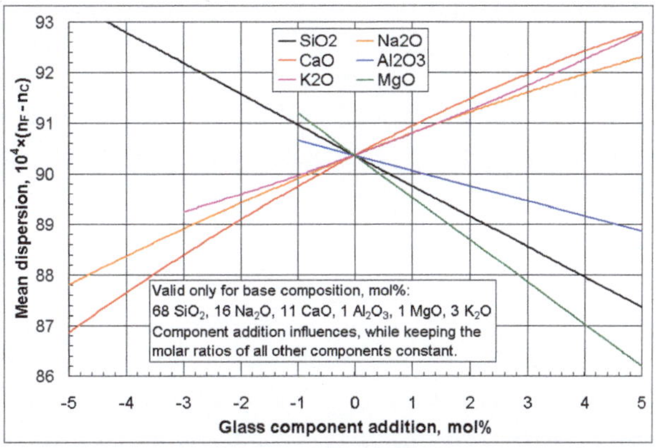

Influences of selected glass component additions on the mean dispersion of a specific base glass (n_F valid for λ = 486 nm (blue), n_C valid for λ = 656 nm (red))

Material dispersion can be a desirable or undesirable effect in optical applications. The dispersion of light by glass prisms is used to construct spectrometers and spectroradiometers. Holographic gratings are also used, as they allow more accurate discrimination of wavelengths. However, in lenses, dispersion causes chromatic aberration, an undesired effect that may degrade images in microscopes, telescopes, and photographic objectives.

The *phase velocity*, *v*, of a wave in a given uniform medium is given by

$$v = \frac{c}{n}$$

where *c* is the speed of light in a vacuum and *n* is the refractive index of the medium.

In general, the refractive index is some function of the frequency *f* of the light, thus $n = n(f)$, or alternatively, with respect to the wave's wavelength $n = n(\lambda)$. The wavelength dependence of a material's refractive index is usually quantified by its Abbe number or its coefficients in an empirical formula such as the Cauchy or Sellmeier equations.

Because of the Kramers–Kronig relations, the wavelength dependence of the real part of the refractive index is related to the material absorption, described by the imaginary part of the refractive index (also called the extinction coefficient). In particular, for non-magnetic materials ($\mu = \mu_o$), the susceptibility χ that appears in the Kramers–Kronig relations is the electric susceptibility $\chi_e = n^2 - 1$.

The most commonly seen consequence of dispersion in optics is the separation of white light into a color spectrum by a prism. From Snell's law it can be seen that the angle of refraction of light in a prism depends on the refractive index of the prism material. Since that refractive index varies with wavelength, it follows that the angle that the light is refracted by will also vary with wavelength, causing an angular separation of the colors known as *angular dispersion*.

For visible light, refraction indices *n* of most transparent materials (e.g., air, glasses) decrease with increasing wavelength λ:

$$1 < n(\lambda_{\text{red}}) < n(\lambda_{\text{yellow}}) < n(\lambda_{\text{blue}}),$$

or alternatively:

$$\frac{dn}{d\lambda} < 0.$$

In this case, the medium is said to have *normal dispersion.* Whereas, if the index increases with increasing wavelength (which is typically the case for X-rays), the medium is said to have *anomalous dispersion.*

At the interface of such a material with air or vacuum (index of ~1), Snell's law predicts that light incident at an angle θ to the normal will be refracted at an angle arcsin(sin θ/n). Thus, blue light, with a higher refractive index, will be bent more strongly than red light, resulting in the well-known rainbow pattern.

Group and Phase Velocity

Another consequence of dispersion manifests itself as a temporal effect. The formula $v = c/n$ calculates the *phase velocity* of a wave; this is the velocity at which the *phase* of any one frequency component of the wave will propagate. This is not the same as the *group velocity* of the wave, that is the rate at which changes in amplitude (known as the *envelope* of the wave) will propagate. For a homogeneous medium, the group velocity v_g is related to the phase velocity v by (here λ is the wavelength in vacuum, not in the medium):

$$v_g = \frac{c}{n - \lambda \frac{dn}{d\lambda}} = \frac{v}{1 - \frac{\lambda}{n}\frac{dn}{d\lambda}}.$$

The group velocity v_g is often thought of as the velocity at which energy or information is conveyed along the wave. In most cases this is true, and the group velocity can be thought of as the *signal velocity* of the waveform. In some unusual circumstances, called cases of anomalous dispersion, the rate of change of the index of refraction with respect to the wavelength changes sign (becoming negative), in which case it is possible for the group velocity to exceed the speed of light ($v_g > c$). Anomalous dispersion occurs, for instance, where the wavelength of the light is close to an absorption resonance of the medium. When the dispersion is anomalous, however, group velocity is no longer an indicator of signal velocity. Instead, a signal travels at the speed of the wavefront, which is c irrespective of the index of refraction. Recently, it has become possible to create gases in which the group velocity is not only larger than the speed of light, but even negative. In these cases, a pulse can appear to exit a medium before it enters. Even in these cases, however, a signal travels at, or less than, the speed of light, as demonstrated by Stenner, et al.

The group velocity itself is usually a function of the wave's frequency. This results in group velocity dispersion (GVD), which causes a short pulse of light to spread in time as a result of different frequency components of the pulse travelling at different velocities. GVD is often quantified as the *group delay dispersion parameter* (again, this formula is for a uniform medium only):

$$D = -\frac{\lambda}{c}\frac{d^2n}{d\lambda^2}.$$

If D is less than zero, the medium is said to have *positive dispersion*. If D is greater than zero, the medium has *negative dispersion*. If a light pulse is propagated through a normally dispersive medium, the result is the higher frequency components travel slower than the lower frequency components. The pulse therefore becomes *positively chirped*, or *up-chirped*, increasing in frequency with time. Conversely, if a pulse travels through an anomalously dispersive medium, high frequency components travel faster than the lower ones, and the pulse becomes *negatively chirped*, or *down-chirped*, decreasing in frequency with time.

The result of GVD, whether negative or positive, is ultimately temporal spreading of the pulse. This makes dispersion management extremely important in optical communications systems based on optical fiber, since if dispersion is too high, a group of pulses representing a bit-stream will spread in time and merge, rendering the bit-stream unintelligible. This limits the length of fiber that a signal can be sent down without regeneration. One possible answer to this problem is to send signals down the optical fibre at a wavelength where the GVD is zero (e.g., around 1.3–1.5 μm in silica fibres), so pulses at this wavelength suffer minimal spreading from dispersion. In practice, however, this approach causes more problems than it solves because zero GVD unacceptably amplifies other nonlinear effects (such as four wave mixing). Another possible option is to use soliton pulses in the regime of anomalous dispersion, a form of optical pulse which uses a nonlinear optical effect to self-maintain its shape. Solitons have the practical problem, however, that they require a certain power level to be maintained in the pulse for the nonlinear effect to be of the correct strength. Instead, the solution that is currently used in practice is to perform dispersion compensation, typically by matching the fiber with another fiber of opposite-sign dispersion so that the dispersion effects cancel; such compensation is ultimately limited by nonlinear effects such as self-phase modulation, which interact with dispersion to make it very difficult to undo.

Dispersion control is also important in lasers that produce short pulses. The overall dispersion of the optical resonator is a major factor in determining the duration of the pulses emitted by the laser. A pair of prisms can be arranged to produce net negative dispersion, which can be used to balance the usually positive dispersion of the laser medium. Diffraction gratings can also be used to produce dispersive effects; these are often used in high-power laser amplifier systems. Recently, an alternative to prisms and gratings has been developed: chirped mirrors. These dielectric mirrors are coated so that different wavelengths have different penetration lengths, and therefore different group delays. The coating layers can be tailored to achieve a net negative dispersion.

Dispersion in Waveguides

Waveguides are highly dispersive due to their geometry (rather than just to their material composition). Optical fibers are a sort of waveguide for optical frequencies (light) widely used in modern telecommunications systems. The rate at which data can be transported on a single fiber is limited by pulse broadening due to chromatic dispersion among other phenomena.

In general, for a waveguide mode with an angular frequency $\omega(\beta)$ at a propagation constant β (so that the electromagnetic fields in the propagation direction z oscillate proportional to $e^{i(\beta z-\omega t)}$), the group-velocity dispersion parameter D is defined as:

$$D = -\frac{2\pi c}{\lambda^2}\frac{d^2\beta}{d\omega^2} = \frac{2\pi c}{v_g^2\lambda^2}\frac{dv_g}{d\omega}$$

where $\lambda = 2\pi c/\omega$ is the vacuum wavelength and $v_g = d\omega/d\beta$ is the group velocity. This formula generalizes the one in the previous section for homogeneous media, and includes both waveguide dispersion and material dispersion. The reason for defining the dispersion in this way is that $|D|$ is the (asymptotic) temporal pulse spreading Δt per unit bandwidth $\Delta\lambda$ per unit distance travelled, commonly reported in ps / nm km for optical fibers.

In the case of multi-mode optical fibers, so-called modal dispersion will also lead to pulse broadening. Even in single-mode fibers, pulse broadening can occur as a result of polarization mode dispersion (since there are still two polarization modes). These are *not* examples of chromatic dispersion as they are not dependent on the wavelength or bandwidth of the pulses propagated.

Higher-order Dispersion over Broad Bandwidths

When a broad range of frequencies (a broad bandwidth) is present in a single wavepacket, such as in an ultrashort pulse or a chirped pulse or other forms of spread spectrum transmission, it may not be accurate to approximate the dispersion by a constant over the entire bandwidth, and more complex calculations are required to compute effects such as pulse spreading.

In particular, the dispersion parameter D defined above is obtained from only one derivative of the group velocity. Higher derivatives are known as *higher-order dispersion*. These terms are simply a Taylor series expansion of the dispersion relation $\beta(\omega)$ of the medium or waveguide around some particular frequency. Their effects can be computed via numerical evaluation of Fourier transforms of the waveform, via integration of higher-order slowly varying envelope approximations, by a split-step method (which can use the exact dispersion relation rather than a Taylor series), or by direct simulation of the full Maxwell's equations rather than an approximate envelope equation.

Dispersion in Gemology

Dispersion values of minerals		
Name	**B–G**	**C–F**
Cinnabar (HgS)	0.40	—
Synth. rutile	0.330	0.190
Rutile (TiO_2)	0.280	0.120–0.180
Anatase (TiO_2)	0.213–0.259	—
Wulfenite	0.203	0.133
Vanadinite	0.202	—
Fabulite	0.190	0.109
Sphalerite (ZnS)	0.156	0.088
Sulfur (S)	0.155	—
Stibiotantalite	0.146	—
Goethite (FeO(OH))	0.14	—
Brookite (TiO_2)	0.131	0.12–1.80

Zincite (ZnO)	0.127	–
Linobate	0.13	0.075
Synthetic moissanite (SiC)	0.104	–
Cassiterite (SnO_2)	0.071	0.035
Zirconia (ZrO_2)	0.060	0.035
Powellite ($CaMoO_4$)	0.058	–
Andradite	0.057	–
Demantoid	0.057	0.034
Cerussite	0.055	0.033–0.050
Titanite	0.051	0.019–0.038
Benitoite	0.046	0.026
Anglesite	0.044	0.025
Diamond (C)	0.044	0.025
Flint glass	0.041	–
Hyacinth	0.039	–
Jargoon	0.039	–
Starlite	0.039	–
Zircon ($ZrSiO_4$)	0.039	0.022
GGG	0.038	0.022
Scheelite	0.038	0.026
Dioptase	0.036	0.021
Whewellite	0.034	–
Alabaster	0.033	–
Gypsum	0.033	0.008
Epidote	0.03	0.012–0.027
Achroite	0.017	–
Cordierite	0.017	0.009
Danburite	0.017	0.009
Dravite	0.017	–
Elbaite	0.017	–
Herderite	0.017	0.008–0.009
Hiddenite	0.017	0.010
Indicolite	0.017	–
Liddicoatite	0.017	–
Kunzite	0.017	0.010
Rubellite	0.017	0.008–0.009
Schorl	0.017	–
Scapolite	0.017	–
Spodumene	0.017	0.010
Tourmaline	0.017	0.009–0.011
Verdelite	0.017	–
Andalusite	0.016	0.009

Baryte ($BaSO_4$)	0.016	0.009
Euclase	0.016	0.009
Alexandrite	0.015	0.011
Chrysoberyl	0.015	0.011
Hambergite	0.015	0.009–0.010
Phenakite	0.01	0.009
Rhodochrosite	0.015	0.010–0.020
Sillimanite	0.015	0.009–0.012
Smithsonite	0.014–0.031	0.008–0.017
Amblygonite	0.014–0.015	0.008
Aquamarine	0.014	0.009–0.013
Beryl	0.014	0.009–0.013
Brazilianite	0.014	0.008
Celestine	0.014	0.008
Goshenite	0.014	–
Heliodor	0.014	0.009–0.013
Morganite	0.014	0.009–0.013
Pyroxmangite	0.015	–
Synth. scheelite	0.015	–
Dolomite	0.013	–
Magnesite ($MgCO_3$)	0.012	–
Synth. emerald	0.012	–
Synth. alexandrite	0.011	–
Synth. sapphire (Al_2O_3)	0.011	–
Phosphophyllite	0.010–0.011	–
Enstatite	0.010	–
Anorthite	0.009–0.010	–
Actinolite	0.009	–
Jeremejevite	0.009	–
Nepheline	0.008–0.009	–
Apophyllite	0.008	–
Hauyne	0.008	–
Natrolite	0.008	–
Synth. quartz (SiO_2)	0.008	–
Aragonite	0.007–0.012	–
Augelite	0.007	–
Tanzanite	0.030	0.011
Thulite	0.03	0.011
Zoisite	0.03	–
YAG	0.028	0.015
Almandine	0.027	0.013–0.016
Hessonite	0.027	0.013–0.015

Spessartine	0.027	0.015
Uvarovite	0.027	0.014–0.021
Willemite	0.027	–
Pleonaste	0.026	–
Rhodolite	0.026	–
Boracite	0.024	0.012
Cryolite	0.024	–
Staurolite	0.023	0.012–0.013
Pyrope	0.022	0.013–0.016
Diaspore	0.02	–
Grossular	0.020	0.012
Hemimorphite	0.020	0.013
Kyanite	0.020	0.011
Peridot	0.020	0.012–0.013
Spinel	0.020	0.011
Vesuvianite	0.019–0.025	0.014
Clinozoisite	0.019	0.011–0.014
Labradorite	0.019	0.010
Axinite	0.018–0.020	0.011
Ekanite	0.018	0.012
Kornerupine	0.018	0.010
Corundum (Al_2O_3)	0.018	0.011
Rhodizite	0.018	–
Ruby (Al_2O_3)	0.018	0.011
Sapphire (Al_2O_3)	0.018	0.011
Sinhalite	0.018	0.010
Sodalite	0.018	0.009
Synth. corundum	0.018	0.011
Diopside	0.018–0.020	0.01
Emerald	0.014	0.009–0.013
Topaz	0.014	0.008
Amethyst (SiO_2)	0.013	0.008
Anhydrite	0.013	–
Apatite	0.013	0.010
Apatite	0.013	0.008
Aventurine	0.013	0.008
Citrine	0.013	0.008
Morion	0.013	–
Prasiolite	0.013	0.008
Quartz (SiO_2)	0.013	0.008
Smoky quartz (SiO_2)	0.013	0.008
Rose quartz (SiO_2)	0.013	0.008

Albite	0.012	—
Bytownite	0.012	—
Feldspar	0.012	0.008
Moonstone	0.012	0.008
Orthoclase	0.012	0.008
Pollucite	0.012	0.007
Sanidine	0.012	—
Sunstone	0.012	—
Beryllonite	0.010	0.007
Cancrinite	0.010	0.008–0.009
Leucite	0.010	0.008
Obsidian	0.010	—
Strontianite	0.008–0.028	—
Calcite ($CaCO_3$)	0.008–0.017	0.013–0.014
Fluorite (CaF_2)	0.007	0.004
Hematite	0.500	—
Synthetic cassiterite (SnO_2)	0.041	—
Gahnite	0.019–0.021	—
Datolite	0.016	—
Tremolite	0.006–0.007	—

In the technical terminology of gemology, *dispersion* is the difference in the refractive index of a material at the B and G (686.7 nm and 430.8 nm) or C and F (656.3 nm and 486.1 nm) Fraunhofer wavelengths, and is meant to express the degree to which a prism cut from the gemstone demonstrates "fire." Fire is a colloquial term used by gemologists to describe a gemstone's dispersive nature or lack thereof. Dispersion is a material property. The amount of fire demonstrated by a given gemstone is a function the gemstone's facet angles, the polish quality, the lighting environment, the material's refractive index, the saturation of color, and the orientation of the viewer relative to the gemstone.

Dispersion in Imaging

In photographic and microscopic lenses, dispersion causes chromatic aberration, which causes the different colors in the image not to overlap properly. Various techniques have been developed to counteract this, such as the use of achromats, multielement lenses with glasses of different dispersion. They are constructed in such a way that the chromatic aberrations of the different parts cancel out.

Dispersion of Pulsar Emissions

Pulsars are spinning neutron stars that emit pulses at very regular intervals ranging from milliseconds to seconds. Astronomers believe that the pulses are emitted simultaneously over a wide range of frequencies. However, as observed on Earth, the components of each pulse emitted at higher radio frequencies arrive before those emitted at lower frequencies. This dispersion occurs because of the ionized component of the interstellar medium, mainly the free electrons, which make the group velocity frequency dependent. The extra delay added at a frequency ν is

$$t = k_{\mathrm{DM}} \times \left(\frac{\mathrm{DM}}{\nu^2} \right)$$

where the dispersion constant k_{DM} is given by

$$k_{\mathrm{DM}} = \frac{e^2}{2\pi m_e c} \simeq 4.149 \mathrm{GHz}^2 \mathrm{pc}^{-1} \mathrm{cm}^3 \mathrm{ms},$$

and the dispersion measure (DM) is the column density of free electrons (total electron content) — i.e. the number density of electrons n_e (electrons/cm^3) integrated along the path traveled by the photon from the pulsar to the Earth — and is given by

$$\mathrm{DM} = \int_0^d n_e \, dl$$

with units of parsecs per cubic centimetre (1 pc/cm^3 = 30.857 × 10^{21} m^{-2}).

Typically for astronomical observations, this delay cannot be measured directly, since the emission time is unknown. What *can* be measured is the difference in arrival times at two different frequencies. The delay Δt between a high frequency ν_{hi} and a low frequency ν_{lo} component of a pulse will be

$$\Delta t = k_{\mathrm{DM}} \times \mathrm{DM} \times \left(\frac{1}{\nu_{\mathrm{lo}}^2} - \frac{1}{\nu_{\mathrm{hi}}^2} \right)$$

Rewriting the above equation in terms of Δt allows one to determine the DM by measuring pulse arrival times at multiple frequencies. This in turn can be used to study the interstellar medium, as well as allow for observations of pulsars at different frequencies to be combined.

Liquidus

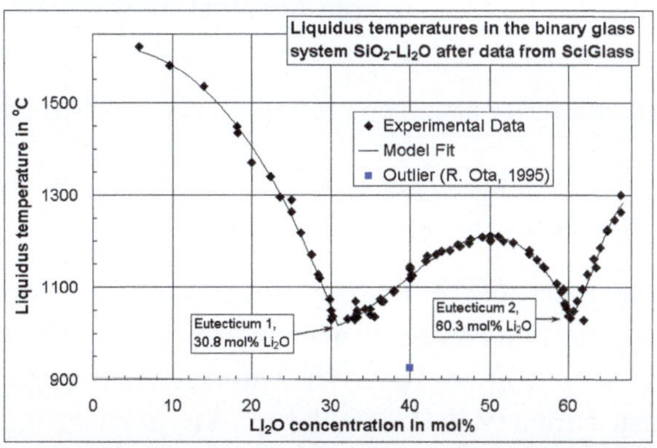

Liquidus temperature curve in the binary glass system SiO_2-Li_2O, based on 91 published data collected in SciGlass; model fit from Glassproperties.com

The liquidus temperature, T_L or T_{liq} specifies the temperature above which a material is completely liquid, and the maximum temperature at which crystals can co-exist with the melt in thermodynamic equilibrium. It is mostly used for impure substances (mixtures) such as glasses, alloys and rocks.

Above the liquidus temperature the material is homogeneous and liquid at equilibrium. Below the liquidus temperature, more and more crystals will form in the melt if one waits a sufficiently long time, depending on the material. Alternately, homogeneous glasses can be obtained through sufficiently fast cooling, i.e., through kinetic inhibition of the crystallization process.

The crystal phase that crystallizes first on cooling a substance to its liquidus temperature is termed primary crystalline phase or primary phase. The composition range within which the primary phase remains constant is known as primary crystalline phase field.

The liquidus temperature is important in the glass industry because crystallization can cause severe problems during the glass melting and forming processes, and it also may lead to product failure.

The liquidus temperature can be contrasted to the solidus temperature. The solidus temperature quantifies the point at which a material completely solidifies (crystallizes). The liquidus and solidus temperatures do not necessarily coincide; if a gap exists between the liquidus and solidus temperatures, then within that gap, the material consists of solid and liquid phases simultaneously (like a slurry).

For pure substances, e.g. pure metal, pure water, etc. the liquidus and solidus are at the same temperature, and the term "melting point" may be used. For impure substances, e.g. alloys, tap water, coca cola, ice cream, etc. the melting point broadens into a melting interval instead. If the temperature is within the melting interval, one may see "slurries" at equilibrium, i.e. the slurry will neither fully solidify nor melt. This is why new snow of high purity either melts or stays solid, while dirty snow on the ground tend to become slushy at certain temperatures. Weld melt pools containing high levels of sulfur, either from melted impurities from the base metal or from the welding electrode, typically have very broad melting intervals, which leads to increased risk of hot cracking.

Heat Capacity

Heat capacity or thermal capacity is a measurable physical quantity equal to the ratio of the heat added to (or removed from) an object to the resulting temperature change. The unit of heat capacity is joule per kelvin $\frac{J}{K}$, or kilogram metre squared per kelvin second squared $\frac{kgm^2}{Ks^2}$ in the International System of Units (SI). The dimensional form is $L^2MT^{-2}\Theta^{-1}$. Specific heat is the amount of heat needed to raise the temperature of one kilogram of mass by 1 kelvin.

Heat capacity is an extensive property of matter, meaning it is proportional to the size of the system. When expressing the same phenomenon as an intensive property, the heat capacity is divided by the amount of substance, mass, or volume, thus the quantity is independent of the size or extent of the sample. The molar heat capacity is the heat capacity per unit amount (SI unit: mole) of a pure substance and the specific heat capacity, often simply called specific heat, is the heat capacity per unit mass of a material. Nonetheless some authors state the term specific heat to refer to the ratio of the specific heat capacity of a substance at any given temperature, to the specific heat capacity of another substance at a reference temperature, much in the fashion of a specific gravity. Occasionally, in engineering contexts, the volumetric heat capacity is used.

Temperature reflects the average randomized kinetic energy of constituent particles of matter (e.g. atoms or molecules) relative to the centre of mass of the system, while heat is the transfer of energy across a system boundary into the body other than by work or matter transfer. Translation, rotation, and vibration of atoms represent the degrees of freedom of motion which classically contribute to the heat capacity of gases, while only vibrations are needed to describe the heat capacities of most solids , as shown by the Dulong–Petit law. Other contributions can come from magnetic and electronic degrees of freedom in solids, but these rarely make substantial contributions.

For quantum mechanical reasons, at any given temperature, some of these degrees of freedom may be unavailable, or only partially available, to store thermal energy. In such cases, the specific heat capacity is a fraction of the maximum. As the temperature approaches absolute zero, the specific capacity of a system approaches zero, due to loss of available degrees of freedom. Quantum theory can be used to quantitatively predict the specific heat capacity of simple systems.

History

In a previous theory of heat common in the early modern period, heat was thought to be a measurement of an invisible fluid, known as the *caloric*. Bodies were capable of holding a certain amount of this fluid, hence the term *heat capacity*, named and first investigated by Scottish chemist Joseph Black in the 1750s.

Since the development of thermodynamics during the 18th and 19th centuries, scientists have abandoned the idea of a physical caloric, and instead understand heat as changes in a system's internal energy. That is, heat is no longer considered a fluid; rather, heat is a transfer of disordered energy. Nevertheless, at least in English, the term "heat capacity" survives. In some other languages, the term *thermal capacity* is preferred, and it is also sometimes used in English.

Units

Extensive Properties

In the International System of Units, heat capacity has the unit joules per kelvin (J/K). An object's heat capacity (symbol C) is defined as the ratio of the amount of heat energy transferred to an object and the resulting increase in temperature of the object,

$$C = \frac{Q}{\Delta T},$$

assuming that the temperature range is sufficiently small thus the heat capacity is constant. More generally, because heat capacity does depend upon temperature, it should be written as

$$C(T) = \frac{\delta Q}{dT},$$

where the symbol δ is used to imply that heat is a path function. Heat capacity is an extensive property, meaning it depends on the extent or size of the physical system in question. A sample containing twice the amount of substance as another sample requires the transfer of twice the amount of heat ( ) to achieve the same change in temperature ($\ddot{A}T$).

Intensive Properties

For many experimental and theoretical purposes it is more convenient to report heat capacity as an intensive property – an intrinsic characteristic of a particular substance. This is most often accomplished by expressing the property in relation to a unit of mass. In science and engineering, such properties are often prefixed with the term *specific*. International standards now recommend that specific heat capacity always refer to division by mass. The units for the specific heat capacity are $[c]=\frac{\mathrm{J}}{\mathrm{kg\cdot K}}$..

In chemistry, heat capacity is often specified relative to one mole, the unit of amount of substance, and is called the molar heat capacity. It has the unit $[C_{\mathrm{mol}}]=\frac{\mathrm{J}}{\mathrm{mol\cdot K}}$..

For some considerations it is useful to specify the volume-specific heat capacity, commonly called volumetric heat capacity, which is the heat capacity per unit volume and has SI units $[s]=\frac{\mathrm{J}}{\mathrm{m}^3\cdot\mathrm{K}}$. This is used almost exclusively for liquids and solids, since for gases it may be confused with specific heat capacity *at constant volume.*

Alternative Unit Systems

While SI units are the most widely used, some countries and industries also use other systems of measurement. One older unit of heat is the kilogram-calorie (Cal), originally defined as the energy required to raise the temperature of one kilogram of water by one degree Celsius, typically from 14.5 to 15.5 °C. The specific average heat capacity of water on this scale would therefore be exactly 1 Cal/(C°·kg). However, due to the temperature-dependence of the specific heat, a large number of different definitions of the calorie came into being. Whilst once it was very prevalent, especially its smaller cgs variant the gram-calorie (cal), defined thus the specific heat of water would be 1 cal/(K·g), in most fields the use of the calorie is now archaic.

In the United States other units of measure for heat capacity may be quoted in disciplines such as construction, civil engineering, and chemical engineering. A still common system is the English Engineering Units in which the mass reference is pound mass and the temperature is specified in degrees Fahrenheit or Rankine. One (rare) unit of heat is the pound calorie (lb-cal), defined as the amount of heat required to raise the temperature of one pound of water by one degree Celsius. On this scale the specific heat of water would be 1 lb-cal/(K·lbm). More common is the British thermal unit, the standard unit of heat in the U.S. construction industry. This is defined such that the specific heat of water is 1 BTU/(F°·lb). The path integral Monte Carlo method is a numerical approach for determining the values of heat capacity, based on quantum dynamical principles. However, good approximations can be made for gases in many states using simpler methods outlined below. For many solids composed of relatively heavy atoms (atomic number > iron), at non-cryogenic temperatures, the heat capacity at room temperature approaches 3R = 24.94 joules per kelvin per mole of atoms (Dulong–Petit law, R is the gas constant). Low temperature approximations for both gases and solids at temperatures less than their characteristic Einstein temperatures or Debye temperatures can be made by the methods of Einstein and Debye discussed below. Water (liquid): CP = 4185.5 J/(kg·K) (15 °C, 101.325 kPa) Water (liquid): CVH = 74.539 J/(mol·K) (25 °C) For liquids and gases, it is important to know the pressure to which given heat capacity data refer. Most published data are given for standard pressure. However, different standard conditions for temperature and pressure have been defined by different organizations. The International Union of Pure and Applied Chemistry (IUPAC) changed its recommendation from one atmosphere to the round value 100 kPa (≈750.062 Torr).

Measurement of Heat Capacity

It may appear that the way to measure heat capacity is to add a known amount of heat to an object, and measure the change in temperature. This works reasonably well for many solids. However, for precise measurements, and especially for gases, other aspects of measurement become critical.

The heat capacity can be affected by many of the state variables that describe the thermodynamic system under study. These include the starting and ending temperature, as well as the pressure and the volume of the system before and after heat is added. So rather than a single way to measure heat capacity, there are actually several slightly different measurements of heat capacity. The most commonly used methods for measurement are to hold the object either at constant pressure (C_P) or at constant volume (C_V). Gases and liquids are typically also measured at constant volume. Measurements under constant pressure produce larger values than those at constant volume because the constant pressure values also include heat energy that is used to do work to expand the substance against the constant pressure as its temperature increases. This difference is particularly notable in gases where values under constant pressure are typically 30% to 66.7% greater than those at constant volume. Hence the heat capacity ratio of gases is typically between 1.3 and 1.67.

The specific heat capacities of substances comprising molecules (as distinct from monatomic gases) are not fixed constants and vary somewhat depending on temperature. Accordingly, the temperature at which the measurement is made is usually also specified. Examples of two common ways to cite the specific heat of a substance are as follows:

- Water (liquid): C_P = 4185.5 J/(kg·K) (15 °C, 101.325 kPa)
- Water (liquid): $C_V H$ = 74.539 J/(mol·K) (25 °C)

For liquids and gases, it is important to know the pressure to which given heat capacity data refer. Most published data are given for standard pressure. However, quite different standard conditions for temperature and pressure have been defined by different organizations. The International Union of Pure and Applied Chemistry (IUPAC) changed its recommendation from one atmosphere to the round value 100 kPa (≈750.062 Torr).

Calculation from First Principles

The path integral Monte Carlo method is a numerical approach for determining the values of heat capacity, based on quantum dynamical principles. However, good approximations can be made for gases in many states using simpler methods outlined below. For many solids composed of relatively heavy atoms (atomic number > iron), at non-cryogenic temperatures, the heat capacity at room temperature approaches $3R = 24.94$ joules per kelvin per mole of atoms (Dulong–Petit law, R is the gas constant). Low temperature approximations for both gases and solids at temperatures less than their characteristic Einstein temperatures or Debye temperatures can be made by the methods of Einstein and Debye discussed below.

Thermodynamic Relations and Definition of Heat Capacity

The internal energy of a closed system changes either by adding heat to the system or by the system performing work. Written mathematically we have

$$\Delta e_{\text{system}} = e_{\text{in}} - e_{\text{out}}$$

Or

$$\mathrm{d}U = \delta Q - \delta W.$$

For work as a result of an increase of the system volume we may write,

$$\mathrm{d}U = \delta Q - P\mathrm{d}V.$$

If the heat is added at constant volume, then the second term of this relation vanishes and one readily obtains

$$\left(\frac{\partial U}{\partial T}\right)_{\mathrm{V}} = \left(\frac{\partial Q}{\partial T}\right)_{\mathrm{V}} = C_{\mathrm{V}}.$$

This defines the *heat capacity at constant volume*, C_V, which is also related to changes in internal energy. Another useful quantity is the *heat capacity at constant pressure*, C_P. This quantity refers to the change in the *enthalpy* of the system, which is given by

$$H = U + PV.$$

A small change in the enthalpy can be expressed as

$$\mathrm{d}H = \delta Q + V\mathrm{d}P,$$

and therefore, at constant pressure, we have

$$\left(\frac{\partial H}{\partial T}\right)_{\mathrm{P}} = \left(\frac{\partial Q}{\partial T}\right)_{\mathrm{P}} = C_{\mathrm{P}}.$$

These two equations:

$$\left(\frac{\partial U}{\partial T}\right)_{\mathrm{V}} = \left(\frac{\partial Q}{\partial T}\right)_{\mathrm{V}} = C_{\mathrm{V}}.$$

$$\left(\frac{\partial H}{\partial T}\right)_{\mathrm{P}} = \left(\frac{\partial Q}{\partial T}\right)_{\mathrm{P}} = C_{\mathrm{P}}.$$

are property relations and are therefore independent of the type of process. In other words, they are valid for any substance going through any process. Both the internal energy and enthalpy of a substance can change with the transfer of energy in many forms i.e., heat.

Relation between Heat Capacities

Measuring the heat capacity, sometimes referred to as specific heat, at constant volume can be prohibitively difficult for liquids and solids. That is, small temperature changes typically require large pressures to maintain a liquid or solid at constant volume implying the containing vessel must be nearly rigid or at least very strong. Instead it is easier to measure the heat capacity at constant pressure (allowing the material to expand or contract freely) and solve for the heat capacity at constant volume using mathematical relationships derived from the basic thermodynamic laws. Starting from the fundamental thermodynamic relation one can show

$$C_{\mathrm{P}} - C_{\mathrm{V}} = T\left(\frac{\partial P}{\partial T}\right)_{\mathrm{V,n}}\left(\frac{\partial V}{\partial T}\right)_{\mathrm{P,n}}$$

where the partial derivatives are taken at constant volume and constant number of particles, and constant pressure and constant number of particles, respectively.

This can also be rewritten

$$C_{\mathrm{P}} - C_{\mathrm{V}} = VT\frac{\alpha^2}{\beta_{\mathrm{T}}}$$

where

α is the coefficient of thermal expansion,

β_{T} is the isothermal compressibility.

The heat capacity ratio or adiabatic index is the ratio of the heat capacity at constant pressure to heat capacity at constant volume. It is sometimes also known as the isentropic expansion factor.

Ideal Gas

For an ideal gas, evaluating the partial derivatives above according to the equation of state where R is the gas constant for an ideal gas

$$PV = nRT$$

$$C_{\mathrm{P}} - C_{\mathrm{V}} = T\left(\frac{\partial P}{\partial T}\right)_{\mathrm{V,n}}\left(\frac{\partial V}{\partial T}\right)_{\mathrm{P,n}}$$

$$P = \frac{nRT}{V} \Rightarrow \left(\frac{\partial P}{\partial T}\right)_{\mathrm{V,n}} = \frac{nR}{V}$$

$$V = \frac{nRT}{P} \Rightarrow \left(\frac{\partial V}{\partial T}\right)_{\mathrm{P,n}} = \frac{nR}{P}$$

substituting

$$T\left(\frac{\partial P}{\partial T}\right)_{\mathrm{V,n}}\left(\frac{\partial V}{\partial T}\right)_{\mathrm{P,n}} = T\left(\frac{nR}{V}\right)\left(\frac{nR}{P}\right) = \left(\frac{nRT}{V}\right)\left(\frac{nR}{P}\right) = P\left(\frac{nR}{P}\right) = nR$$

this equation reduces simply to Mayer's relation,

$$C_{\mathrm{P,m}} - C_{\mathrm{V,m}} = R$$

Specific Heat Capacity

The specific heat capacity of a material on a per mass basis is

$$c = \frac{\partial C}{\partial m},$$

which in the absence of phase transitions is equivalent to

$$c = E_{\mathrm{m}} = \frac{C}{m} = \frac{C}{\rho V},$$

where

C is the heat capacity of a body made of the material in question,

m is the mass of the body,

V is the volume of the body, and

$\rho = \frac{m}{V}$ is the density of the material.

For gases, and also for other materials under high pressures, there is need to distinguish between different boundary conditions for the processes under consideration (since values differ significantly between different conditions). Typical processes for which a heat capacity may be defined include isobaric (constant pressure, $\mathrm{d}P = 0$) or isochoric (constant volume, $\mathrm{d}V = 0$) processes. The corresponding specific heat capacities are expressed as

$$\mathrm{c}_P = \left(\frac{\partial \mathrm{C}}{\partial \mathrm{m}}\right)_P,$$

$$c_{\mathrm{V}} = \left(\frac{\partial C}{\partial m}\right)_{\mathrm{V}}.$$

From the results of the previous section, dividing through by the mass gives the relation

$$c_P - c_V = \frac{\alpha^2 T}{\rho \beta_T}.$$

A related parameter to c is CV^{-1}, the volumetric heat capacity. In engineering practice, c_V for solids or liquids often signifies a volumetric heat capacity, rather than a constant-volume one. In such cases, the mass-specific heat capacity (specific heat) is often explicitly written with the subscript m, as c_m. Of course, from the above relationships, for solids one writes

$$c_m = \frac{C}{m} = \frac{c_{volumetric}}{\rho}.$$

For pure homogeneous chemical compounds with established molecular or molar mass or a molar quantity is established, heat capacity as an intensive property can be expressed on a per mole basis instead of a per mass basis by the following equations analogous to the per mass equations:

$$C_{P,m} = \left(\frac{\partial C}{\partial n}\right)_P = \text{molar heat capacity at constant pressure}$$

$$C_{V,m} = \left(\frac{\partial C}{\partial n}\right)_V = \text{molar heat capacity at constant volume}$$

where n is the number of moles in the body or thermodynamic system. One may refer to such a *per mole* quantity as molar heat capacity to distinguish it from specific heat capacity on a per mass basis.

Polytropic Heat Capacity

The polytropic heat capacity is calculated at processes if all the thermodynamic properties (pressure, volume, temperature) change

$$C_{i,m} = \left(\frac{\partial C}{\partial n}\right) = \text{molar heat capacity at polytropic process}$$

The most important polytropic processes run between the adiabatic and the isotherm functions, the polytropic index is between 1 and the adiabatic exponent (γ or κ)

Dimensionless Heat Capacity

The dimensionless heat capacity of a material is

$$C^* = \frac{C}{nR} = \frac{C}{Nk}$$

where

C is the heat capacity of a body made of the material in question (J/K)

n is the amount of substance in the body (mol)

R is the gas constant (J/(K·mol))

N is the number of molecules in the body. (dimensionless)

k is Boltzmann's constant (J/(K·molecule))

In the ideal gas article, dimensionless heat capacity C^* is expressed as $\hat{c}$, and is related there directly to half the number of degrees of freedom per particle. This holds true for quadratic degrees of freedom, a consequence of the equipartition theorem.

More generally, the dimensionless heat capacity relates the logarithmic increase in temperature to the increase in the dimensionless entropy per particle $S^* = S / Nk$, measured in nats.

$$C^* = \frac{\mathrm{d}S^*}{\mathrm{d}\ln T}$$

Alternatively, using base 2 logarithms, C^* relates the base-2 logarithmic increase in temperature to the increase in the dimensionless entropy measured in bits.

Heat Capacity at Absolute Zero

From the definition of entropy

$$T\,\mathrm{d}S = \delta Q$$

the absolute entropy can be calculated by integrating from zero kelvins temperature to the final temperature T_f

$$S(T_f) = \int_{T=0}^{T_f} \frac{\delta Q}{T} = \int_0^{T_f} \frac{\delta Q}{\mathrm{d}T}\frac{\mathrm{d}T}{T} = \int_0^{T_f} C(T)\frac{\mathrm{d}T}{T}.$$

The heat capacity must be zero at zero temperature in order for the above integral not to yield an infinite absolute entropy, which would violate the third law of thermodynamics. One of the strengths of the Debye model is that (unlike the preceding Einstein model) it predicts the proper mathematical form of the approach of heat capacity toward zero, as absolute zero temperature is approached.

Negative Heat Capacity (Stars)

Most physical systems exhibit a positive heat capacity. However, even though it can seem paradoxical at first, there are some systems for which the heat capacity is *negative*. These are inhomogeneous systems which do not meet the strict definition of thermodynamic equilibrium. They include gravitating objects such as stars, galaxies; and also sometimes some nano-scale clusters of a few tens of atoms, close to a phase transition. A negative heat capacity can result in a negative temperature.

According to the virial theorem, for a self-gravitating body like a star or an interstellar gas cloud, the average potential energy U_{Pot} and the average kinetic energy U_{Kin} are locked together in the relation

$$U_{Pot} = -2U_{Kin},$$

The total energy U (= $U_{Pot} + U_{Kin}$) therefore obeys

$$U = -U_{Kin},$$

If the system loses energy, for example by radiating energy away into space, the average kinetic energy actually increases. If a temperature is defined by the average kinetic energy, then the system therefore can be said to have a negative heat capacity.

A more extreme version of this occurs with black holes. According to black hole thermodynamics, the more mass and energy a black hole absorbs, the colder it becomes. In contrast, if it is a net emitter of energy, through Hawking radiation, it will become hotter and hotter until it boils away.

Theory of Heat Capacity

Factors that Affect Specific Heat Capacity

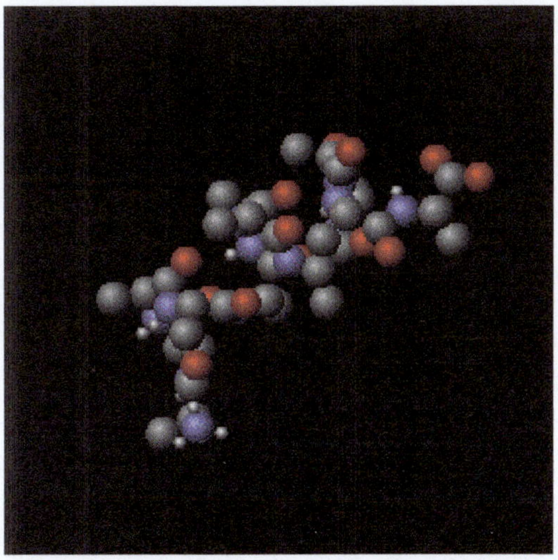

Molecules undergo many characteristic internal vibrations. Potential energy stored in these internal degrees of freedom contributes to a sample's energy content, but not to its temperature. More internal degrees of freedom tend to increase a substance's specific heat capacity, so long as temperatures are high enough to overcome quantum effects.

For any given substance, the heat capacity of a body is directly proportional to the amount of substance it contains (measured in terms of mass or moles or volume). Doubling the amount of substance in a body doubles its heat capacity, etc.

However, when this effect has been corrected for, by dividing the heat capacity by the quantity of substance in a body, the resulting specific heat capacity is a function of the structure of the substance itself. In particular, it depends on the number of degrees of freedom that are available to the particles in the substance; each independent degree of freedom allows the particles to store thermal energy. The translational kinetic energy of substance particles which manifests as *temperature change* is only one of the many possible degrees of freedom, and thus the larger the number of degrees of freedom available to the particles of a substance *other* than translational kinetic energy, the larger will be the specific heat capacity for the substance. For example, rotational kinetic energy of gas molecules stores heat energy in a way that increases heat capacity, since this energy does not contribute to temperature.

In addition, quantum effects require that whenever energy be stored in any mechanism associated with a bound system which confers a degree of freedom, it must be stored in certain minimal-sized deposits (quanta) of energy, or else not stored at all. Such effects limit the full ability of some degrees of freedom to store energy when their lowest energy storage quantum amount is not easily supplied at the average energy of particles at a given temperature. In general, for this reason, specific heat capacities tend to fall at lower temperatures where the average thermal energy available to each particle degree of freedom is smaller, and thermal energy storage begins to be limited by these quantum effects. Due to this process, as temperature falls toward absolute zero, so also does heat capacity.

Degrees of Freedom

Molecules are quite different from the monatomic gases like helium and argon. With monatomic gases, thermal energy comprises only translational motions. Translational motions are ordinary, whole-body movements in 3D space whereby particles move about and exchange energy in collisions—like rubber balls in a vigorously shaken container. These simple movements in the three dimensions of space mean individual atoms have three translational degrees of freedom. A degree of freedom is any form of energy in which heat transferred into an object can be stored. This can be in translational kinetic energy, rotational kinetic energy, or other forms such as potential energy in vibrational modes. Only three translational degrees of freedom (corresponding to the three independent directions in space) are available for any individual atom, whether it is free, as a monatomic molecule, or bound into a polyatomic molecule.

As to rotation about an atom's axis (again, whether the atom is bound or free), its energy of rotation is proportional to the moment of inertia for the atom, which is extremely small compared to moments of inertia of collections of atoms. This is because almost all of the mass of a single atom is concentrated in its nucleus, which has a radius too small to give a significant moment of inertia. In contrast, the *spacing* of quantum energy levels for a rotating object is inversely proportional to its moment of inertia, and so this spacing becomes very large for objects with very small moments of inertia. For these reasons, the contribution from rotation of atoms on their axes is essentially zero in monatomic gases, because the energy spacing of the associated quantum levels is too large for significant thermal energy to be stored in rotation of systems with such small moments of inertia. For similar reasons, axial rotation around bonds joining atoms in diatomic gases (or along the linear axis in a linear molecule of any length) can also be neglected as a possible "degree of freedom" as well, since such rotation is similar to rotation of monatomic atoms, and so occurs about an axis with a moment of inertia too small to be able to store significant heat energy.

In polyatomic molecules, other rotational modes may become active, due to the much higher moments of inertia about certain axes which do not coincide with the linear axis of a linear molecule. These modes take the place of some translational degrees of freedom for individual atoms, since the atoms are moving in 3-D space, as the molecule rotates. The narrowing of quantum mechanically determined energy spacing between rotational states results from situations where atoms are rotating around an axis that does not connect them, and thus form an assembly that has a large moment of inertia. This small difference between energy states allows the kinetic energy of this type of rotational motion to store heat energy at ambient temperatures. Furthermore, internal vibrational degrees of freedom also may become active (these are also a type of translation, as seen from the view of each atom). In summary, molecules are complex objects with a population of atoms that may move about within the molecule in a number of different ways, and each of these ways of moving is capable of storing energy if the temperature is sufficient.

The heat capacity of molecular substances (on a "per-atom" or atom-molar, basis) does not exceed the heat capacity of monatomic gases, unless vibrational modes are brought into play. The reason for this is that vibrational modes allow energy to be stored as potential energy in inter-atomic bonds in a molecule, which are not available to atoms in monatomic gases. Up to about twice as much energy (on a per-atom basis) per unit of temperature increase can be stored in a solid as in a monatomic gas, by this mechanism of storing energy in the potentials of interatomic bonds. This gives many solids about twice the atom-molar heat capacity at room temperature of monatomic gases.

However, quantum effects heavily affect the actual ratio at lower temperatures (i.e., much lower than the melting temperature of the solid), especially in solids with light and tightly bound atoms (e.g., beryllium metal or diamond). Polyatomic gases store intermediate amounts of energy, giving them a "per-atom" heat capacity that is between that of monatomic gases ($3/2$ R per mole of atoms, where R is the ideal gas constant), and the maximum of fully excited warmer solids (3 R per mole of atoms). For gases, heat capacity never falls below the minimum of $3/2$ R per mole (of molecules), since the kinetic energy of gas molecules is always available to store at least this much thermal energy. However, at cryogenic temperatures in solids, heat capacity falls toward zero, as temperature approaches absolute zero.

Example of Temperature-dependent Specific Heat Capacity, in a Diatomic Gas

To illustrate the role of various degrees of freedom in storing heat, we may consider nitrogen, a diatomic molecule that has five active degrees of freedom at room temperature: the three comprising translational motions plus two rotational degrees of freedom internally. Although the constant-volume molar heat capacity of nitrogen at this temperature is five-thirds that of monatomic gases, on a per-mole of atoms basis, it is five-sixths that of a monatomic gas. The reason for this is the loss of a degree of freedom due to the bond when it does not allow storage of thermal energy. Two separate nitrogen atoms would have a total of six degrees of freedom—the three translational degrees of freedom of each atom. When the atoms are bonded the molecule will still only have three translational degrees of freedom, as the two atoms in the molecule move as one. However, the molecule cannot be treated as a point object, and the moment of inertia has increased sufficiently about two axes to allow two rotational degrees of freedom to be active at room temperature to give five degrees of freedom. The moment of inertia about the third axis remains small, as this is the axis passing through the centres of the two atoms, and so is similar to the small moment of inertia for atoms of a monatomic gas. Thus, this degree of freedom does not act to store heat, and does not contribute to the heat capacity of nitrogen. The heat capacity *per atom* for nitrogen (5/2 R per mole molecules = 5/4 R per mole atoms) is therefore less than for a monatomic gas (3/2 R per mole molecules or atoms), so long as the temperature remains low enough that no vibrational degrees of freedom are activated.

At higher temperatures, however, nitrogen gas gains one more degree of internal freedom, as the molecule is excited into higher vibrational modes that store thermal energy. A vibrational degree of freedom contributes a heat capacity of 1/2 R each for kinetic and potential energy, for a total of R. Now the bond is contributing heat capacity, and (because of storage of energy in potential energy) is contributing more than if the atoms were not bonded. With full thermal excitation of bond vibration, the heat capacity per volume, or per mole of gas *molecules* approaches seven-thirds that of monatomic gases. Significantly, this is seven-sixths of the monatomic gas value on a mole-of-atoms basis, so this is now a *higher* heat capacity *per atom* than the monatomic figure, because the vibrational mode enables for diatomic gases allows an extra degree of *potential energy* freedom per pair of atoms, which monatomic gases cannot possess. See thermodynamic temperature for more information on translational motions, kinetic (heat) energy, and their relationship to temperature.

However, even at these large temperatures where gaseous nitrogen is able to store $7/6^{ths}$ of the energy *per atom* of a monatomic gas (making it more efficient at storing energy on an atomic basis), it still only stores $7/12^{ths}$ of the maximal per-atom heat capacity of a *solid,* meaning it is not nearly as efficient at storing thermal energy on an atomic basis, as solid substances can be. This is typical of gases, and results because many of the potential bonds which might be storing potential energy in gaseous nitrogen (as opposed to solid nitrogen) are lacking, because only one of the spatial dimensions for each nitrogen atom offers a bond into which potential energy can be stored without increasing the kinetic energy of the atom. In general, solids are most efficient, on an atomic basis, at storing thermal energy (that is, they have the highest per-atom or per-mole-of-atoms heat capacity).

Per Mole of Different Units

Per Mole of Molecules

When the specific heat capacity, *c*, of a material is measured (lowercase *c* means the unit quantity is in terms of mass), different values arise because different substances have different molar masses (essentially, the weight of the individual atoms or molecules). In solids, thermal energy arises due to the number of atoms that are vibrating. “Molar” heat capacity *per mole of molecules*, for both gases and solids, offer figures which are arbitrarily large, since molecules may be arbitrarily large. Such heat capacities are thus not intensive quantities for this reason, since the quantity of mass being considered can be increased without limit.

Per Mole of Atoms

Conversely, for *molecular-based* substances (which also absorb heat into their internal degrees of freedom), massive, complex molecules with high atomic count—like octane—can store a great deal of energy per mole and yet are quite unremarkable on a mass basis, or on a per-atom basis. This is because, in fully excited systems, heat is stored independently by each atom in a substance, not primarily by the bulk motion of molecules.

Thus, it is the heat capacity per-mole-of-atoms, not per-mole-of-molecules, which is the intensive quantity, and which comes closest to being a constant for all substances at high temperatures. This relationship was noticed empirically in 1819, and is called the Dulong–Petit law, after its two discoverers. Historically, the fact that specific heat capacities are approximately equal when corrected by the presumed weight of the atoms of solids, was an important piece of data in favor of the atomic theory of matter.

Because of the connection of heat capacity to the number of atoms, some care should be taken to specify a mole-of-molecules basis vs. a mole-of-atoms basis, when comparing specific heat capacities of molecular solids and gases. Ideal gases have the same numbers of molecules per volume, so increasing molecular complexity adds heat capacity on a per-volume and per-mole-of-molecules basis, but may lower or raise heat capacity on a per-atom basis, depending on whether the temperature is sufficient to store energy as atomic vibration.

In solids, the quantitative limit of heat capacity in general is about 3 *R* per mole of atoms, where *R* is the ideal gas constant. This 3 *R* value is about 24.9 J/mole.K. Six degrees of freedom (three kinetic and three potential) are available to each atom. Each of these six contributes $\frac{1}{2}$R specific heat capacity per mole of atoms. This limit of 3 *R* per mole specific heat capacity is approached at room temperature for most solids, with significant departures at this temperature only for solids composed of the lightest atoms which are bound very strongly, such as beryllium (where the value is only of 66% of 3 *R*), or diamond (where it is only 24% of 3 *R*). These large departures are due to quantum effects which prevent full distribution of heat into all vibrational modes, when the energy difference between vibrational quantum states is very large compared to the average energy available to each atom from the ambient temperature.

For monatomic gases, the specific heat is only half of 3 *R* per mole, i.e. ($\frac{3}{2}$R per mole) due to loss of all potential energy degrees of freedom in these gases. For polyatomic gases, the heat capacity will be intermediate between these values on a per-mole-of-atoms basis, and (for heat-stable molecules) would approach the limit of 3 *R* per mole of atoms, for gases composed of complex molecules, and at higher temperatures at which all vibrational modes accept excitational energy. This is because very large and complex gas molecules may be thought of as relatively large blocks of solid matter which have lost only a relatively small fraction of degrees of freedom, as compared to a fully integrated solid.

For a list of heat capacities per atom-mole of various substances, in terms of R.

Corollaries of these Considerations for Solids (Volume-specific Heat Capacity)

Since the bulk density of a solid chemical element is strongly related to its molar mass (usually about 3 *R* per mole, as noted above), there exists a noticeable inverse correlation between a solid's density and its specific heat capacity on a per-mass basis. This is due to a very approximate tendency of atoms of most elements to be about the same size (and constancy of mole-specific heat capacity) resulting in a good correlation between the *volume* of any given solid chemical element and its total heat capacity. Another way of stating this, is that the volume-specific heat capacity (volumetric heat capacity) of solid elements is roughly a constant. The molar volume of solid elements is very roughly constant, and (even more reliably) so also is the molar heat capacity for most solid substances. These two factors determine the volumetric heat capacity, which as a bulk property may be striking in consistency. For example, the element uranium is a metal which has a density almost 36 times that of the metal lithium, but uranium's specific heat capacity on a volumetric basis (i.e. per given volume of metal) is only 18% larger than lithium's.

Since the volume-specific corollary of the Dulong–Petit specific heat capacity relationship requires that atoms of all elements take up (on average) the same volume in solids, there are many departures from it, with most of these due to variations in atomic size. For instance, arsenic, which is only 14.5% less dense than antimony, has nearly 59% more specific heat capacity on a mass basis. In other words; even though an ingot of arsenic is only about 17% larger than an antimony one of the same mass, it absorbs about 59% more heat for a given temperature rise. The heat capacity ratios of the two substances closely follows the ratios of their molar volumes (the ratios of numbers of atoms in the same volume of each substance); the departure from the correlation to simple volumes in this case is due to lighter arsenic atoms being significantly more closely packed than antimony atoms, instead of similar size. In other words, similar-sized atoms would cause a mole of arsenic to be 63% larger than a mole of antimony, with a correspondingly lower density, allowing its volume to more closely mirror its heat capacity behavior.

Other Factors

Hydrogen Bonds

Hydrogen-containing polar molecules like ethanol, ammonia, and water have powerful, intermolecular hydrogen bonds when in their liquid phase. These bonds provide another place where heat may be stored as potential energy of vibration, even at comparatively low temperatures. Hydrogen bonds account for the fact that liquid water stores nearly the theoretical limit of 3 R per mole of atoms, even at relatively low temperatures (i.e. near the freezing point of water).

Impurities

In the case of alloys, there are several conditions in which small impurity concentrations can greatly affect the specific heat. Alloys may exhibit marked difference in behaviour even in the case of small amounts of impurities being one element of the alloy; for example impurities in semiconducting ferromagnetic alloys may lead to quite different specific heat properties.

The Simple Case of the Monatomic Gas

In the case of a monatomic gas such as helium under constant volume, if it is assumed that no electronic or nuclear quantum excitations occur, each atom in the gas has only 3 degrees of freedom, all of a translational type. No energy dependence is associated with the degrees of freedom which define the position of the atoms. While, in fact, the degrees of freedom corresponding to the momenta of the atoms are quadratic, and thus contribute to the heat capacity. There are N atoms, each of which has 3 components of momentum, which leads to $3N$ total degrees of freedom. This gives:

$$C_V = \left(\frac{\partial U}{\partial T}\right)_V = \frac{3}{2} N k_B = \frac{3}{2} nR$$

$$C_{V,m} = \frac{C_V}{n} = \frac{3}{2} R$$

where

C_V is the heat capacity at constant volume of the gas

$C_{V,m}$ is the *molar heat capacity* at constant volume of the gas

N is the total number of atoms present in the container

n is the number of moles of atoms present in the container (n is the ratio of N and Avogadro's number)

R is the ideal gas constant, (8.3144621 J/(mol·K). R is equal to the product of Boltzmann's constant k_B and Avogadro's number

The following table shows experimental molar constant volume heat capacity measurements taken for each noble monatomic gas (at 1 atm and 25 °C):

Monatomic gas	$C_{V,m}$ (J/(mol·K))	$C_{V,m}/R$
He	12.5	1.50
Ne	12.5	1.50
Ar	12.5	1.50
Kr	12.5	1.50
Xe	12.5	1.50

It is apparent from the table that the experimental heat capacities of the monatomic noble gases agrees with this simple application of statistical mechanics to a very high degree.

The molar heat capacity of a monatomic gas at constant pressure is then

$$C_{p,m} = C_{V,m} + R = \frac{5}{2}R$$

Diatomic Gas

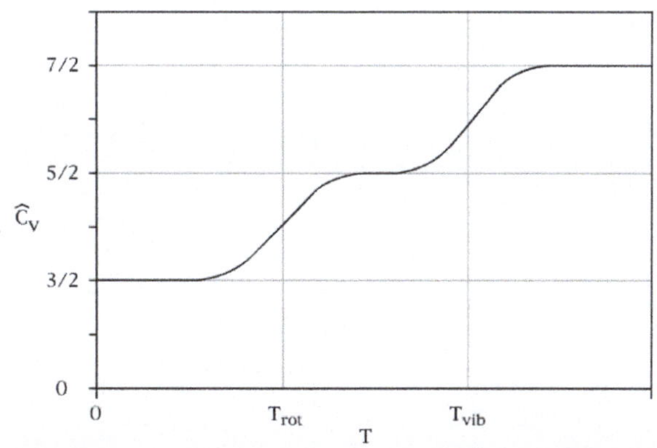

Constant volume specific heat capacity of a diatomic gas (idealised). As temperature increases, heat capacity goes from 3/2 R (translation contribution only), to 5/2 R (translation plus rotation), finally to a maximum of 7/2 R (translation + rotation + vibration)

In the somewhat more complex case of an ideal gas of diatomic molecules, the presence of internal degrees of freedom are apparent. In addition to the three translational degrees of freedom, there are rotational and vibrational degrees of freedom. In general, the number of degrees of freedom, f, in a molecule with n_a atoms is $3n_a$:

$$f = 3n_a$$

Mathematically, there are a total of three rotational degrees of freedom, one corresponding to rotation about each of the axes of three-dimensional space. However, in practice only the existence of two degrees of rotational freedom for linear molecules will be considered. This approximation is valid because the moment of inertia about the internuclear axis is vanishingly small with respect to other moments of inertia in the molecule (this is due to the very small rotational moments of single atoms, due to the concentration of almost all their mass at their centers; compare also the extremely small radii of the atomic nuclei compared to the distance between them in a diatomic molecule). Quantum mechanically, it can be shown that the interval between successive rotational energy eigenstates is inversely proportional to the moment of inertia about that axis. Because the moment of inertia about the internuclear axis is vanishingly small relative to the other two rotational axes, the energy spacing can be considered so high that no excitations of the rotational state can occur unless the temperature is extremely high. It is easy to calculate the expected number of vibrational degrees of freedom (or vibrational modes). There are three degrees of translational freedom, and two degrees of rotational freedom, therefore

$$f_{\text{vib}} = f - f_{\text{trans}} - f_{\text{rot}} = 6 - 3 - 2 = 1$$

Each rotational and translational degree of freedom will contribute $R/2$ in the total molar heat capacity of the gas. Each vibrational mode will contribute R to the total molar heat capacity, however. This is because for each vibrational mode, there is a potential and kinetic energy component. Both the potential and kinetic components will contribute $R/2$ to the total molar heat capacity of the gas. Therefore, a diatomic molecule would be expected to have a molar constant-volume heat capacity of

$$C_{V,m} = \frac{3R}{2} + R + R = \frac{7R}{2} = 3.5R$$

where the terms originate from the translational, rotational, and vibrational degrees of freedom, respectively.

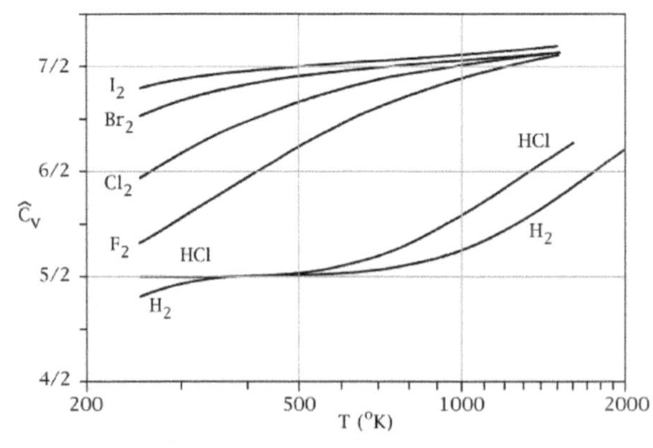

Constant volume specific heat capacity of diatomic gases (real gases) between about 200 K and 2000 K. This temperature range is not large enough to include both quantum transitions in all gases. Instead, at 200 K, all but hydrogen are fully rotationally excited, so all have at least 5/2 R heat capacity. (Hydrogen is already below 5/2, but it will require cryogenic conditions for even H2 to fall to 3/2 R). Further, only the heavier gases fully reach 7/2 R at the highest temperature, due to the relatively small vibrational energy spacing of these molecules. HCl and H2 begin to make the transition above 500 K, but have not achieved it by 1000 K, since their vibrational energy level spacing is too wide to fully participate in heat capacity, even at this temperature.

The following is a table of some molar constant-volume heat capacities of various diatomic gases at standard temperature (25 °C = 298 K)

Diatomic gas	$C_{V,m}$ (J/(mol·K))	$C_{V,m}/R$
H_2	20.18	2.427
CO	20.2	2.43
N_2	19.9	2.39
Cl_2	24.1	3.06
Br_2 (vapour)	28.2	3.39

From the above table, clearly there is a problem with the above theory. All of the diatomics examined have heat capacities that are lower than those predicted by the equipartition theorem, except Br_2. However, as the atoms composing the molecules become heavier, the heat capacities move closer to their expected values. One of the reasons for this phenomenon is the quantization of vibrational, and to a lesser extent, rotational states. In fact, if it is assumed that the molecules remain in their lowest energy vibrational state because the inter-level energy spacings for vibration-energies are large, the predicted molar constant volume heat capacity for a diatomic molecule becomes just that from the contributions of translation and rotation:

$$C_{V,m} = \frac{3R}{2} + R = \frac{5R}{2} = 2.5R$$

which is a fairly close approximation of the heat capacities of the lighter molecules in the above table. If the quantum harmonic oscillator approximation is made, it turns out that the quantum vibrational energy level spacings are actually inversely proportional to the square root of the reduced mass of the atoms composing the diatomic molecule. Therefore, in the case of the heavier diatomic molecules such as chlorine or bromine, the quantum vibrational energy level spacings become finer, which allows more excitations into higher vibrational levels at lower temperatures. This limit for storing heat capacity in vibrational modes, as discussed above, becomes 7R/2 = 3.5 R per mole of gas molecules, which is fairly consistent with the measured value for Br_2 at room temperature. As temperatures rise, all diatomic gases approach this value.

General Gas Phase

The specific heat of the gas is best conceptualized in terms of the degrees of freedom of an individual molecule. The different degrees of freedom correspond to the different ways in which the molecule may store energy. The molecule may store energy in its translational motion according to the formula:

$$E = \frac{1}{2}m\left(v_x^2 + v_y^2 + v_z^2\right)$$

where m is the mass of the molecule and $[v_x, v_y, v_z]$ is velocity of the center of mass of the molecule. Each direction of motion constitutes a degree of freedom, thus there are three translational degrees of freedom.

In addition, a molecule may have rotational motion. The kinetic energy of rotational motion is generally expressed as

$$E = \frac{1}{2}\left(I_1\omega_1^2 + I_2\omega_2^2 + I_3\omega_3^2\right)$$

where I is the moment of inertia tensor of the molecule, and $[\omega_1, \omega_2, \omega_3]$ is the angular velocity pseudo-vector (in a coordinate system aligned with the principal axes of the molecule). In general, then, there will be three additional degrees of freedom corresponding to the rotational motion of the molecule, (For linear molecules one of the inertia tensor terms vanishes and there are only two rotational degrees of freedom). The degrees of freedom corresponding to translations and rotations are called the rigid degrees of freedom, since they do not involve any deformation of the molecule.

The motions of the atoms in a molecule which are not part of its gross translational motion or rotation may be classified as vibrational motions. It can be shown that if there are n atoms in the molecule, there will be as many as $v = 3n - 3 - n_r$ vibrational degrees of freedom, where n_r is the number of rotational degrees of freedom. A vibrational degree of freedom corresponds to a specific way in which all the atoms of a molecule can vibrate. The actual number of possible vibrations may be less than this maximal one, due to various symmetries.

For example, triatomic nitrous oxide N_2O will have only 2 degrees of rotational freedom (since it is a linear molecule) and contains n=3 atoms: thus the number of possible vibrational degrees of freedom will be v = (3·3) – 3 – 2 = 4. There are four ways or "modes" in which the three atoms can vibrate, corresponding to *1)* A mode in which an atom at each end of the molecule moves away from, or towards, the center atom at the same time, *2)* a mode in which either end atom moves asynchronously with regard to the other two, and *3)* and *4)* two modes in which the molecule bends out of line, from the center, in the two possible planar directions that are orthogonal to its axis. Each vibrational degree of freedom confers TWO total degrees of freedom, since vibrational energy mode partitions into 1 kinetic and 1 potential mode. This would give nitrous oxide 3 translational, 2 rotational, and 4 vibrational modes (but these last giving 8 vibrational degrees of freedom), for storing energy. This is a total of f = 3 + 2 + 8 = 13 total energy-storing degrees of freedom, for N_2O.

For a bent molecule like water H_2O, a similar calculation gives 9 – 3 – 3 = 3 modes of vibration, and 3 (translational) + 3 (rotational) + 6 (vibrational) = 12 degrees of freedom.

The Storage of Energy into Degrees of Freedom

If the molecule could be entirely described using classical mechanics, then the theorem of equipartition of energy could be used to predict that each degree of freedom would have an average energy in the amount of $(1/2)kT$ where k is Boltzmann's constant and T is the temperature. Our calculation of the constant-volume heat capacity would be straightforward. Each molecule would be holding, on average, an energy of $(f/2)kT$ where f is the total number of degrees of freedom in the molecule. Note that $Nk = R$ if N is Avogadro's number, which is the case in considering the heat capacity of a mole of molecules. Thus, the total internal energy of the gas would be $(f/2)NkT$ where N is the total number of molecules. The heat capacity (at constant volume) would then be a constant $(f/2)Nk$ the mole-specific heat capacity would be $(f/2)R$ the molecule-specific heat capacity would be $(f/2)k$ and the dimensionless heat capacity would be just $f/2$. Here again, each vibrational degree of freedom contributes 2f. Thus, a mole of nitrous oxide would have a total constant-volume heat capacity (including vibration) of $(13/2)R$ by this calculation.

In summary, the molar heat capacity (mole-specific heat capacity) of an ideal gas with f degrees of freedom is given by

$$C_{V,m} = \frac{f}{2}R$$

This equation applies to all polyatomic gases, if the degrees of freedom are known.

The constant-pressure heat capacity for any gas would exceed this by an extra factor of R. As example C_p would be a total of (15/2)R/mole for nitrous oxide.

The Effect of Quantum Energy Levels in Storing Energy in Degrees of Freedom

The various degrees of freedom cannot generally be considered to obey classical mechanics, however. Classically, the energy residing in each degree of freedom is assumed to be continuous—it can take on any positive value, depending on the temperature. In reality, the amount of energy that may reside in a particular degree of freedom is quantized: It may only be increased and decreased in finite amounts. A good estimate of the size of this minimum amount is the energy of the first excited state of that degree of freedom above its ground state. For example, the first vibrational state of the hydrogen chloride (HCl) molecule has an energy of about 5.74×10^{-20} joule. If this amount of energy were deposited in a classical degree of freedom, it would correspond to a temperature of about 4156 K.

If the temperature of the substance is so low that the equipartition energy of $(1/2)kT$ is much smaller than this excitation energy, then there will be little or no energy in this degree of freedom. This degree of freedom is then said to be "frozen out». As mentioned above, the temperature corresponding to the first excited vibrational state of HCl is about 4156 K. For temperatures well below this value, the vibrational degrees of freedom of the HCl molecule will be frozen out. They will contain little energy and will not contribute to the thermal energy or the heat capacity of HCl gas.

Energy Storage Mode "Freeze-out" Temperatures

It can be seen that for each degree of freedom there is a critical temperature at which the degree of freedom "unfreezes" and begins to accept energy in a classical way. In the case of translational degrees of freedom, this temperature is that temperature at which the thermal wavelength of the molecules is roughly equal to the size of the container. For a container of macroscopic size (e.g.

10 cm) this temperature is extremely small and has no significance, since the gas will certainly liquify or freeze before this low temperature is reached. For any real gas translational degrees of freedom may be considered to always be classical and contain an average energy of $(3/2)kT$ per molecule.

The rotational degrees of freedom are the next to "unfreeze". In a diatomic gas, for example, the critical temperature for this transition is usually a few tens of kelvins, although with a very light molecule such as hydrogen the rotational energy levels will be spaced so widely that rotational heat capacity may not completely "unfreeze" until considerably higher temperatures are reached. Finally, the vibrational degrees of freedom are generally the last to unfreeze. As an example, for diatomic gases, the critical temperature for the vibrational motion is usually a few thousands of kelvins, and thus for the nitrogen in our example at room temperature, no vibration modes would be excited, and the constant-volume heat capacity at room temperature is $(5/2)R$/mole, not $(7/2)$ R/mole. As seen above, with some unusually heavy gases such as iodine gas I_2, or bromine gas Br_2, some vibrational heat capacity may be observed even at room temperatures.

It should be noted that it has been assumed that atoms have no rotational or internal degrees of freedom. This is in fact untrue. For example, atomic electrons can exist in excited states and even the atomic nucleus can have excited states as well. Each of these internal degrees of freedom are assumed to be frozen out due to their relatively high excitation energy. Nevertheless, for sufficiently high temperatures, these degrees of freedom cannot be ignored. In a few exceptional cases, such molecular electronic transitions are of sufficiently low energy that they contribute to heat capacity at room temperature, or even at cryogenic temperatures. One example of an electronic transition degree of freedom which contributes heat capacity at standard temperature is that of nitric oxide (NO), in which the single electron in an anti-bonding molecular orbital has energy transitions which contribute to the heat capacity of the gas even at room temperature.

An example of a nuclear magnetic transition degree of freedom which is of importance to heat capacity, is the transition which converts the spin isomers of hydrogen gas (H_2) into each other. At room temperature, the proton spins of hydrogen gas are aligned 75% of the time, resulting in *orthohydrogen* when they are. Thus, some thermal energy has been stored in the degree of freedom available when *parahydrogen* (in which spins are anti-aligned) absorbs energy, and is converted to the higher energy ortho form. However, at the temperature of liquid hydrogen, not enough heat energy is available to produce orthohydrogen (that is, the transition energy between forms is large enough to "freeze out" at this low temperature), and thus the parahydrogen form predominates. The heat capacity of the transition is sufficient to release enough heat, as orthohydrogen converts to the lower-energy parahydrogen, to boil the hydrogen liquid to gas again, if this evolved heat is not removed with a catalyst after the gas has been cooled and condensed. This example also illustrates the fact that some modes of storage of heat may not be in constant equilibrium with each other in substances, and heat absorbed or released from such phase changes may "catch up" with temperature changes of substances, only after a certain time. In other words, the heat evolved and absorbed from the ortho-para isomeric transition contributes to the heat capacity of hydrogen on long time-scales, but not on *short* time-scales. These time scales may also depend on the presence of a catalyst.

Less exotic phase-changes may contribute to the heat-capacity of substances and systems, as well, as (for example) when water is converted back and forth from solid to liquid or gas form.

Phase changes store heat energy entirely in breaking the bonds of the potential energy interactions between molecules of a substance. As in the case of hydrogen, it is also possible for phase changes to be hindered as the temperature drops, thus they do not catch up and become apparent, without a catalyst. For example, it is possible to supercool liquid water to below the freezing point, and not observe the heat evolved when the water changes to ice, so long as the water remains liquid. This heat appears instantly when the water freezes.

Solid Phase

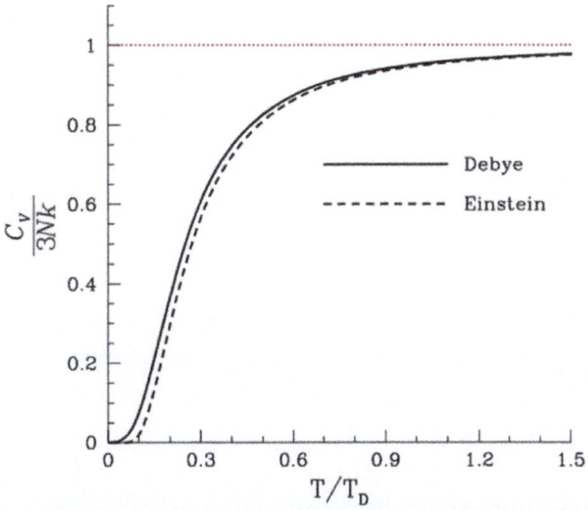

The dimensionless heat capacity divided by three, as a function of temperature as predicted by the Debye model and by Einstein's earlier model. The horizontal axis is the temperature divided by the Debye temperature. Note that, as expected, the dimensionless heat capacity is zero at absolute zero, and rises to a value of three as the temperature becomes much larger than the Debye temperature. The red line corresponds to the classical limit of the Dulong–Petit law

For matter in a crystalline solid phase, the Dulong–Petit law, which was discovered empirically, states that the molar heat capacity assumes the value 3 *R*. Indeed, for solid metallic chemical elements at room temperature, molar heat capacities range from about 2.8 *R* to 3.4 *R*. Large exceptions at the lower end involve solids composed of relatively low-mass, tightly bonded atoms, such as beryllium at 2.0 *R*, and diamond at only 0.735 *R*. The latter conditions create larger quantum vibrational energy spacing, thus many vibrational modes have energies too high to be populated (and thus are "frozen out") at room temperature. At the higher end of possible heat capacities, heat capacity may exceed *R* by modest amounts, due to contributions from anharmonic vibrations in solids, and sometimes a modest contribution from conduction electrons in metals. These are not degrees of freedom treated in the Einstein or Debye theories.

The theoretical maximum heat capacity for multi-atomic gases at higher temperatures, as the molecules become larger, also approaches the Dulong–Petit limit of 3 *R*, so long as this is calculated per mole of atoms, not molecules. The reason for this behavior is that, in theory, gases with very large molecules have almost the same high-temperature heat capacity as solids, lacking only the (small) heat capacity contribution that comes from potential energy that cannot be stored between separate molecules in a gas.

The Dulong–Petit limit results from the equipartition theorem, and as such is only valid in the classical limit of a microstate continuum, which is a high temperature limit. For light and non-metallic elements, as well as most of the common molecular solids based on carbon compounds at standard ambient temperature, quantum effects may also play an important role, as they do in multi-atomic gases. These effects usually combine to give heat capacities lower than 3 *R* per mole of *atoms* in the solid, although in molecular solids, heat capacities calculated *per mole of molecules* in molecular solids may be more than 3 *R*. For example, the heat capacity of water ice at the melting point is about 4.6 *R* per mole of molecules, but only 1.5 *R* per mole of atoms. As noted, heat capacity values far lower than 3 *R* "per atom" (as is the case with diamond and beryllium) result from "freezing out" of possible vibration modes for light atoms at suitably low temperatures, just as happens in many low-mass-atom gases at room temperatures (where vibrational modes are all frozen out). Because of high crystal binding energies, the effects of vibrational mode freezing are observed in solids more often than liquids: for example the heat capacity of liquid water is twice that of ice at near the same temperature, and is again close to the 3 *R* per mole of atoms of the Dulong–Petit theoretical maximum.

Liquid Phase

A general theory of the heat capacity of liquids has not yet been achieved, and is still an active area of research. It was long thought that phonon theory is not able to explain the heat capacity of liquids, since liquids only sustain longitudinal, but not transverse phonons, which in solids are responsible for 2/3 of the heat capacity. However, Brillouin scattering experiments with neutrons and with X-rays, confirming an intuition of Yakov Frenkel, have shown that transverse phonons do exist in liquids, albeit restricted to frequencies above a threshold called the Frenkel frequency. Since most energy is contained in these high-frequency modes, a simple modification of the Debye model is sufficient to yield a good approximation to experimental heat capacities of simple liquids.

Amorphous materials can be considered a type of liquid. The specific heat of amorphous materials has characteristic discontinuities at the glass transition temperature. These discontinuities are frequently used to detect the glass transition temperature where a supercooled liquid transforms to a glass.

Table of Specific Heat Capacities

Note that the especially high molar values, as for paraffin, gasoline, water and ammonia, result from calculating specific heats in terms of moles of *molecules*. If specific heat is expressed per mole of *atoms* for these substances, none of the constant-volume values exceed, to any large extent, the theoretical Dulong–Petit limit of 25 $J \cdot mol^{-1} \cdot K^{-1}$ = 3 R per mole of atoms. Paraffin, for example, has very large molecules and thus a high heat capacity per mole, but as a substance it does not have remarkable heat capacity in terms of volume, mass, or atom-mol (which is just 1.41 R per mole of atoms, or less than half of most solids, in terms of heat capacity per atom).

In the last column, major departures of solids at standard temperatures from the Dulong–Petit law value of 3 R, are usually due to low atomic weight plus high bond strength (as in diamond) causing some vibration modes to have too much energy to be available to store thermal energy at the measured temperature. For gases, departure from 3 R per mole of atoms in this table is generally due to two factors: (1) failure of the higher quantum-energy-spaced vibration modes in gas molecules to be excited at room temperature, and (2) loss of potential energy degree of freedom for small gas molecules, simply because most of their atoms are not bonded maximally in space to other atoms, as happens in many solids.

Notable minima and maxima are shown in maroon						
Table of specific heat capacities at 25 °C (298 K) unless otherwise noted. **Substance**	**Phase**	**Isobaric mass heat capacity** c_P **J·g⁻¹·K⁻¹**	**Isobaric molar heat capacity** $C_{P,m}$ **J·mol⁻¹·K⁻¹**	**Isochore molar heat capacity** $C_{V,m}$ **J·mol⁻¹·K⁻¹**	**Isobaric volumetric heat capacity** $C_{P,v}$ **J·cm⁻³·K⁻¹**	**Isochore atom-molar heat capacity in units of R** $C_{V,am}$ **atom-mol⁻¹**
Air (Sea level, dry, 0 °C (273.15 K))	gas	1.0035	29.07	20.7643	0.001297	~ 1.25 R
Air (typical room conditions[A])	gas	1.012	29.19	20.85	0.00121	~ 1.25 R
Aluminium	solid	0.897	24.2		2.422	2.91 R
Ammonia	liquid	4.700	80.08		3.263	3.21 R
Animal tissue (incl. human)	mixed	3.5			3.7*	
Antimony	solid	0.207	25.2		1.386	3.03 R
Argon	gas	0.5203	20.7862	12.4717		1.50 R
Arsenic	solid	0.328	24.6		1.878	2.96 R
Beryllium	solid	1.82	16.4		3.367	1.97 R
Bismuth	solid	0.123	25.7		1.20	3.09 R
Cadmium	solid	0.231	26.02			3.13 R
Carbon dioxide CO_2	gas	0.839*	36.94	28.46		1.14 R
Chromium	solid	0.449	23.35			2.81 R
Copper	solid	0.385	24.47		3.45	2.94 R
Diamond	solid	0.5091	6.115		1.782	0.74 R
Ethanol	liquid	2.44	112		1.925	1.50 R
Gasoline (octane)	liquid	2.22	228		1.64	1.05 R
Glass	solid	0.84			2.1	
Gold	solid	0.129	25.42		2.492	3.05 R
Granite	solid	0.790			2.17	
Graphite	solid	0.710	8.53		1.534	1.03 R
Helium	gas	5.1932	20.7862	12.4717		1.50 R
Hydrogen	gas	14.30	28.82			1.23 R
Hydrogen sulfide H_2S	gas	1.015*	34.60			1.05 R
Iron	solid	0.412	25.09		3.537	3.02 R

Lead	solid	0.129	26.4		1.44	3.18 R
Lithium	solid	3.58	24.8		1.912	2.98 R
Lithium at 181 °C	liquid	4.379	30.33		2.242	3.65 R
Magnesium	solid	1.02	24.9		1.773	2.99 R
Mercury	liquid	0.1395	27.98		1.888	3.36 R
Methane at 2 °C	gas	2.191	35.69			0.85 R
Methanol	liquid	2.14	68.62			1.38 R
Molten salt (142–540 °C)	liquid	1.56			2.62	
Nitrogen	gas	1.040	29.12	20.8		1.25 R
Neon	gas	1.0301	20.7862	12.4717		1.50 R
Oxygen	gas	0.918	29.38	21.0		1.26 R
Paraffin wax $C_{25}H_{52}$	solid	2.5 (ave)	900		2.325	1.41 R
Polyethylene (rotomolding grade)	solid	2.3027				
Silica (fused)	solid	0.703	42.2		1.547	1.69 R
Silver	solid	0.233	24.9		2.44	2.99 R
Sodium	solid	1.230	28.23			3.39 R
Steel	solid	0.466			3.756	
Tin	solid	0.227	27.112		1.659	3.26 R
Titanium	solid	0.523	26.060		2.6384	3.13 R
Tungsten	solid	0.134	24.8		2.58	2.98 R
Uranium	solid	0.116	27.7		2.216	3.33 R
Water at 100 °C (steam)	gas	2.080	37.47	28.03		1.12 R
Water at 25 °C	liquid	4.1813	75.327	74.53	4.1796	3.02 R
Water at 100 °C	liquid	4.1813	75.327	74.53	4.2160	3.02 R
Water at −10 °C (ice)	solid	2.05	38.09		1.938	1.53 R
Zinc	solid	0.387	25.2		2.76	3.03 R
Substance	Phase	Isobaric mass heat capacity c_P $J \cdot g^{-1} \cdot K^{-1}$	Isobaric molar heat capacity $C_{P,m}$ $J \cdot mol^{-1} \cdot K^{-1}$	Isochore molar heat capacity $C_{V,m}$ $J \cdot mol^{-1} \cdot K^{-1}$	Isobaric volumetric heat capacity $C_{P,v}$ $J \cdot cm^{-3} \cdot K^{-1}$	Isochore atom-molar heat capacity in units of R $C_{V,am}$ atom-mol^{-1}

[A] Assuming an altitude of 194 metres above mean sea level (the world–wide median altitude of human habitation), an indoor temperature of 23 °C, a dewpoint of 9 °C (40.85% relative humidity), and 760 mm–Hg sea level–corrected barometric pressure (molar water vapor content = 1.16%).
*Derived data by calculation. This is for water-rich tissues such as brain. The whole-body average figure for mammals is approximately 2.9 J·cm^{-3}·K^{-1}

Mass Heat Capacity of Building Materials

(Usually of interest to builders and solar designers)

Mass heat capacity of building materials		
Substance	**Phase**	c_P J·g^{-1}·K^{-1}
Asphalt	solid	0.920
Brick	solid	0.840
Concrete	solid	0.880
Glass, silica	solid	0.840
Glass, crown	solid	0.670
Glass, flint	solid	0.503
Glass, pyrex	solid	0.753
Granite	solid	0.790
Gypsum	solid	1.090
Marble, mica	solid	0.880
Sand	solid	0.835
Soil	solid	0.800
Water	liquid	4.1813
Wood	solid	1.7 (1.2 to 2.9)
Substance	**Phase**	c_P J g^{-1} K^{-1}

Surface Tension

Surface tension is the elastic tendency of a fluid surface which makes it acquire the least surface area possible. Surface tension allows insects (e.g. water striders), usually denser than water, to float and stride on a water surface.

Surface tension and hydrophobicity interact in this attempt to cut a water droplet.

Surface tension experimental demonstration with soap

At liquid-air interfaces, surface tension results from the greater attraction of liquid molecules to each other (due to cohesion) than to the molecules in the air (due to adhesion). The net effect is an inward force at its surface that causes the liquid to behave as if its surface were covered with a stretched elastic membrane. Thus, the surface becomes under tension from the imbalanced forces, which is probably where the term "surface tension" came from. Because of the relatively high attraction of water molecules for each other through a web of hydrogen bonds, water has a higher surface tension (72.8 millinewtons per meter at 20 °C) compared to that of most other liquids. Surface tension is an important factor in the phenomenon of capillarity.

Surface tension has the dimension of force per unit length, or of energy per unit area. The two are equivalent, but when referring to energy per unit of area, it is common to use the term surface energy, which is a more general term in the sense that it applies also to solids.

In materials science, surface tension is used for either surface stress or surface free energy.

Causes

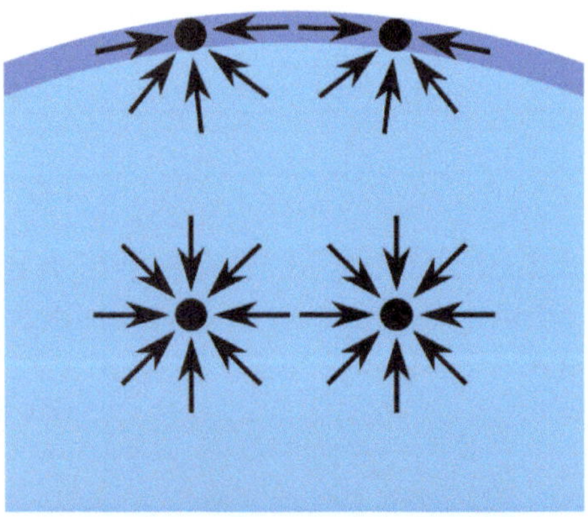

Diagram of the forces on molecules of a liquid

The cohesive forces among liquid molecules are responsible for the phenomenon of surface tension. In the bulk of the liquid, each molecule is pulled equally in every direction by neighboring liquid molecules, resulting in a net force of zero. The molecules at the surface do not have the *same* molecules on all sides of them and therefore are pulled inwards. This creates some internal pressure and forces liquid surfaces to contract to the minimal area.

Surface tension is responsible for the shape of liquid droplets. Although easily deformed, droplets of water tend to be pulled into a spherical shape by the imbalance in cohesive forces of the surface layer. In the absence of other forces, including gravity, drops of virtually all liquids would be approximately spherical. The spherical shape minimizes the necessary "wall tension" of the surface layer according to Laplace's law.

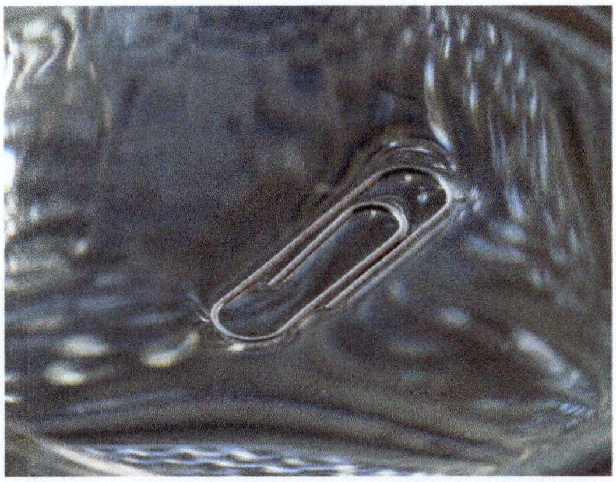

Surface tension preventing a paper clip from submerging.

Another way to view surface tension is in terms of energy. A molecule in contact with a neighbor is in a lower state of energy than if it were alone (not in contact with a neighbor). The interior molecules have as many neighbors as they can possibly have, but the boundary molecules are missing neighbors (compared to interior molecules) and therefore have a higher energy. For the liquid to minimize its energy state, the number of higher energy boundary molecules must be minimized. The minimized quantity of boundary molecules results in a minimal surface area. As a result of surface area minimization, a surface will assume the smoothest shape it can (mathematical proof that "smooth" shapes minimize surface area relies on use of the Euler–Lagrange equation). Since any curvature in the surface shape results in greater area, a higher energy will also result. Consequently, the surface will push back against any curvature in much the same way as a ball pushed uphill will push back to minimize its gravitational potential energy.

Effects of Surface Tension

Water

Several effects of surface tension can be seen with ordinary water:

A. Beading of rain water on a waxy surface, such as a leaf. Water adheres weakly to wax and strongly to itself, so water clusters into drops. Surface tension gives them their near-spherical shape, because a sphere has the smallest possible surface area to volume ratio.

B. Formation of drops occurs when a mass of liquid is stretched. The animation shows water adhering to the faucet gaining mass until it is stretched to a point where the surface tension can no longer keep the drop linked to the faucet. It then separates and surface tension forms the drop into a sphere. If a stream of water was running from the faucet, the stream would break up into drops during its fall. Gravity stretches the stream, then surface tension pinches it into spheres.

C. Flotation of objects denser than water occurs when the object is nonwettable and its weight is small enough to be borne by the forces arising from surface tension. For example, water striders use surface tension to walk on the surface of a pond by the following way. Nonwettability of leg of the water strider means no attraction between molecules of the leg and molecules of the water, so when the leg pushes down the water, the surface tension of the water only tries to recover its flatness from its deformation due to the leg. This behavior of the water push the water strider upward so it can stand on the surface of the water as long as its mass is small enough so that the water can support it. The surface of the water behaves like an elastic film: the insect's feet cause indentations in the water's surface, increasing its surface area and tendency of minimization of surface curvature (so area) of the water pushes the insect's feet upward.

D. Separation of oil and water (in this case, water and liquid wax) is caused by a tension in the surface between dissimilar liquids. This type of surface tension is called "interface tension", but its chemistry is the same.

E. Tears of wine is the formation of drops and rivulets on the side of a glass containing an alcoholic beverage. Its cause is a complex interaction between the differing surface tensions of water and ethanol; it is induced by a combination of surface tension modification of water by ethanol together with ethanol evaporating faster than water.

A. Water beading on a leaf

B. Water dripping from a tap

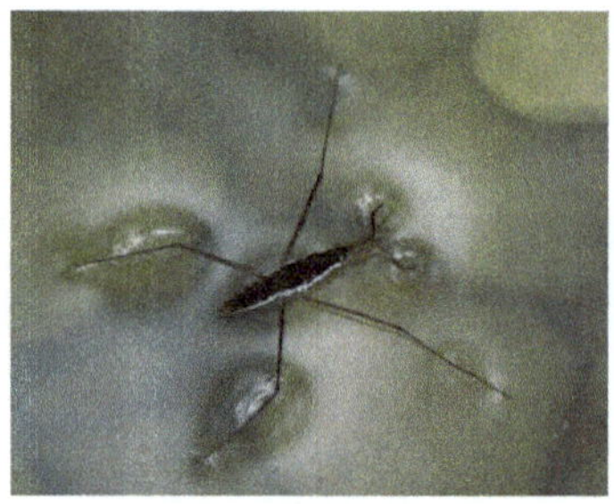

C. Water striders stay atop the liquid because of surface tension

D. Lava lamp with interaction between dissimilar liquids: water and liquid wax

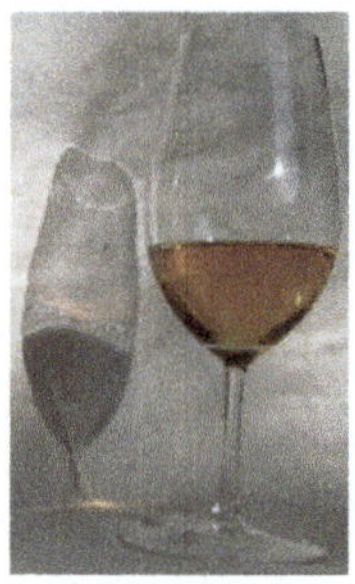

E. Photo showing the "tears of wine" phenomenon.

Surfactants

Surface tension is visible in other common phenomena, especially when surfactants are used to decrease it:

- Soap bubbles have very large surface areas with very little mass. Bubbles in pure water are unstable. The addition of surfactants, however, can have a stabilizing effect on the bubbles.

Note that surfactants actually reduce the surface tension of water by a factor of three or more.

- Emulsions are a type of colloid in which surface tension plays a role. Tiny fragments of oil suspended in pure water will spontaneously assemble themselves into much larger masses. But the presence of a surfactant provides a decrease in surface tension, which permits stability of minute droplets of oil in the bulk of water (or vice versa).

Physics

Physical Units

Surface tension, usually represented by the symbol γ, is measured in force per unit length. Its SI unit is newton per meter but the cgs unit of dyne per cm is also used.

$$\gamma = 1\frac{dyn}{cm} = 1\frac{erg}{cm^2} = 1\frac{mN}{m} = 0.001\frac{N}{m} = 0.001\frac{J}{m^2}$$

Surface Area Growth

This diagram illustrates the force necessary to increase the surface area.
This force is proportional to the surface tension.

Surface tension can be defined in terms of force or energy. In terms of force: surface tension γ of a liquid is the force per unit length. In the illustration on the right, the rectangular frame, composed of three unmovable sides (black) that form a "U" shape, and a fourth movable side (blue) that can slide to the right. Surface tension will pull the blue bar to the left; the force F required to hold immobile the movable side is proportional to the length L of the movable side. Thus the ratio $F >$ depends only on the intrinsic properties of the liquid (composition, temperature, etc.), not on its geometry. For example, if the frame had a more complicated shape, the ratio F / L, with L the length of the movable side and F the force required to stop it from sliding, is found to be the same for all shapes. We therefore define the surface tension as

$$\gamma = \frac{1}{2}\frac{F}{L}..$$

The reason for the $1/2$ is that the film has two sides, each of which contributes equally to the force; so the force contributed by a single side is $\gamma L = F / 2$.

In terms of energy: surface tension γ of a liquid is the ratio of 1) the change in the energy of the liquid, and 2) the change in the surface area of the liquid (that led to the change in energy). This can be easily related to the previous definition in terms of force: if F is the force required to stop the side from *starting* to slide, then this is also the force that would keep the side in the state of *sliding at a constant speed* (by Newton's Second Law). But if the side is moving to the right (in the direction the force is applied), then the surface area of the stretched liquid is increasing while the applied force is doing work on the liquid. This means that increasing the surface area increases the energy of the film. The work done by the force F in moving the side by distance Äx is $W = F$Äx; at the same time the total area of the film increases by Ä$A = 2L$Äx (the factor of 2 is here because the liquid has two sides, two surfaces). Thus, multiplying both the numerator and the denominator of $\gamma = (1/2)F/L$ by Äx, we get

$$\gamma = \frac{F}{2L} = \frac{F\Delta x}{2L\Delta x} = \frac{W}{\Delta A}.$$

This work W is, by the usual arguments, interpreted as being stored as potential energy. Consequently, surface tension can be also measured in SI system as joules per square meter and in the cgs system as ergs per cm^2. Since mechanical systems try to find a state of minimum potential energy, a free droplet of liquid naturally assumes a spherical shape, which has the minimum surface area for a given volume. The equivalence of measurement of energy per unit area to force per unit length can be proven by dimensional analysis.

Surface Curvature and Pressure

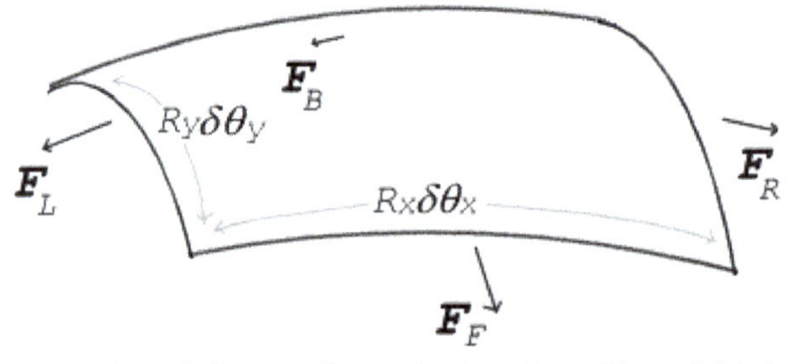

Surface tension forces acting on a tiny (differential) patch of surface. $\delta\theta_x$ and $\delta\theta_y$ indicate the amount of bend over the dimensions of the patch. Balancing the tension forces with pressure leads to the Young–Laplace equation

If no force acts normal to a tensioned surface, the surface must remain flat. But if the pressure on one side of the surface differs from pressure on the other side, the pressure difference times surface area results in a normal force. In order for the surface tension forces to cancel the force due to pressure, the surface must be curved. The diagram shows how surface curvature of a tiny patch of surface leads to a net component of surface tension forces acting normal to the center of the patch. When all the forces are balanced, the resulting equation is known as the Young–Laplace equation:

$$\Delta p = \gamma\left(\frac{1}{R_x} + \frac{1}{R_y}\right)$$

where:

- Δp is the pressure difference, known as the Laplace pressure.
- is surface tension.
- R_x and R_y are radii of curvature in each of the axes that are parallel to the surface.

The quantity in parentheses on the right hand side is in fact (twice) the mean curvature of the surface (depending on normalisation). Solutions to this equation determine the shape of water drops, puddles, menisci, soap bubbles, and all other shapes determined by surface tension (such as the shape of the impressions that a water strider's feet make on the surface of a pond). The table below shows how the internal pressure of a water droplet increases with decreasing radius. For not very small drops the effect is subtle, but the pressure difference becomes enormous when the drop sizes approach the molecular size. (In the limit of a single molecule the concept becomes meaningless.)

Δp for water drops of different radii at STP				
Droplet radius	1 mm	0.1 mm	1 μm	10 nm
Δp (atm)	0.0014	0.0144	1.436	143.6

Floating Objects

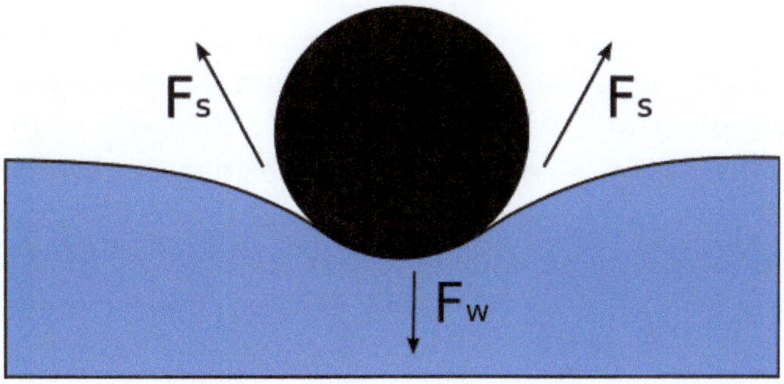

Cross-section of a needle floating on the surface of water. F_w is the weight and F_s are surface tension resultant forces.

When an object is placed on a liquid, its weight F_w depresses the surface, and if surface tension and downward force becomes equal than is balanced by the surface tension forces on either side F_s, which are each parallel to the water's surface at the points where it contacts the object. Notice that small moment in the body may cause the object sink as the angle of contact decreases surface tension decreaseases the horizontal components of the two F_s arrows point in opposite directions, so they cancel each other, but the vertical components point in the same direction and therefore add up to balance F_w. The object's surface must not be wettable for this to happen, and its weight must be low enough for the surface tension to support it.

$$F_w = 2F_s \cos\theta \Leftrightarrow \rho A_s Lg = 2\gamma Lz\cos\theta$$

Liquid Surface

To find the shape of the minimal surface bounded by some arbitrary shaped frame using strictly mathematical means can be a daunting task. Yet by fashioning the frame out of wire and dipping it in soap-solution, a locally minimal surface will appear in the resulting soap-film within seconds.

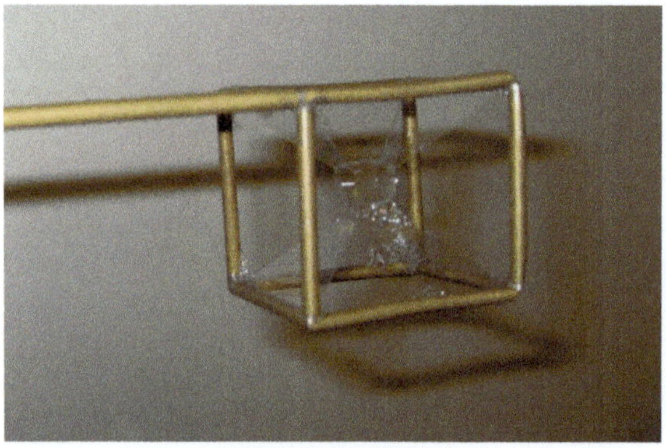

Minimal surface

The reason for this is that the pressure difference across a fluid interface is proportional to the mean curvature, as seen in the Young-Laplace equation. For an open soap film, the pressure difference is zero, hence the mean curvature is zero, and minimal surfaces have the property of zero mean curvature.

Contact Angles

The surface of any liquid is an interface between that liquid and some other medium. The top surface of a pond, for example, is an interface between the pond water and the air. Surface tension, then, is not a property of the liquid alone, but a property of the liquid's interface with another medium. If a liquid is in a container, then besides the liquid/air interface at its top surface, there is also an interface between the liquid and the walls of the container. The surface tension between the liquid and air is usually different (greater than) its surface tension with the walls of a container. And where the two surfaces meet, their geometry must be such that all forces balance.

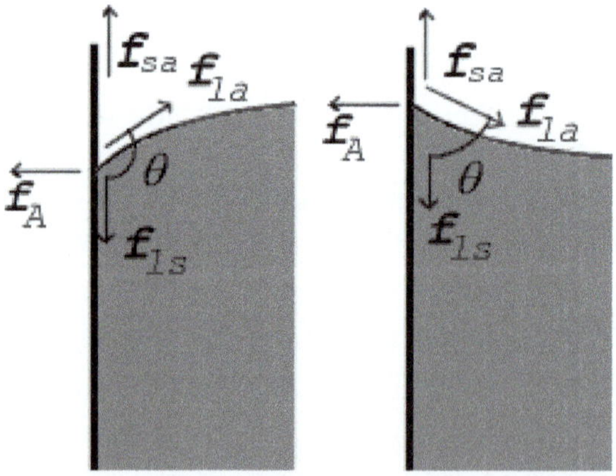

Forces at contact point shown for contact angle greater than 90° (left) and less than 90° (right)

Where the two surfaces meet, they form a contact angle, θ, which is the angle the tangent to the surface makes with the solid surface. The diagram to the right shows two examples. Tension forces are shown for the liquid-air interface, the liquid-solid interface, and the solid-air interface. The example on the left is where the difference between the liquid-solid and solid-air surface tension, $\gamma_{ls} - \gamma_{sa}$, is less than the liquid-air surface tension, γ_{la}, but is nevertheless positive, that is

$$\gamma_{la} > \gamma_{ls} - \gamma_{sa} > 0$$

In the diagram, both the vertical and horizontal forces must cancel exactly at the contact point, known as equilibrium. The horizontal component of f_{la} is canceled by the adhesive force, f_A.

$$f_A = f_{la} \sin\theta$$

The more telling balance of forces, though, is in the vertical direction. The vertical component of f_{la} must exactly cancel the force, f_{ls}.

$$f_{ls} - f_{sa} = -f_{la} \cos\theta$$

<table>
<tr><th>Liquid</th><th>Solid</th><th>Contact angle</th></tr>
<tr><td>water</td><td rowspan="6">soda-lime glass
lead glass
fused quartz</td><td rowspan="6">0°</td></tr>
<tr><td>ethanol</td></tr>
<tr><td>diethyl ether</td></tr>
<tr><td>carbon tetrachloride</td></tr>
<tr><td>glycerol</td></tr>
<tr><td>acetic acid</td></tr>
<tr><td rowspan="2">water</td><td>paraffin wax</td><td>107°</td></tr>
<tr><td>silver</td><td>90°</td></tr>
<tr><td rowspan="3">methyl iodide</td><td>soda-lime glass</td><td>29°</td></tr>
<tr><td>lead glass</td><td>30°</td></tr>
<tr><td>fused quartz</td><td>33°</td></tr>
<tr><td>mercury</td><td>soda-lime glass</td><td>140°</td></tr>
<tr><td colspan="3">Some liquid-solid contact angles</td></tr>
</table>

Since the forces are in direct proportion to their respective surface tensions, we also have:

$$\gamma_{ls} - \gamma_{sa} = -\gamma_{la} \cos\theta$$

where

- γ_{ls} is the liquid-solid surface tension,
- γ_{la} is the liquid-air surface tension,
- γ_{sa} is the solid-air surface tension,
- θ is the contact angle, where a concave meniscus has contact angle less than 90°

and a convex meniscus has contact angle of greater than 90°.

This means that although the difference between the liquid-solid and solid-air surface tension, $\gamma_{\text{ls}} - \gamma_{\text{sa}}$, is difficult to measure directly, it can be inferred from the liquid-air surface tension, γ_{la}, and the equilibrium contact angle, θ, which is a function of the easily measurable advancing and receding contact angles.

This same relationship exists in the diagram on the right. But in this case we see that because the contact angle is less than 90°, the liquid-solid/solid-air surface tension difference must be negative:

$$\gamma_{\text{la}} > 0 > \gamma_{\text{ls}} - \gamma_{\text{sa}}$$

Special Contact Angles

Observe that in the special case of a water-silver interface where the contact angle is equal to 90°, the liquid-solid/solid-air surface tension difference is exactly zero.

Another special case is where the contact angle is exactly 180°. Water with specially prepared Teflon approaches this. Contact angle of 180° occurs when the liquid-solid surface tension is exactly equal to the liquid-air surface tension.

$$\gamma_{\text{la}} = \gamma_{\text{ls}} - \gamma_{\text{sa}} > 0 \qquad \theta = 180^{\circ}$$

Methods of Measurement

Surface tension can be measured using the pendant drop method on a goniometer.

Because surface tension manifests itself in various effects, it offers a number of paths to its measurement. Which method is optimal depends upon the nature of the liquid being measured, the conditions under which its tension is to be measured, and the stability of its surface when it is deformed.

- du Noüy ring method: The traditional method used to measure surface or interfacial tension. Wetting properties of the surface or interface have little influence on this measuring technique. Maximum pull exerted on the ring by the surface is measured.

- Du Noüy-Padday method: A minimized version of Du Noüy method uses a small diameter metal needle instead of a ring, in combination with a high sensitivity microbalance to record maximum pull. The advantage of this method is that very small sample volumes (down to few tens of microliters) can be measured with very high precision, without the need to correct for buoyancy (for a needle or rather, rod, with proper geometry). Further, the measurement can be performed very quickly, minimally in about 20 seconds. First commercial multichannel tensiometers [CMCeeker] were recently built based on this principle.
- Wilhelmy plate method: A universal method especially suited to check surface tension over long time intervals. A vertical plate of known perimeter is attached to a balance, and the force due to wetting is measured.
- Spinning drop method: This technique is ideal for measuring low interfacial tensions. The diameter of a drop within a heavy phase is measured while both are rotated.
- Pendant drop method: Surface and interfacial tension can be measured by this technique, even at elevated temperatures and pressures. Geometry of a drop is analyzed optically.
- Bubble pressure method (Jaeger's method): A measurement technique for determining surface tension at short surface ages. Maximum pressure of each bubble is measured.
- Drop volume method: A method for determining interfacial tension as a function of interface age. Liquid of one density is pumped into a second liquid of a different density and time between drops produced is measured.
- Capillary rise method: The end of a capillary is immersed into the solution. The height at which the solution reaches inside the capillary is related to the surface tension by the equation discussed below.
- Stalagmometric method: A method of weighting and reading a drop of liquid.
- Sessile drop method: A method for determining surface tension and density by placing a drop on a substrate and measuring the contact angle.
- Vibrational frequency of levitated drops: The natural frequency of vibrational oscillations of magnetically levitated drops has been used to measure the surface tension of superfluid ^4He. This value is estimated to be 0.375 dyn/cm at T = 0 K.
- Resonant oscillations of spherical and hemispherical liquid drop: The technique is based on measuring the resonant frequency of spherical and hemispherical pendant droplets driven in oscillations by a modulated electric field. The surface tension and viscosity can be evaluated from the obtained resonant curves.

Effects

Liquid in a Vertical Tube

An old style mercury barometer consists of a vertical glass tube about 1 cm in diameter partially filled with mercury, and with a vacuum (called Torricelli's vacuum) in the unfilled volume. Notice that the mercury level at the center of the tube is higher than at the edges, making the upper surface of the mercury dome-shaped. The center of mass of the entire column of mercury would be slightly lower if the top surface of the mercury were flat over the entire crossection of the tube. But the dome-shaped top gives slightly less surface area to the entire mass of mercury. Again the two effects combine to minimize the total potential energy. Such a surface shape is known as a convex meniscus.

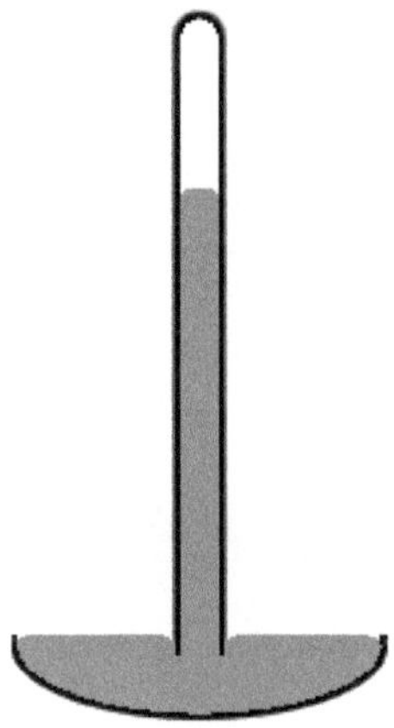

Diagram of a mercury barometer

We consider the surface area of the entire mass of mercury, including the part of the surface that is in contact with the glass, because mercury does not adhere to glass at all. So the surface tension of the mercury acts over its entire surface area, including where it is in contact with the glass. If instead of glass, the tube was made out of copper, the situation would be very different. Mercury aggressively adheres to copper. So in a copper tube, the level of mercury at the center of the tube will be lower than at the edges (that is, it would be a concave meniscus). In a situation where the liquid adheres to the walls of its container, we consider the part of the fluid's surface area that is in contact with the container to have *negative* surface tension. The fluid then works to maximize the contact surface area. So in this case increasing the area in contact with the container decreases rather than increases the potential energy. That decrease is enough to compensate for the increased potential energy associated with lifting the fluid near the walls of the container.

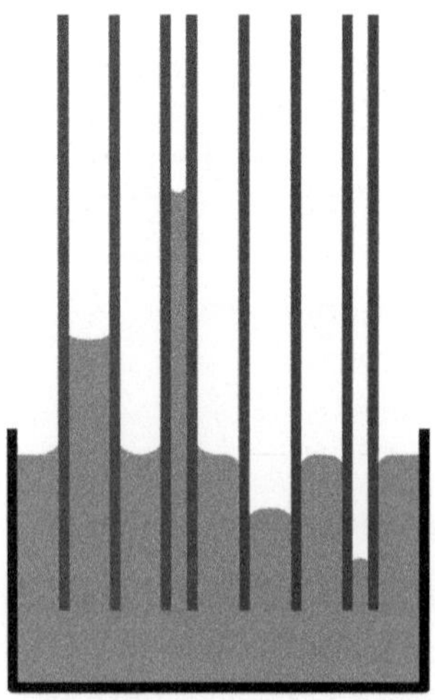

Illustration of capillary rise and fall. Red=contact angle less than 90°; blue=contact angle greater than 90°

If a tube is sufficiently narrow and the liquid adhesion to its walls is sufficiently strong, surface tension can draw liquid up the tube in a phenomenon known as capillary action. The height to which the column is lifted is given by:

$$h = \frac{2\gamma_{la} \cos\theta}{\rho g r}$$

where

- h is the height the liquid is lifted,
- γ_{la} is the liquid-air surface tension,
- ρ is the density of the liquid,
- r is the radius of the capillary,
- g is the acceleration due to gravity,
- θ is the angle of contact described above. If θ is greater than 90°, as with mercury in a glass container, the liquid will be depressed rather than lifted.

Puddles on a Surface

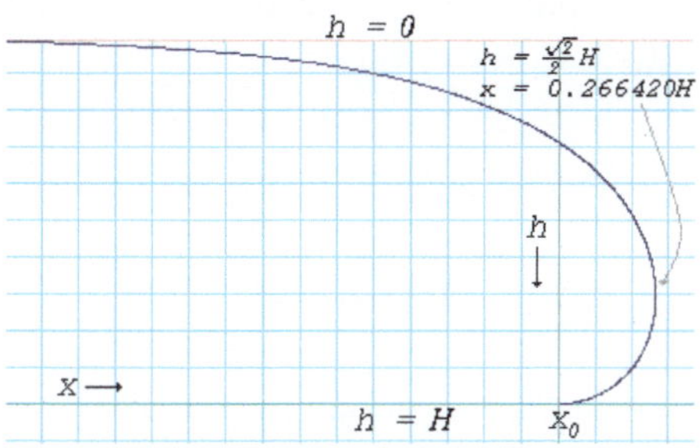

Profile curve of the edge of a puddle where the contact angle is 180°. The curve is given by the formula: where

Small puddles of water on a smooth clean surface have perceptible thickness.

Pouring mercury onto a horizontal flat sheet of glass results in a puddle that has a perceptible thickness. The puddle will spread out only to the point where it is a little under half a centimetre thick, and no thinner. Again this is due to the action of mercury's strong surface tension. The liquid mass flattens out because that brings as much of the mercury to as low a level as possible, but the surface tension, at the same time, is acting to reduce the total surface area. The result of the compromise is a puddle of a nearly fixed thickness.

The same surface tension demonstration can be done with water, lime water or even saline, but only on a surface made of a substance to which water does not adhere. Wax is such a substance. Water poured onto a smooth, flat, horizontal wax surface, say a waxed sheet of glass, will behave similarly to the mercury poured onto glass.

The thickness of a puddle of liquid on a surface whose contact angle is 180° is given by:

$$h = 2\sqrt{\frac{\gamma}{g\rho}}$$

where

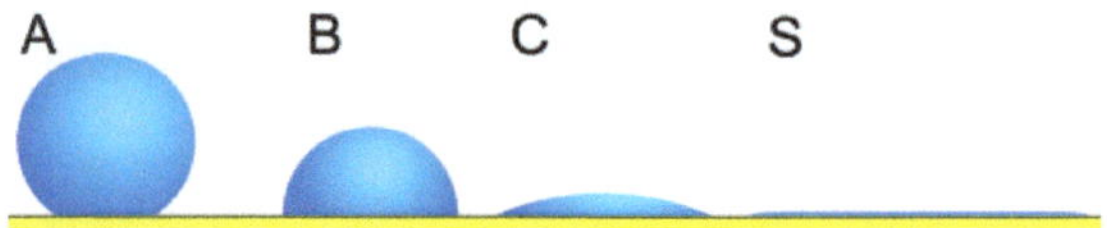

Illustration of how lower contact angle leads to reduction of puddle depth

In reality, the thicknesses of the puddles will be slightly less than what is predicted by the above formula because very few surfaces have a contact angle of 180° with any liquid. When the contact angle is less than 180°, the thickness is given by:

$$h = \sqrt{\frac{2\gamma_{la}\left(1-\cos\theta\right)}{g\rho}}.$$

For mercury on glass, γ_{Hg} = 487 dyn/cm, ρ_{Hg} = 13.5 g/cm³ and θ = 140°, which gives h_{Hg} = 0.36 cm. For water on paraffin at 25 °C, γ = 72 dyn/cm, ρ = 1.0 g/cm³, and θ = 107° which gives h_{H2O} = 0.44 cm.

The formula also predicts that when the contact angle is 0°, the liquid will spread out into a micro-thin layer over the surface. Such a surface is said to be fully wettable by the liquid.

The Breakup of Streams into Drops

In day-to-day life all of us observe that a stream of water emerging from a faucet will break up into droplets, no matter how smoothly the stream is emitted from the faucet. This is due to a phenomenon called the Plateau–Rayleigh instability, which is entirely a consequence of the effects of surface tension.

The explanation of this instability begins with the existence of tiny perturbations in the stream. These are always present, no matter how smooth the stream is. If the perturbations are resolved into sinusoidal components, we find that some components grow with time while others decay with time. Among those that grow with time, some grow at faster rates than others. Whether a component decays or grows, and how fast it grows is entirely a function of its wave number (a measure of how many peaks and troughs per centimeter) and the radii of the original cylindrical stream.

Surface Tension of Water and of Seawater

The two most abundant liquids on Earth are fresh water and seawater.

Surface Tension of Water

The surface tension of pure liquid water in contact with its vapor has been given by IAPWS as

$$\gamma_w = 235.8\left(1-\frac{T}{T_C}\right)^{1.256}\left[1-0.625\left(1-\frac{T}{T_C}\right)\right]\frac{\text{mN}}{\text{m}}$$

where both T and the critical temperature, T_C = 647.098 K, are expressed in kelvin. The region of validity the entire vapor-liquid saturation curve, from the triple point (0.01 °C) to the critical point. It also provides reasonable results when extrapolated to metastable (supercooled) conditions, down to at least −25 °C. This formulation was originally adopted by IAPWS in 1976, and was adjusted in 1994 to conform to the International Temperature Scale of 1990.

The uncertainty of this formulation is given over the full range of temperature by IAPWS. For temperatures below 100 °C, the uncertainty is ± 0.5%.

Surface Tension of Seawater

Nayar et al. published reference data for the surface tension of seawater over the salinity range of $20 \le S \le 131$ g/kg and a temperature range of $1 \le t \le 92$ °C at atmospheric pressure. The uncertainty of the measurements varied from 0.18 to 0.37 mN/m with the average uncertainty being 0.22 mN/m. This data is correlated by the following equation

$$\gamma_{sw} = \gamma_w\left[1+3.766\times10^{-4}S+2.347\times10^{-6}St\right]$$

where γ_{sw} is the surface tension of seawater in mN/m, γ_w is the surface tension of water in mN/m, S is the Reference Salinity in g/kg, and t is temperature in degrees Celsius. The average absolute percentage deviation between measurements and the correlation was 0.19% while the maximum deviation is 0.60%.

The range of temperature and salinity encompasses both the oceanographic range and the range of conditions encountered in thermal desalination technologies.

Data Table

Surface tension of various liquids in dyn/cm against air Mixture compositions denoted "%" are by mass dyn/cm is equivalent to the SI units of mN/m (millinewton per meter)		
Liquid	**Temperature °C**	**Surface tension, γ**
Acetic acid	20	27.60
Acetic acid (40.1%) + Water	30	40.68

Acetic acid (10.0%) + Water	30	54.56
Acetone	20	23.70
Diethyl ether	20	17.00
Ethanol	20	22.27
Ethanol (40%) + Water	25	29.63
Ethanol (11.1%) + Water	25	46.03
Glycerol	20	63.00
n-Hexane	20	18.40
Hydrochloric acid 17.7M aqueous solution	20	65.95
Isopropanol	20	21.70
Liquid helium II	−273	0.37
Liquid nitrogen	−196	8.85
Mercury	15	487.00
Methanol	20	22.60
n-Octane	20	21.80
Sodium chloride 6.0M aqueous solution	20	82.55
Sucrose (55%) + water	20	76.45
Water	0	75.64
Water	25	71.97
Water	50	67.91
Water	100	58.85
Toluene	25	27.73

Fused Quartz

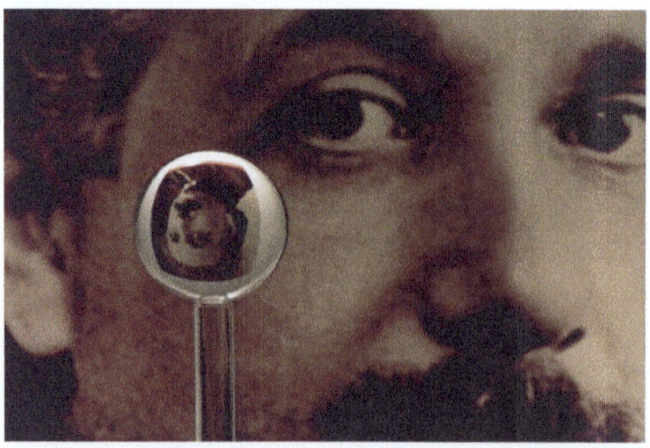

A fused quartz sphere manufactured for use in a gyroscope in the Gravity Probe B experiment. It is one of the most accurate spheres ever manufactured, deviating from a perfect sphere by no more than 40 atoms of thickness. It is thought that only neutron stars are smoother.

Fused quartz or fused silica is glass consisting of silica in amorphous (non-crystalline) form. It differs from traditional glasses in containing no other ingredients, which are typically added to glass to lower the melt temperature. Fused silica, therefore, has high working and melting temperatures. The optical and thermal properties of fused quartz are superior to those of other types of glass due to its purity. For these reasons, it finds use in situations such as semiconductor fabrication and laboratory equipment. It has better ultraviolet transmission than most other glasses, and so is used to make lenses and other optics for the ultraviolet spectrum. Its low coefficient of thermal expansion also makes it a useful material for precision mirror substrates.

Manufacture

Fused quartz is produced by fusing (melting) high-purity silica sand, which consists of quartz crystals. Quartz contains only silicon and oxygen, although commercial quartz glass often contains impurities. The most dominant impurities are aluminium and titanium.

Fusion

Melting is effected at approximately 1650 °C (3000 F) using either an electrically heated furnace (electrically fused) or a gas/oxygen-fuelled furnace (flame fused). Fused silica can be made from almost any silicon-rich chemical precursor, usually using a continuous process which involves flame oxidation of volatile silicon compounds to silicon dioxide, and thermal fusion of the resulting dust (although there are alternative processes). This results in a transparent glass with an ultra-high purity and improved optical transmission in the deep ultraviolet. One common method involves adding silicon tetrachloride to a hydrogen–oxygen flame, however use of this precursor results in environmentally unfriendly by-products including chlorine and hydrochloric acid.

Product Quality

Fused quartz is normally transparent, the process of fusion results in a material that is translucent. The material can however appear opaque owing to the presence of small air bubbles trapped within the material. The water content (and therefore infrared transmission of fused quartz and fused silica) is determined by the manufacturing process. Flame fused material always has a higher water content due to the combination of the hydrocarbons and oxygen fuelling the furnace forming hydroxyl [OH] groups within the material. An IR grade material typically has an [OH] content of <10 parts per million.

Applications

Most of the applications of fused silica exploit its wide transparency range, which extends from the UV to the near IR. Fused silica is the key starting material for optical fiber, used for telecommunications.

Because of its strength and high melting point (compared to ordinary glass), fused silica is used as an envelope for halogen lamps and high-intensity discharge lamps, which must operate at a high envelope temperature to achieve their combination of high brightness and long life. Vacuum tubes with silica envelopes allowed for radiation cooling by incandescent anodes.

Because of its strength fused silica was used in deep diving vessels such as the Bathysphere and Benthoscope.

The combination of strength, thermal stability, and UV transparency makes it an excellent substrate for projection masks for photolithography.

An EPROM with fused quartz window in the top of the package

Its UV transparency also finds uses in the semiconductor industry; an EPROM, or *erasable programmable read only memory,* is a type of memory chip that retains its data when its power supply is switched off, but which can be erased by exposure to strong ultraviolet light. EPROMs are recognizable by the transparent fused quartz window which sits on top of the package, through which the silicon chip is visible, and which permits exposure to UV light during erasing.

Due to the thermal stability and composition it is used in semiconductor fabrication furnaces.

Fused quartz has nearly ideal properties for fabricating first surface mirrors such as those used in telescopes. The material behaves in a predictable way and allows the optical fabricator to put a very smooth polish onto the surface and produce the desired figure with fewer testing iterations. In some instances, a high-purity UV grade of fused quartz has been used to make several of the individual uncoated lens elements of special purpose lenses including the Zeiss 105mm f/4.3 UV Sonnar, a lens formerly made for the Hasselblad camera, and the Nikon UV-Nikkor 105mm f/4.5 (presently sold as the Nikon PF10545MF-UV) lens. These lenses are used for UV photography, as the quartz glass has a lower extinction rate than lens made with more common flint or crown glass formulas.

Fused quartz can be metallised and etched for use as a substrate for high precision microwave circuits, the thermal stability making it a good choice for narrowband filters and similar demanding applications. The lower dielectric constant than alumina allows higher impedance tracks or thinner substrates.

Fused quartz is also the material used for modern glass instruments such as the glass harp and the verrophone, and is also used for new builds of the historical glass harmonica. Here, the superior strength and structure of fused quartz gives it a greater dynamic range than the historically used lead crystal, and a clearer sound.

Refractory Material Applications

Fused silica as an industrial raw material is used to make various refractory shapes such as crucibles, trays, shrouds, and rollers for many high-temperature thermal processes including steelmaking, investment casting, and glass manufacture. Refractory shapes made from fused silica have excellent thermal shock resistance and are chemically inert to most elements and compounds including virtually all acids, regardless of concentration, except hydrofluoric acid which is very reactive even in fairly low concentrations. Translucent fused silica tubes are commonly used to sheathe electric elements in room heaters, industrial furnaces and other similar applications.

Owing to its low mechanical damping at ordinary temperatures, it is used for high-Q resonators, in particular, for wine-glass resonator of hemispherical resonator gyro (HRG).

Quartz glassware is occasionally used in chemistry laboratories when standard borosilicate glass cannot withstand high temperatures; it is more commonly found as a very basic element, such as a tube in a furnace, or as a flask, the elements in direct exposure to the heat.

Physical Properties

The extremely low coefficient of thermal expansion, about 5.5×10^{-7}/°C (20–320 °C), accounts for its remarkable ability to undergo large, rapid temperature changes without cracking .

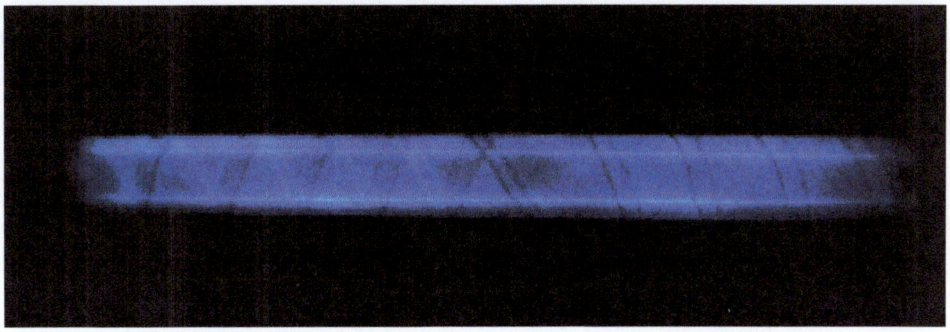

Phosphorescence in fused quartz from an extremely intense pulse of ultraviolet light, centered at 170 nm, in a flashtube

Fused quartz is prone to phosphorescence and "solarisation" (purplish discoloration) under intense UV illumination, as is often seen in flashtubes. "UV grade" synthetic fused silica (sold under various tradenames including "HPFS", "Spectrosil" and "Suprasil") has a very low metallic impurity content making it transparent deeper into the ultraviolet. An optic with a thickness of 1 cm has a transmittance of about 50% at a wavelength of 170 nm, which drops to only a few percent at 160 nm. However, its infrared transmission is limited by strong water absorptions at 2.2 μm and 2.7 μm.

"Infrared grade" fused quartz (tradenames "Infrasil", "Vitreosil IR" and others), which is electrically fused, has a greater presence of metallic impurities, limiting its UV transmittance wavelength to around 250 nm, but a much lower water content, leading to excellent infrared transmission up to 3.6 μm wavelength. All grades of transparent fused quartz/fused silica have nearly identical physical properties.

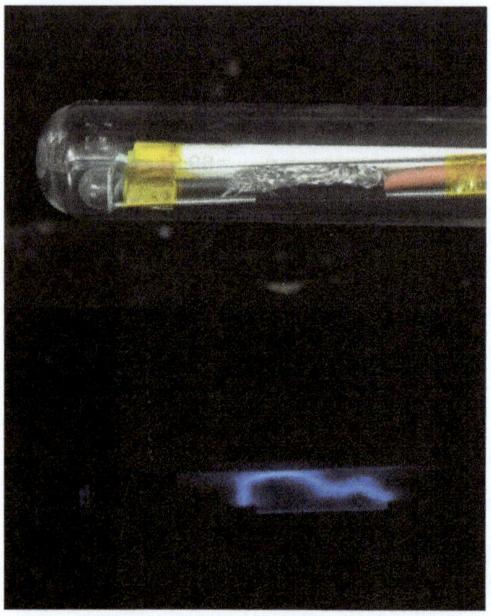

Phosphorescence of the quartz ignition tube of an air-gap flash

Optical Properties

The optical dispersion of fused silica can be approximated by the following Sellmeier equation:

$$\varepsilon = n^2 = 1 + \frac{0.6961663\lambda^2}{\lambda^2 - 0.0684043^2} + \frac{0.4079426\lambda^2}{\lambda^2 - 0.1162414^2} + \frac{0.8974794\lambda^2}{\lambda^2 - 9.896161^2},$$

where the wavelength λ is measured in micrometers. This equation is valid between 0.21 and 3.71 micrometers and at 20 °C. Its validity was confirmed for wavelengths up to 6.7 μm. Experimental data for the real (refractive index) and imaginary (absorption index) parts of the complex refractive index of fused quartz reported in the literature over the spectral range from 30 nm to 1000 μm has been reviewed by Kitamura et al. and are available online.

Typical Properties of Clear Fused Silica

- Density: 2.203 g/cm^3
- Hardness: 5.3–6.5 (Mohs scale), 8.8 GPa
- Tensile strength: 48.3 MPa
- Compressive strength: >1.1 GPa
- Bulk modulus: ~37 GPa
- Rigidity modulus: 31 GPa
- Young's modulus: 71.7 GPa
- Poisson's ratio: 0.17
- Lamé elastic constants: λ = 15.87 GPa, μ = 31.26 GPa

- Coefficient of thermal expansion: 5.5×10^{-7}/°C (average from 20 °C to 320 °C)
- Thermal conductivity: 1.3 W/(m·K)
- Specific heat capacity: 45.3 J/(mol·K)
- Softening point: ≈ 1665 °C
- Annealing point: ≈ 1140 °C
- Strain point: 1070 °C
- Electrical resistivity: $>10^{18}$ Ω·m
- Dielectric constant: 3.75 at 20 °C 1 MHz
- Dielectric loss factor: less than 0.0004 at 20 °C 1 MHz typically 6E-5 at 10 GHz
- Index of refraction: n_d = 1.4585 (at 587.6 nm)
- Change of refractive index with temperature (0 to 700 °C): 1.28×10^{-5}/°C (between 20 and 30 °C)
- Strain-optic coefficients: p_{11} = 0.113, p_{12} = 0.252.
- Hamaker constant: A = 6.5×10^{-20} J.
- Dielectric strength: 250–400 kV/cm at 20 °C
- Surface tension: 0.300 N/m at 1800–2400 °C

Borosilicate Glass

Guitar slide made of borosilicate glass

Borosilicate glass is a type of glass with silica and boron trioxide as the main glass-forming constituents. Borosilicate glasses are known for having very low coefficients of thermal expansion ($\sim 3 \times 10^{-6}$ K^{-1} at 20 °C), making them resistant to thermal shock, more so than any other common glass. Such glass is less subject to thermal stress and is commonly used for the construction of reagent bottles. Borosilicate glass is sold under such trade names as Simax, Borcam, Borosil, Suprax, Kimax, Heatex, Endural, Schott, or Refmex, Kimble, and some (but not all) items sold under the trade name Pyrex.

History

Borosilicate glass was first developed by German glassmaker Otto Schott in the late 19th century. Otto Schott is also founder of today's SCHOTT AG, which sells borosilicate glass under the brand name DURAN since 1893. Another manufacturer of DURAN is the DURAN Group. After Corning Glass Works introduced Pyrex in 1915, the name became a synonym for borosilicate glass in the English-speaking world. However, borosilicate glass is the name of a glass family with various members tailoring completely different purposes. Most common today is borosilicate 3.3 glass like SCHOTT Duran and Pyrex by Corning.

The European manufacturer of Pyrex, Arc International, uses borosilicate glass in its Pyrex glass kitchen products; however, the U.S. manufacturer of Pyrex kitchenware uses tempered soda-lime glass. Thus Pyrex can refer to either soda-lime glass or borosilicate glass when discussing kitchen glassware, while Pyrex, Bomex, Duran, TGI and Simax all refer to borosilicate glass when discussing laboratory glassware. The real difference is the trademark and the company that owns the Pyrex name. The original Corning ware made of borosilicate glass was trademarked in capital letters (PYREX). When the kitchenware division was sold, the trademark was changed to lowercase (pyrex) and switched to low thermal-expansion soda-lime glass. The bottom of new kitchenware and old kitchenware can be inspected for an immediate difference. The scientific division of Pyrex has always been using borosilicate glass.

In addition to quartz, sodium carbonate and aluminium oxide traditionally used in glassmaking, boron is used in the manufacture of borosilicate glass. The composition of low-expansion borosilicate glass, such as those laboratory glasses mentioned above, is approximately 80% silica, 13% boric oxide, 4% sodium oxide and 2–3% aluminium oxide. Though more difficult to make than traditional glass due to the high melting temperature required (Corning conducted a major revamp of their operations to manufacture it), it is economical to produce. Its superior durability, chemical and heat resistance finds excellent use in chemical laboratory equipment, cookware, lighting and, in certain cases, windows.

Manufacturing Process

Borosilicate glass is created by adding boric oxide to the traditional glassmaker's frit of silica sand, soda, and ground lime. Since borosilicate glass melts at a higher temperature than ordinary silicate glass, some new techniques were required for industrial production. Borrowing from the welding trade, burners combining oxygen with natural gas were required. The manufacturing process depends on the product geometry and can be differentiated between different methods like floating, tube drawing or moulding.

Physical Characteristics

The common type of borosilicate glass used for laboratory glassware has a very low thermal expansion coefficient (3.3×10^{-6} K^{-1}), about one-third that of ordinary soda-lime glass. This reduces material stresses caused by temperature gradients, which makes borosilicate a more suitable type of glass for certain applications. Fused quartzware is even better in this respect (having a fifteen times lower thermal expansion than soda-lime glass), however the difficulty of working with fused quartz makes quartzware much more expensive; borosilicate glass is a low-cost compromise. While more resistant to thermal shock than other types of glass, borosilicate glass can still crack or shatter when subjected to rapid or uneven temperature variations. When broken, borosilicate glass tends to crack into large pieces rather than shattering (it will snap rather than splinter).

The softening point (temperature at which viscosity is approximately $10^{7.6}$ poise) of type 7740 Pyrex is 820 °C (1,510 °F).

Borosilicate glass is less dense (about 2.23 g/cm^3) than typical soda-lime glass due to the low atomic mass of boron.

The temperature differential that borosilicate glass can withstand before fracturing is about 165 °C (329 °F). This compares well with soda lime glass, which can withstand only a 37 °C (99 °F) change in temperature and is why typical kitchenware made from traditional soda-lime glass will shatter if a vessel containing boiling water is placed on ice, but Pyrex or other borosilicate glass laboratory will not.

Optically, borosilicate glasses are crown glasses with low dispersion (Abbe numbers around 65) and relatively low refractive indices (1.51–1.54 across the visible range).

Glass Families

For the purposes of classification, borosilicate glass can be roughly arranged in the following groups, according to their oxide composition (in mass fractions). Characteristic of borosilicate glasses is the presence of substantial amounts of silica (SiO_2) and boric oxide (B_2O_3, >8%) as glass network formers. The amount of boric oxide affects the glass properties in a particular way. Apart from the highly resistant varieties (B_2O_3 up to a maximum of 13%), there are others that – due to the different way in which the boric oxide is incorporated into the structural network – have only low chemical resistance (B_2O_3 content over 15%). Hence we differentiate between the following subtypes.

Non-alkaline-earth borosilicate glass (borosilicate glass 3.3)

The B_2O_3 content for borosilicate glass is typically 12–13% and the SiO_2 content over 80%. High chemical durability and low thermal expansion (3.3×10^{-6} K^{-1}) – the lowest of all commercial glasses for large-scale technical applications – make this a multitalented glass material. High-grade borosilicate flat glasses are used in a wide variety of industries, mainly for technical applications that require either good thermal resistance, excellent chemical durability, or high light transmission in combination with a pristine surface quality. Other typical applications for different forms of borosilicate glass include glass tubing, glass piping, glass containers, etc. especially for the chemical industry.

Alkaline-earth-containing borosilicate glasses

In addition to about 75% SiO_2 and 8–12% B_2O_3, these glasses contain up to 5% alkaline earths and alumina (Al_2O_3). This is a subtype of slightly softer glasses (as compared with non-alkaline-earth borosilicate glass), which have thermal expansions in the range (4.0–5.0) × 10^{-6} K^{-1}.

High-borate borosilicate glasses

Glasses containing 15–25% B_2O_3, 65–70% SiO_2, and smaller amounts of alkalis and Al_2O_3 as additional components, have low softening points and low thermal expansion. Sealability to metals in the expansion range of tungsten and molybdenum and high electrical insulation are their most important features. The increased B_2O_3 content reduces the chemical resistance; in this respect, high-borate borosilicate glasses differ widely from non-alkaline-earth and alkaline-earth borosilicate glasses. Among this glasses are also borosilicate glasses that transmit UV-ray down to 180 nm, which combine the best of the borosilicate glass- and the quartz world.

Usage

Borosilicate glass has a wide variety of uses ranging from cookware to lab equipment, as well as a component of high-quality products such as implantable medical devices and devices used in space exploration

Health and Science

Virtually all modern laboratory glassware is made of borosilicate glass. It is widely used in this application due to its chemical and thermal resistance and good optical clarity, but the glass can react with sodium hydride upon heating to produce sodium borohydride, a common laboratory reducing agent. Fused quartz is also found in some laboratory equipment when its higher melting point and transmission of UV are required (e.g. for tube furnace liners and UV cuvettes), but the cost and difficulty of working with quartz make it excessive for the majority of laboratory equipment.

Borosilicate beakers

Additionally, borosilicate tubing is used as the feedstock for the production of parenteral drug packaging, such as vials and pre-filled syringes, as well as ampoules and dental cartridges. The chemical resistance of borosilicate glass minimizes the migration of sodium ions from the glass matrix, thus making it well suited for injectable-drug applications. This type of glass is typically referred to as USP / EP JP Type I.

Borosilicate is widely used in implantable medical devices such as prosthetic eyes, artificial hip joints, bone cements, dental composite materials (white fillings) and even in breast implants.

Many implantable devices benefit from the unique advantages of borosilicate glass encapsulation. Applications include veterinary tracking devices, neurostimulators for the treatment of epilepsy, implantable drug pumps, cochlear implants, and physiological sensors.

Electronics

During the mid-twentieth century, borosilicate glass tubing was used to pipe coolants (often distilled water) through high-power vacuum-tube–based electronic equipment, such as commercial broadcast transmitters.

Borosilicate glasses also have an application in the semiconductor industry in the development of microelectromechanical systems (MEMS), as part of stacks of etched silicon wafers bonded to the etched borosilicate glass.

Cookware

Glass cookware is another common usage. Borosilicate glass is used for measuring cups, featuring screen printed markings providing graduated measurements. Borosilicate glass is sometimes used for high-quality beverage glassware. Borosilicate glass is thin and durable, microwave- and dishwasher-safe.

Arc International bakeware

Lighting

Many high-quality flashlights use borosilicate glass for the lens. This increases light transmittance through the lens compared to plastics and lower-quality glass.

Several types of high-intensity discharge (HID) lamps, such as mercury-vapor and metal-halide lamps, use borosilicate glass as the outer envelope material.

New lampworking techniques led to artistic applications such as contemporary glass marbles. The modern studio glass movement has responded to color. Borosilicate is commonly used in the glassblowing form of lampworking and the artists create a range of products such as jewelry, kitchenware, sculpture, as well as for artistic glass smoking pipes.

Lighting manufacturers use borosilicate glass in their refractors.

Organic light-emitting diode (for display and lighting purposes) also uses borosilicate glass (BK7). The thicknesses of the BK7 glass substrates are usually less than 1 millimeter for the OLED fabrication. Due to its optical and mechanical characteristics in relation with cost, BK7 is a common substrate in OLEDs. However, depending on the application, soda-lime glass substrates of similar thicknesses are also used in OLED fabrication.

Optics

Most astronomical reflecting telescope use glass mirror components made of borosilicate glass because of its low coefficient of thermal expansion. This makes very precise optical surfaces possible that change very little with temperature, and matched glass mirror components that "track" across temperature changes and retain the optical system's characteristics.

The optical glass most often used for making instrument lenses is Schott BK-7 (or the equivalent from other makers), a very finely made borosilicate crown glass. It is also designated as 517642 glass after its 1.517 refractive index and 64.2 Abbe number. Other less costly borosilicate glasses, such as Schott B270 or the equivalent, are used to make "crown-glass" eyeglass lenses. Ordinary lower-cost borosilicate glass, like that used to make kitchenware and even reflecting telescope mirrors, cannot be used for high-quality lenses because of the striations and inclusions common to lower grades of this type of glass. The maximal working temperature is 268 °C (514 °F). While it transitions to a liquid starting at 288 °C (550 °F) (just before it turns red-hot), it is not workable until it reaches over 538 °C (1,000 °F). That means that in order to industrially produce this glass, oxygen/fuel torches must be used. Glassblowers borrowed technology and techniques from welders.

Rapid Prototyping

Borosilicate glass has become the material of choice for fused deposition modeling (FDM), or fused filament fabrication (FFF), build plates. Its low coefficient of expansion makes borosilicate glass, when used in combination with resistance-heating plates and pads, an ideal material for the heated build platform onto which plastic materials are extruded one layer at a time. The initial layer of build must be placed onto a substantially flat, heated surface to minimize shrinkage of some build materials (ABS, polycarbonate, polyamide, etc.) due to cooling after deposition. The build plate will cycle from room temperature to between 100 °C and 130 °C for each prototype that is built. The temperature, along with various coatings (Kapton tape, painter tape, hair spray, ABS+acetone slurry, etc.), ensure that the first layer may be adhered to and remain adhered to the plate, without warping, as the first and subsequent layers cool following extrusion. Subsequently, following the build, the heating elements and plate are allowed to cool. The resulting residual stress formed when the plastic contracts as it cools, while the glass remains relatively dimensionally unchanged due to the low coefficient of thermal expansion, provides a convenient aid in removing the otherwise mechanically bonded plastic from the build plate. In some cases the parts self-separate as the developed stresses overcome the adhesive bond of the build material to the coating material and underlying plate.

Other

Aquarium heaters are sometimes made of borosilicate glass. Due to its high heat resistance, it can tolerate the significant temperature difference between the water and the nichrome heating element.

Specialty glass smoking pipes for cannabis and tobacco are made from borosilicate glass. The high heat resistance makes the pipes more durable.

Most premanufactured glass guitar slides are also made of borosilicate glass.

Borosilicate is also a material of choice for evacuated-tube solar thermal technology, because of its high strength and heat resistance.

The thermal insulation tiles on the Space Shuttle were coated with a borosilicate glass.

Borosilicate glasses are used for immobilisation and disposal of radioactive wastes. In most countries high-level radioactive waste has been incorporated into alkali borosilicate or phosphate vitreous waste forms for many years, and vitrification is an established technology. Vitrification is a particularly attractive immobilization route because of the high chemical durability of the vitrified glass product. This characteristic has been used by industry for centuries. The chemical resistance of glass can allow it to remain in a corrosive environment for many thousands and even millions of years.

Borosilicate glass tubing is used in specialty TIG welding torch nozzles in place of standard alumina nozzles. This allows a clear view of the arc in situations where visibility is limited.

Trade Names

Borosilicate glass is offered in slightly different compositions under different trade names:

- Borofloat of Schott AG, a borosilicate glass, which is produced to flat glass in a float process.
- BK7 of Schott, a borosilicate glass with a high level of purity. Main use in lens and mirrors for laser, cameras and telescopes.
- Duran of DURAN Group, similar to Pyrex, Simax or Jenaer Glas.
- Fiolax of Schott, main use for containers for pharmaceutical applications.
- Ilmabor of TGI (2014 insolvency), main use for containers and equipment in laboratories and medicine.
- Jenaer Glas of Zwiesel Kristallglas, formerly Schott AG. Main use for kitchenware.
- Pyrex of Arc International Cookware, formerly Corning. Main use for kitchenware
- Rasotherm of VEB Jenaer Glaswerk Schott & Genossen, for technical glass

- Simax of Pegasus Industrial Specialties or Kavalier Glaswerke, similar to Pyrex or Jenaer Glas
- Willow Glass is an alkali free, thin and flexible borosilicate glass of Corning

Borosilicate Nanoparticles

It was initially thought that borosilicate glass could not be formed into nanoparticles, since an unstable boron oxide precursor prevented successful forming of these shapes. However, in 2008 a team of researchers from the Swiss Federal Institute of Technology at Lausanne were successful in forming borosilicate nanoparticles of 100 to 500 nanometers in diameter. The researchers formed a gel of tetraethylorthosilicate and trimethoxyboroxine. When this gel is exposed to water under proper conditions, a dynamic reaction ensues which results in the nanoparticles.

In Lampworking

Borosilicate (or "boro", as it is often called) is used extensively in the glassblowing process lampworking; the glassworker uses a burner torch to melt and form glass, using a variety of metal and graphite tools to shape it. Borosilicate is referred to as "hard glass" and has a higher melting point (approximately 3,000 °F / 1648 °C) than "soft glass", which is preferred for glassblowing by beadmakers. Raw glass used in lampworking comes in glass rods for solid work and glass tubes for hollow work tubes and vessels/containers. Lampworking is used to make complex and custom scientific apparatus; most major universities have a lampworking shop to manufacture and repair their glassware. For this kind of "scientific glassblowing", the specifications must be exact and the glassblower must be highly skilled and able to work with precision. Lampworking is also done as art, and common items made include goblets, paper weights, pipes, pendants, compositions and figurines.

In 1968, English metallurgist John Burton brought his hobby of hand-mixing metallic oxides into borosilicate glass to Los Angeles. Burton began a glass workshop at Pepperdine College, with instructor Margaret Youd. A few of the students in the classes, including Suellen Fowler, discovered that a specific combination of oxides made a glass that would shift from amber to purples and blues, depending on the heat and flame atmosphere. Fowler shared this combination with Paul Trautman, who formulated the first small-batch colored borosilicate recipes. He then founded Northstar Glassworks in the mid-1980s, the first factory in the world to devoted solely to producing colored borosilicate glass rods and tubes for use by artists in the flame. Trautman also developed the techniques and technology to make the small-batch colored boro that is used by a number of similar companies. By the time Trautman sold Northstar in 2002, he had composed hundreds of his own recipes for colored borosilicate that are still in production today. In 2004, Trautman came back to making glass, founding Trautman Art Glass. At TAG he developed a number of newer colors that are also industry standards for the boro art community. Fowler continues to teach lampworking and hand-mixing of color in the "Burton style" around the US and Europe.

Beadmaking

In recent years, with the resurgence of lampworking as a technique to make handmade glass beads, borosilicate has become a popular material in many glass artists' studios. Borosilicate for beadmaking comes in thin, pencil-like rods. Northstar, Trautman Art Glass, Glass Alchemy and Momka's Glass are popular manufacturers, although there are other brands available. The metals used to color borosilicate glass, particularly silver, often create strikingly beautiful and unpredictable results when melted in an oxygen-gas torch flame. Because it is more shock-resistant and stronger than soft glass, borosilicate is particularly suited for pipe making, as well as sculpting figures and creating large beads. The tools used for making glass beads from borosilicate glass are the same as those used for making glass beads from soft glass.

Soda-lime Glass

Reusable soda-lime glass milk bottles

Old window made from soda-lime flat glass, Jena, Germany: The distorted reflections of a tree indicates that the flat glass was possibly not made by the float glass process.

Soda-lime glass, also called soda-lime-silica glass, is the most prevalent type of glass, used for windowpanes and glass containers (bottles and jars) for beverages, food, and some commodity items. Glass bakeware is often made of tempered soda-lime glass. Soda-lime glass accounts for about 90% of manufactured glass.

Soda-lime glass is relatively inexpensive, chemically stable, reasonably hard, and extremely workable. Because it can be resoftened and remelted numerous times, it is ideal for glass recycling.

Soda-lime glass is prepared by melting the raw materials, such as sodium carbonate (soda), lime, dolomite, silicon dioxide (silica), aluminium oxide (alumina), and small quantities of fining agents (e.g., sodium sulfate, sodium chloride) in a glass furnace at temperatures locally up to 1675 °C. The temperature is only limited by the quality of the furnace superstructure material and by the glass composition. Relatively inexpensive minerals such as trona, sand, and feldspar are usually used instead of pure chemicals. Green and brown bottles are obtained from raw materials containing iron oxide. The mix of raw materials is termed batch.

Soda-lime glass is divided technically into glass used for windows, called flat glass, and glass for containers, called container glass. The two types differ in the application, production method (float process for windows, blowing and pressing for containers), and chemical composition. Flat glass has a higher magnesium oxide and sodium oxide content than container glass, and a lower silica, calcium oxide, and aluminium oxide content. From the lower content of highly water-soluble ions (sodium and magnesium) in container glass comes its slightly higher chemical durability against water, which is required especially for storage of beverages and food.

Typical Compositions and Properties

The following table lists some physical properties of soda-lime glasses. Unless otherwise stated, the glass compositions and many experimentally determined properties are taken from one large study. Those values marked in *italic* font have been interpolated from similar glass compositions due to the lack of experimental data.

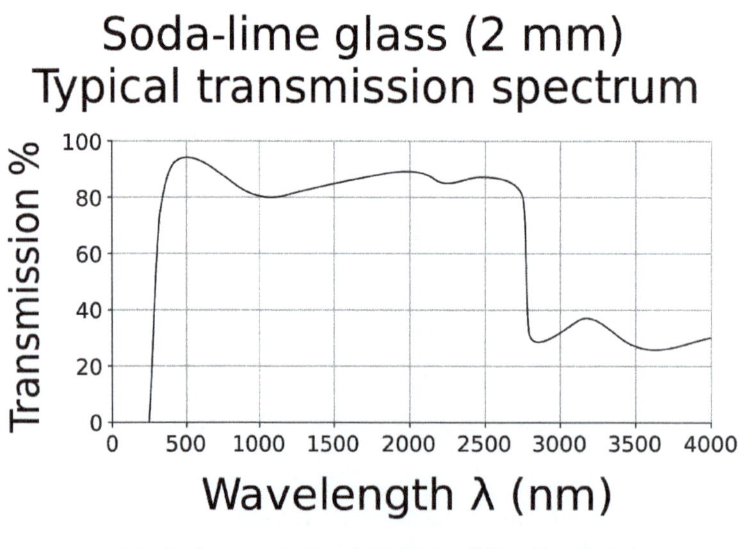

Typical transmission spectrum of a 2-mm glass.

Properties	Container glass	Flat glass
Chemical composition, wt%	74 SiO_2 13 Na_2O 10.5 CaO 1.3 Al_2O_3 0.3 K_2O 0.2 MgO 0.04 Fe_2O_3 0.01 TiO_2 0.2 SO_3	73 SiO_2 14 Na_2O 9 CaO 0.15 Al_2O_3 0.03 K_2O 4 MgO 0.1 Fe_2O_3 0.02 TiO_2
Viscosity log(η, dPa·s or Poise) = A + B / (T in °C – T_o)	550 °C (1,022 °F) | 1,450 °C (2,640 °F) A: −2.309 B: 3922 T_o: 291	550 °C (1,022 °F) | 1,450 °C (2,640 °F) A: −2.585 B: 4215 T_o: 263
Glass transition temperature, T_g	573 °C (1,063 °F)	564 °C (1,047 °F)
Coefficient of thermal expansion, ppm/K, ~100–300 °C (212–572 °F)	9	9.5
Density at 20 °C (68 °F), g/cm³	2.52	2.53
Refractive index n_D at 20 °C (68 °F)	*1.518*	1.520
Dispersion at 20 °C (68 °F), $10^4\times(n_F-n_C)$	*86.7*	*87.7*
Young's modulus at 20 °C (68 °F), GPa	*72*	*74*
Shear modulus at 20 °C (68 °F), GPa	*29.8*	*29.8*
Liquidus temperature	*1,040 °C (1,900 °F)*	*1,000 °C (1,830 °F)*
Heat capacity at 20 °C (68 °F), J/(mol·K)	*49*	*48*
Surface tension, at ~1,300 °C (2,370 °F), mJ/m²	315	
Chemical durability, Hydrolytic class, after ISO 719	*3*	*3...4*
Critical stress intensity factor, (K_{IC}), MPa.m$^{0.5}$	*?*	*0.75*

- Coefficient of restitution (glass sphere vs. glass wall): 0.97 ± 0.01
- Thermal conductivity: 0.7-1.3 W/m.K
- Hardness (Mohs scale): 6
- Knoop hardness: 585 kg/mm2 + 20

Float Glass

Float glass is a sheet of glass made by floating molten glass on a bed of molten metal, typically tin, although lead and various low melting point alloys were used in the past. This method gives the sheet uniform thickness and very flat surfaces. Modern windows are made from float glass. Most float glass is soda-lime glass, but relatively minor quantities of specialty borosilicate and flat panel display glass are also produced using the float glass process. The float glass process is also known as the Pilkington process, named after the British glass manufacturer Pilkington, which pioneered the technique (invented by Sir Alastair Pilkington) in the 1950s.

Use of float glass at Crystal Palace railway station, London

History

Old window containing a single sheet of float glass in the upper left section, Jena, Germany.
The remaining sections are possibly not float glass as indicated by the distorted reflections of a tree.

Until the 16th century, window glass or flat glass was generally cut from large discs (or rondels) of crown glass. Larger sheets of glass were made by blowing large cylinders which were cut open and flattened, then cut into panes. Most window glass in the early 19th century was made using the cylinder method. The 'cylinders' were 6 to 8 feet (180 to 240 cm) long and 10 to 14 inches (25 to 36 cm) in diameter, limiting the width that panes of glass could be cut, and resulting in windows divided by transoms into rectangular panels.

The first advances in automating glass manufacturing were patented in 1848 by Henry Bessemer, an English engineer. His system produced a continuous ribbon of flat glass by forming the ribbon between rollers. This was an expensive process, as the surfaces of the glass needed polishing. If the glass could be set on a perfectly smooth, flat, body, like the surface of an open pan of calm liquid, this would reduce costs considerably. Attempts were made to form flat glass on a bath of molten tin--one of the few liquids denser than glass that would be calm at the high temperatures needed to make glass--most notably in the US. Several patents were granted, but this process was unworkable at the time.

Before the development of float glass, larger sheets of plate glass were made by casting a large puddle of glass on an iron surface, and then polishing both sides, a costly process. From the early 1920s, a continuous ribbon of plate glass was passed through a lengthy series of inline grinders and polishers, reducing glass losses and cost.

Glass of lower quality, sheet glass, was made by drawing upwards from a pool of molten glass a thin sheet, held at the edges by rollers. As it cooled the rising sheet stiffened and could then be cut. The two surfaces were of lower quality i.e. not as smooth or uniform as those of float glass. This process continued in use for many years after the development of float glass.

Between 1953 and 1957, Sir Alastair Pilkington and Kenneth Bickerstaff of the UK's Pilkington Brothers developed the first successful commercial application for forming a continuous ribbon of glass using a molten tin bath on which the molten glass flows unhindered under the influence of gravity. The success of this process lay in the careful balance of the volume of glass fed onto the bath, where it was flattened by its own weight. Full scale profitable sales of float glass were first achieved in 1960.

In 1964, CH3 Tank, the first purpose-built float line in the world, was built after the conversion of CH4 Tank. CH1 Tank closed down in 1977 as CH2 Float Line started up. The Pilkington's production line at Cowley Hill, the birthplace of float glass, is no longer in operation as of 2014.

Manufacture

Float glass uses common glass-making raw materials, typically consisting of sand, soda ash (sodium carbonate), dolomite, limestone, and salt cake (sodium sulfate) etc. Other materials may be used as colourants, refining agents or to adjust the physical and chemical properties of the glass. The raw materials are mixed in a batch process, then fed together with suitable cullet (waste glass), in a controlled ratio, into a furnace where it is heated to approximately 1500 °C. Common flat glass furnaces are 9 m wide, 45 m long, and contain more than 1200 tons of glass. Once molten, the temperature of the glass is stabilised to approximately 1200 °C to ensure a homogeneous specific gravity.

The molten glass is fed into a "tin bath", a bath of molten tin (about 3–4 m wide, 50 m long, 6 cm deep), from a delivery canal and is poured into the tin bath by a ceramic lip known as the spout lip. The amount of glass allowed to pour onto the molten tin is controlled by a gate called a *tweel*.

Tin is suitable for the float glass process because it has a high specific gravity, is cohesive, and is immiscible with molten glass. Tin, however, oxidises in a natural atmosphere to form tin dioxide (SnO_2). Known in the production process as dross, the tin dioxide adheres to the glass. To prevent oxidation, the tin bath is provided with a positive pressure protective atmosphere of nitrogen and hydrogen.

The glass flows onto the tin surface forming a floating ribbon with perfectly smooth surfaces on both sides and of even thickness. As the glass flows along the tin bath, the temperature is gradually reduced from 1100 °C until at approximately 600 °C the sheet can be lifted from the tin onto rollers. The glass ribbon is pulled off the bath by rollers at a controlled speed. Variation in the flow speed and roller speed enables glass sheets of varying thickness to be formed. Top rollers positioned above the molten tin may be used to control both the thickness and the width of the glass ribbon.

Once off the bath, the glass sheet passes through a lehr kiln for approximately 100 m, where it is cooled gradually so that it anneals without strain and does not crack from the temperature change. On exiting the "cold end" of the kiln, the glass is cut by machines.

Market

As of 2009, the world float glass market, not including China and Russia, is dominated by the four companies: Asahi Glass, NSG/Pilkington, Saint-Gobain, and Guardian Industries. Other companies include Sise Cam AS, PPG, Central Glass, Hankuk (HanGlas), Zeledyne, and Cardinal Glass Industries.

Stained Glass

The north transept rose of Chartres Cathedral donated by Blanche of Castile. It represents the Virgin Mary as Queen of Heaven, surrounded by Biblical kings and prophets. Below is St Anne, mother of the Virgin, with four righteous leaders. The window includes the arms of France and Castile.

Sunlight shining through stained glass onto coloured carpet of Nasir-ol-Molk Mosque

The term stained glass can refer to coloured glass as a material or to works created from it. Throughout its thousand-year history, the term has been applied almost exclusively to the windows of churches, mosques and other significant buildings. Although traditionally made in flat panels and used as windows, the creations of modern stained glass artists also include three-dimensional structures and sculpture. Modern vernacular usage has often extended the term "stained glass" to include domestic leadlight and objets d'art created from came glasswork exemplified in the famous lamps of Louis Comfort Tiffany.

As a material *stained glass* is glass that has been coloured by adding metallic salts during its manufacture. The coloured glass is crafted into *stained glass windows* in which small pieces of glass are arranged to form patterns or pictures, held together (traditionally) by strips of lead and supported by a rigid frame. Painted details and yellow stain are often used to enhance the design. The term *stained glass* is also applied to windows in which the colours have been painted onto the glass and then fused to the glass in a kiln.

Stained glass, as an art and a craft, requires the artistic skill to conceive an appropriate and workable design, and the engineering skills to assemble the piece. A window must fit snugly into the space for which it is made, must resist wind and rain, and also, especially in the larger windows, must support its own weight. Many large windows have withstood the test of time and remained substantially intact since the late Middle Ages. In Western Europe they constitute the major form of pictorial art to have survived. In this context, the purpose of a stained glass window is not to allow those within a building to see the world outside or even primarily to admit light but rather to control it. For this reason stained glass windows have been described as 'illuminated wall decorations'.

The design of a window may be abstract or figurative; may incorporate narratives drawn from the Bible, history, or literature; may represent saints or patrons, or use symbolic motifs, in particular armorial. Windows within a building may be thematic, for example: within a church - episodes from the life of Christ; within a parliament building - shields of the constituencies; within a college hall - figures representing the arts and sciences; or within a home - flora, fauna, or landscape.

Stained glass is still popular today, but often referred to as art glass. It is prevalent in luxury homes, commercial buildings, and places of worship. Artists and companies are contracted to create beautiful art glass ranging from domes, windows, backsplashes, etc.

Manufacture

Glass Production

During the late Medieval period, glass factories were set up where there was a ready supply of silica, the essential material for glass manufacture. Silica requires very high heat to become molten, something not all glass factories were able to achieve. Such materials as potash, soda, and lead can be added to lower the melting temperature. Other substances, such as lime, are added to rebuild the weakened network and make the glass more stable. Glass is coloured by adding metallic oxide powders or finely divided metals while it is in a molten state. Copper oxides produce green or bluish green, cobalt makes deep blue, and gold produces wine red and violet glass. Much modern red glass is produced using copper, which is less expensive than gold and gives a brighter, more vermilion shade of red. Glass coloured while in the clay pot in the furnace is known as pot metal glass, as opposed to flashed glass.

Cylinder glass or Muff Using a blow-pipe, a "gather" (glob) of molten glass is taken from the pot heating in the furnace. The gather is formed to the correct shape and a bubble of air blown into it. Using metal tools, molds of wood that have been soaking in water, and gravity, the gather is manipulated to form a long, cylindrical shape. As it cools, it is reheated so the manipulation can continue. During the process, the bottom of the cylinder is removed. Once brought to the desired size it is left to cool. One side of the cylinder is opened. It is put into another oven to quickly heat and flatten it, and then placed in an annealer to cool at a controlled rate, making the material more stable. "Hand-blown" cylinder (also called muff glass) and crown glass were the types used in ancient stained-glass windows.

Crown glass This hand-blown glass is created by blowing a bubble of air into a gather of molten glass and then spinning it, either by hand or on a table that revolves rapidly like a potter's wheel. The centrifugal force causes the molten bubble to open up and flatten. It can then be cut into small sheets. Glass formed this way can be both coloured and used for stained-glass windows, or uncoloured as seen in small paned windows in 16th- and 17th-century houses. Concentric, curving waves are characteristic of the process. The center of each piece of glass, known as the "bull's-eye", receives less force during spinning, so it remains thicker than the rest of the sheet. It also has the distinctive lump of glass left by the "pontil" rod, which holds the glass as it is spun out. This lumpy, refractive quality means the bulls-eyes are less transparent, but they have still been used for windows, both domestic and ecclesiastical. Crown glass is still made today, but not on a large scale.

Rolled glass Rolled glass (sometimes called "table glass") is produced by pouring molten glass onto a metal or graphite table and immediately rolling it into a sheet using a large metal cylinder, similar to rolling out a pie crust. The rolling can be done by hand or machine. Glass can be "double rolled", which means it is passed through two cylinders at once (similar to the clothes wringers on older washing machines) to yield glass of a specified thickness (typically approximately 1/8"). Glass made this way is never fully transparent, but doesn't necessarily have much texture. It can be pushed and tugged while molten for certain effects. For distinct textures the metal cylinder can be imprinted with a pattern that is pressed into the molten glass as it passes through the rollers. The glass is then annealed. Rolled glass was first commercially produced around the mid-1830s and is widely used today. It is often called cathedral glass, but this has nothing to do with medieval cathedrals, where the glass used was hand-blown.

Flashed glass Architectural glass must be at least 1/8 of an inch thick to survive the push and pull of typical wind loads. However, in the creation of red glass, the colouring ingredients must be of a certain concentration, or the colour will not develop. This results in a colour so intense that at the thickness of 1/8 inch, the red glass transmits little light and appears black. The method employed is to laminate a thin layer of red glass to a thicker body of glass that is clear or lightly tinted, forming "flashed glass".

A lightly coloured molten gather is dipped into a pot of molten red glass, that is then blown into a sheet of laminated glass using either the cylinder (muff) or the crown technique described above.

Once this solution was found for making red glass, other colours were also made this way as well. A great advantage is that the double-layered glass can be engraved or abraded to reveal the clear or tinted glass below. The method allows rich detailing and patterns to be achieved without needing to add more lead-lines, giving artists greater freedom in their designs. A number of artists have embraced the possibilities flashed glass gives them. For instance, 16th-century heraldic windows relied heavily on a variety of flashed colours for their intricate crests and creatures. In the medieval period the glass was "abraded" (ground off); later, hydrofluoric acid was used to remove the flash in a chemical reaction (a very dangerous technique), and in the 19th century sandblasting started to be used for this purpose.

Modern production of traditional glass There are a number of glass factories, notably in Germany, USA, England, France, Poland and Russia, which produce high-quality glass, both hand-blown (cylinder, muff, crown) and rolled (cathedral and opalescent). Modern stained-glass artists have a number of resources to use and the work of centuries of other artists from which to learn as they continue the tradition in new ways. In the late 19th and 20th centuries there have been many innovations in techniques and in the types of glass used. Many new types of glass have been developed for use in stained glass windows, in particular Tiffany glass and dalle de verre (slab glass).

Colours

Stained glass window at Süleymaniye Mosque.

Transparent glass Ordinary soda-lime glass appears colourless to the naked eye when it is thin, although iron oxide impurities produce a green tint which becomes evident in thick pieces or can be seen with the aid of scientific instruments. A number of additives are used to reduce the green tint, particularly if the glass is to be used for plain window glass, rather than stained glass windows. Additives that reduce the green tint include manganese dioxide which produces sodium permanganate, and may result in a slightly mauve tint, characteristic of the glass in older houses in New England, US. Selenium has been used for the same purpose.

Green glass While very pale green is the typical colour of transparent glass, deeper greens can be achieved by the addition of Iron(II) oxide which results in a bluish-green glass. Together with chromium it gives glass of a richer green colour, typical of the glass used to make wine bottles. The addition of Chromium yields dark green glass, suitable for flashed glass. Together with tin oxide and arsenic it yields emerald green glass.

Blue Glass

- In medieval times, blue glass was made by adding cobalt, which at a concentration of 0.025 to 0.1% in soda-lime glass achieves the brilliant blue characteristic of Chartres Cathedral.
- The addition of sulphur to boron-rich borosilicate glasses imparts a blue colour.
- The addition of copper oxide at 2-3% produces a turquoise colour.
- The addition of nickel, at different concentrations, produces blue, violet, or black glass.

Red glass

- Metallic gold, in very small concentrations (around 0.001%), produces a rich ruby-coloured glass ("ruby gold"); in lower concentrations it produces a less intense red, often marketed as "cranberry glass". The colour is caused by the size and dispersion of gold particles. Ruby gold glass is usually made of lead glass with tin added.
- Pure metallic copper produces a very dark red, opaque glass. Glass created in this manner is generally "flashed" (laminated glass). It was used extensively in the late 19th and early 20th centuries and exploited for the decorative effects that could be achieved by sanding and engraving.
- Selenium is an important agent to make pink and red glass. When used together with cadmium sulphide, it yields a brilliant red colour known as "Selenium Ruby".

Yellow glass

- Silver compounds (notably silver nitrate) are used as stain applied to the surface of glass and fired on. They can produce a range of colours from orange-red to yellow. The way the glass is heated and cooled can significantly affect the colours produced by these compounds. The chemistry involved is complex and not well understood.
- The addition of sulphur, together with carbon and iron salts, is used to form iron polysulphides and produce amber glass ranging from yellowish to almost black. With calcium it yields a deep yellow colour.
- Adding titanium produces yellowish-brown glass. Titanium is rarely used on its own and is more often employed to intensify and brighten other additives.
- Cadmium together with sulphur results in deep yellow colour, often used in glazes. However, cadmium is toxic.
- Uranium (0.1 to 2%) can be added to give glass a fluorescent yellow or green colour. Uranium glass is typically not radioactive enough to be dangerous, but if ground into a

powder, such as by polishing with sandpaper, and inhaled it can be carcinogenic. When used with lead glass with a very high proportion of lead, it produces a deep red colour.

Purple glass

- The addition of Manganese gives an amethyst colour. Manganese is one of the oldest glass additives, and purple manganese glass has been used since early Egyptian history.
- Nickel, depending on the concentration, produces blue, or violet, or even black glass. Lead crystal with added nickel acquires a purplish colour.

White Glass

- Tin oxide with antimony and arsenic oxides produce an opaque white glass, first used in Venice to produce an imitation porcelain. White glass was used extensively by Louis Comfort Tiffany to create a range of opalescent, mottled and streaky glasses.

13th-century window from Chartres showing extensive use of the ubiquitous cobalt blue with green and purple-brown glass, details of amber and borders of flashed red glass.

A 19th-century window illustrates the range of colours common in both Medieval and Gothic Revival glass, Lucien Begule, Lyon (1896)

A 16th-century window by Arnold of Nijmegen showing the combination of painted glass and intense colour common in Renaissance windows

A late 20th-century window showing a graded range of colours. Ronald Whiting, Chapel Studios. Tattershall Castle, UK

A window by Tiffany illustrating the development and use of multi-coloured flashed, opalised and streaky glasses at the end of the 19th century

Creating Stained Glass Windows

Design

The first stage in the production of a window is to make, or acquire from the architect or owners of the building, an accurate template of the window opening that the glass is to fit.

The subject matter of the window is determined to suit the location, a particular theme, or the whim of the patron. A small design called a *Vidimus* (from Latin "we saw it") is prepared which can be shown to the patron. A scaled model maquette may also be provided. The designer must take into account the design, the structure of the window, the nature and size of the glass available and his or her own preferred technique.

A traditional narrative window has panels which relate a story. A figurative window could have rows of saints or dignitaries. Scriptural texts or mottoes are sometimes included and perhaps the names of the patrons or the person as whose memorial the window is dedicated. In a window of a traditional type, it is usually left to the discretion of the designer to fill the surrounding areas with borders, floral motifs and canopies.

A full sized cartoon is drawn for every "light" (opening) of the window. A small church window might typically have two lights, with some simple *tracery* lights above. A large window might have four or five lights. The east or west window of a large cathedral might have seven lights in three *tiers* with elaborate tracery. In Medieval times the cartoon was drawn directly on the surface of a whitewashed table, which was then used as a pattern for cutting, painting and assembling the window. The cartoon is then divided into a patchwork, providing a template for each small glass piece. The exact position of the lead which holds the glass in place is also noted as it is part of the calculated visual effect.

Selecting and Painting the Glass

Each piece of glass is selected for the desired colour and cut to match a section of the template. An exact fit is ensured by *grozing* the edges with a tool which can nibble off small pieces.

Details of faces, hair and hands can be painted onto the inner surface of the glass using a special glass paint which contains finely ground lead or copper filings, ground glass, gum arabic and a medium such as wine, vinegar or (traditionally) urine. The art of painting details became increasingly elaborate and reached its height in the early 20th century.

From 1300 onwards, artists started using *silver stain* which was made with silver nitrate. It gave a yellow effect ranging from pale lemon to deep orange. It was usually painted onto the outside of a piece of glass, then fired to make it permanent. This yellow was particularly useful for enhancing borders, canopies and haloes, and turning blue glass into green glass.

By about 1450 a stain known as *Cousin's rose* was used to enhance flesh tones.

In the 16th century a range of glass stains were introduced, most of them coloured by ground glass particles. They were a form of enamel. Painting on glass with these stains was initially used for small heraldic designs and other details. By the 17th century a style of stained glass had evolved that was no longer dependent upon the skilful cutting of coloured glass into sections. Scenes were painted onto glass panels of square format, like tiles. The colours were then annealed to the glass before the pieces were assembled.

A method used for embellishment and gilding is the decoration of one side of each of two pieces of thin glass which are then placed back to back within the lead came. This allows for the use of techniques such as Angel gilding and Eglomise to produce an effect visible from both sides but not exposing the decorated surface to the atmosphere or mechanical damage.

Assembly and Mounting

Once the glass is cut and painted, the pieces are assembled by slotting them into H-sectioned lead cames. All the joints are then soldered together and the glass pieces are prevented from rattling and the window made weatherproof by forcing a soft oily cement or *mastic* between the glass and the cames. In modern windows, copper foil is now sometimes used instead of lead.

Traditionally, when the windows were inserted into the window spaces, iron rods were put across it at various points to support its weight. The window was tied to these rods with copper wire. Some very large early Gothic windows are divided into sections by heavy metal frames called *ferramenta*. This method of support was also favoured for large, usually painted, windows of the Baroque period.

- Technical details

Maquette by Heaton, Butler and Bayne, 19th-century English manufacturers

Exterior of a window at Sé Velha de Coimbra, Portugal, showing a modern steel armature

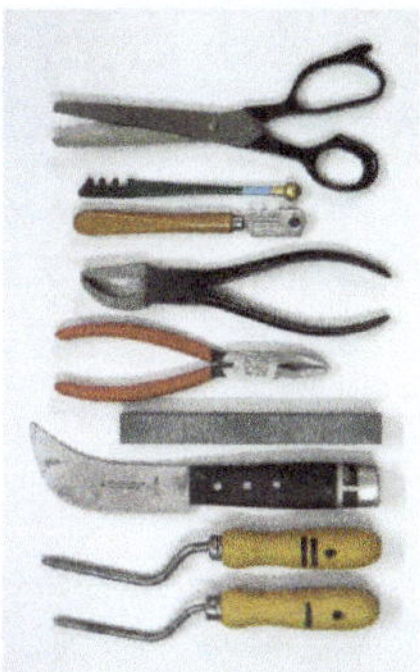

A set of glaziers' tools

Skilled glass cutting and leading in a 19th-century window at Meaux Cathedral, France

Thomas Becket window from Canterbury showing the *pot metal* and painted glass, lead H-sectioned *cames*, modern steel rods and copper wire attachments

Colours and mediums for painting on glass

A palette and tools to prepare colours for painting on glass

A small panel by G. Owen Bonawit at Yale University, c. 1930, demonstrates grisaille glass painting enlivened with silver stain.

Swiss armourial glass of the Arms of Unterwalden, 1564, with typical painted details, extensive *silver stain*, *Cousin's rose* on the face, and flashed *ruby glass* with abraded white motif

Detail from a 19th or 20th-century window in Eyneburg, Belgium, showing detailed polychrome painting of face.

History

Origins

Coloured glass has been produced since ancient times. Both the Egyptians and the Romans excelled at the manufacture of small coloured glass objects. Phoenicia was important in glass manufacture with its chief centres Sidon, Tyre and Antioch. The British Museum holds two of the finest Roman pieces, the Lycurgus Cup, which is a murky mustard colour but glows purple-red to transmitted light, and the Portland vase which is midnight blue, with a carved white overlay.

In early Christian churches of the 4th and 5th centuries, there are many remaining windows which are filled with ornate patterns of thinly-sliced alabaster set into wooden frames, giving a stained-glass like effect.

Evidence of stained glass windows in churches and monasteries in Britain can be found as early as the 7th century. The earliest known reference dates from 675 AD when Benedict Biscop imported workmen from France to glaze the windows of the monastery of St Peter which he was building at Monkwearmouth. Hundreds of pieces of coloured glass and lead, dating back to the late 7th century, have been discovered here and at Jarrow.

In the Middle East, the glass industry of Syria continued during the Islamic period with major centres of manufacture at Ar-Raqqah, Aleppo and Damascus and the most important products being highly transparent colourless glass and gilded glass, rather than coloured glass. The production of coloured glass in Southwest Asia existed by the 8th century, at which time the alchemist Jābir ibn Hayyān, in *Kitab al-Durra al-Maknuna*, gave 46 recipes for producing coloured glass and described the technique of cutting glass into artificial gemstones.

A perfume flask from 100 BC-200 AD

The Portland Vase, a rare example of Roman flashed glass

An alabaster window in Orvieto Cathedral, Italy

Stained glass in the Nasir al-Mulk mosque in Shiraz, Iran

Medieval Glass

Stained glass, as an art form, reached its height in the Middle Ages when it became a major pictorial form used to illustrate the narratives of the Bible to a largely illiterate populace.

In the Romanesque and Early Gothic period, from about 950 AD to 1240 AD, the untraceried windows demanded large expanses of glass which of necessity were supported by robust iron frames, such as may be seen at Chartres Cathedral and at the eastern end of Canterbury Cathedral. As Gothic architecture developed into a more ornate form, windows grew larger, affording greater illumination to the interiors, but were divided into sections by vertical shafts and tracery of stone. This elaboration of form reached its height of complexity in the Flamboyant style in Europe, and windows grew still larger with the development of the Perpendicular style in England.

Integrated with the lofty verticals of Gothic cathedrals and parish churches, glass designs became more daring. The circular form, or rose window, developed in France from relatively simple windows with openings pierced through slabs of thin stone to wheel windows, as exemplified by the West front of Chartres Cathedral, and ultimately to designs of enormous complexity, the tracery being drafted from hundreds of different points, such as those at Sainte-Chapelle, Paris and the "Bishop's Eye" at Lincoln Cathedral.

While stained glass was widely manufactured, Chartres in France was the greatest centre of stained glass manufacture, producing glass of unrivalled quality.

- Medieval glass in France

Charlemagne from a Romanesque window in Strasbourg Cathedral

The Crucifixion window of Poitiers Cathedral

Late Gothic Tree of Jesse window from Evreux Cathedral

- Medieval glass in Germany and Austria

The Crucifixion and *Virgin and Child in Majesty*, Cologne Cathedral, (1340)

Ulm Munster, *The Last Judgement* by Hans Acker (1430)

The windows of the choir of Cologne Cathedral, (early 14th century)

- Medieval glass in England

Detail of a Tree of Jesse from York Minster (c. 1170), the oldest stained glass window in England.

The Poor Man's Bible Window from Canterbury Cathedral

South Transept window at Canterbury Cathedral, 13th century

Renaissance, Reformation and Classical Windows

Probably the earliest scheme of stained glass windows that was created during the Renaissance was that for Florence Cathedral, devised by Lorenzo Ghiberti. The scheme includes three ocular windows for the dome and three for the facade which were designed from 1405-1445 by several of the most renowned artists of this period: Ghiberti, Donatello, Uccello and Andrea del Castagno. Each major ocular window contains a single picture drawn from the Life of Christ or the Life of the Virgin Mary, surrounded by a wide floral border, with two smaller facade windows by Ghiberti showing the martyred deacons, St Stephen and St Lawrence. One of the cupola windows has since been lost, and that by Donatello has lost nearly all of its painted details.

In Europe, stained glass continued to be produced with the style evolving from the Gothic to the Classical, which is well represented in Germany, Belgium and the Netherlands, despite the rise of Protestantism. In France, much glass of this period was produced at the Limoges factory, and in Italy at Murano, where stained glass and faceted lead crystal are often coupled together in the same window. Ultimately, the French Revolution brought about the neglect or destruction of many windows in France. At the Reformation, in England large numbers of Medieval and Renaissance windows were smashed and replaced with plain glass. The Dissolution of the Monasteries under Henry VIII and the injunctions of Thomas Cromwell against "abused images" (the object of veneration) resulted in the loss of thousands of windows. Few remain undamaged; of them the windows in the private chapel at Hengrave Hall in Suffolk are among the finest. With the latter wave of destruction the traditional methods of working with stained glass died and were not to be rediscovered in England until the early 19th century.

In the Netherlands a rare scheme of glass has remained intact at Grote Sint-Jan Church, Gouda. The windows, some of which are 18 metres (60 feet) high date from 1555 to the early 1600s, with the earliest being the work of Dirck Crabeth and his brother Wouter. Many of the original cartoons still exist.

The Resurrection, Paolo Uccello, (1443–45) one of a series in the dome of Florence Cathedral designed by renowned Renaissance artists.

Tree of Jesse window, Church of Saint-Étienne, Beauvais, France, Engrand Le Prince, (1522-1524)

Detail of Adam and Eve from the Cathedral of St-Etienne, Châlons-en-Champagne, France

Glass painting depicting Mordnacht (murder's night) on 23/24 February 1350 and heraldry of the first Meisen guild's Zunfthaus. (c. 1650)

The story of how the *Crown of Thorns*, passed from John of Brienne and Baldwin II of Constantinople to Saint Louis IX of France, Moulins Cathedral (16th century)

The *Death and Assumption of the Virgin Mary*, Church of Ss Ägidius and Koloman, Steyr, Austria

Revival in Britain

The Catholic revival in England, gaining force in the early 19th century with its renewed interest in the medieval church, brought a revival of church building in the Gothic style, claimed by John Ruskin to be "the true Catholic style". The architectural movement was led by Augustus Welby Pugin. Many new churches were planted in large towns and many old churches were restored. This brought about a great demand for the revival of the art of stained glass window making.

Among the earliest 19th-century English manufacturers and designers were William Warrington and John Hardman of Birmingham, whose nephew, John Hardman Powell, had a commercial eye and exhibited works at the Philadelphia Exhibition of 1876, influencing stained glass in the United States of America. Other manufacturers included William Wailes, Ward and Hughes, Clayton and Bell, Heaton, Butler and Bayne and Charles Eamer Kempe. A Scottish designer, Daniel Cottier, opened firms in Australia and the US

Detail, *Apostles John and Paul*, Hardman of Birmingham, 1861–67, typical of Hardman in its elegant arrangement of figures and purity of colour. St. Andrew's Cathedral, Sydney

One of England's largest windows, the east window of Lincoln Cathedral, Ward and Nixon (1855), is a formal arrangement of small narrative scenes in roundels

William Wailes. This window has the bright pastel colour, wealth of inventive ornament, and stereotypical gestures of windows by this firm. St Mary's, Chilham

Clayton and Bell. A narrative window with elegant forms and colour which is both brilliant and subtle in its combinations. Peterborough Cathedral

Revival in France

In France there was a greater continuity of stained glass production than in England. In the early 19th century most stained glass was made of large panes that were extensively painted and fired, the designs often being copied directly from oil paintings by famous artists. In 1824 the Sèvres porcelain factory began producing stained glass to supply the increasing demand. In France many churches and cathedrals suffered despoliation during the French Revolution. During the 19th century a great number of churches were restored by Viollet-le-Duc. Many of France's finest ancient windows were restored at that time. From 1839 onwards much stained glass was produced that very closely imitated medieval glass, both in the artwork and in the nature of the glass itself. The pioneers were Henri Gèrente and André Lusson. Other glass was designed in a more Classical manner, and characterised by the brilliant cerulean colour of the blue backgrounds (as against the purple-blue of the glass of Chartres) and the use of pink and mauve glass.

Detail of a "Tree of Jesse" window in Reims Cathedral designed in the 13th-century style by L. Steiheil and painted by Coffetier for Viollet-le-Duc, (1861)

St Louis administering Justice by Lobin in the painterly style. (19th century) Church of St Medard, Thouars.

A brilliantly-coloured window at Cassagnes-Bégonhès, Aveyron

West window from Saint-Urbain, Troyes, (about 1900)

Revival in Europe

During the mid- to late 19th century, many of Germany's ancient buildings were restored, and some, such as Cologne Cathedral, were completed in the medieval style. There was a great demand for stained glass. The designs for many windows were based directly on the work of famous engravers such as Albrecht Dürer. Original designs often imitate this style. Much 19th-century German glass has large sections of painted detail rather than outlines and details dependent on the lead. The Royal Bavarian Glass Painting Studio was founded by Ludwig I in 1827. A major firm was Mayer of Munich, which commenced glass production in 1860, and is still operating as *Franz Mayer of Munich, Inc.*. German stained glass found a market across Europe, in America and Australia. Stained glass studios were also founded in Italy and Belgium at this time.

In the Austrian Empire and later Austria-Hungary, one of the leading stained glass artists was Carl Geyling, who founded his studio in 1841. His son would continue the tradition as Carl Geyling's Erben, which still exists today. Carl Geyling's Erben completed numerous stained glass windows for major churches in Vienna and elsewhere, and received an Imperial and Royal Warrant of Appointment from emperor Franz Joseph I of Austria.

A painted memorial window, Castle Bodenstein, Germany, early 19th century

One of five windows donated to Cologne Cathedral by Ludwig II

Post-Reformation window in the Memorial Church, Speyer, Germany

A window in the Late Gothic style, St Maurice's Church, Olomouc, Czech Republic, early 20th century

Innovations in the United States

J&R Lamb Studios, established in 1857 in New York City, was the first major decorative arts studio in the United States and for many years a major producer of ecclesiastical stained glass.

Notable American practitioners include John La Farge (1835–1910), who invented opalescent glass and for which he received a U.S. patent on 24 February 1880, and Louis Comfort Tiffany (1848–1933), who received several patents for variations of the same opalescent process in November of the same year and is believed to have invented the copper foil method as an alternative to lead, and used it extensively in windows, lamps and other decorations.

Many of the distinctive types of glass invented by Tiffany are demonstrated within this single small panel including "fracture-streamer glass" and "drapery glass".

John La Farge, *The Angel of Help*, North Easton, MA shows the use of tiny panes contrasting with large areas of flashed or opalescent glass.

Religion Enthroned, J&R Lamb Studios, designer Frederick Stymetz Lamb, c. 1900. Brooklyn Museum. Symmetrical design, "Aesthetic Style", a limited palette and extensive use of mottled glass.

A trompe l'oeil glass c. 1884, Eugène Stanislas Oudinot, design Richard Morris Hunt, for home of Henry Gurdon Marquand, New York City.

Innovations in Britain and Europe

Among the most innovative English designers were the Pre-Raphaelites, William Morris (1834–1898) and Edward Burne-Jones (1833–1898), whose work heralds Art Nouveau. Art Nouveau or Belle Epoch stained glass design flourished in France, and Eastern Europe, where it can be identified by the use of curving, sinuous lines in the lead, and swirling motifs. In France it is seen in the work of Francis Chigot of Limoges. In Britain it appears in the refined and formal leadlight designs of Charles Rennie Macintosh.

- *David's charge to Solomon* shows the strongly linear design and use of flashed glass for which Burne-Jones' designs are famous. Trinity Church, Boston, US, (1882)

God the Creator by Stanisław Wyspiański, this window has no glass painting, but relies entirely on leadlines and skilful placement of colour and tone. Franciscan Church, Kraków (c. 1900)

Art Nouveau by Jacques Grüber, the glass harmonising with the curving architectural forms that surround it, Musée de l'École de Nancy (1904).

20th and 21st Centuries

Many 19th-century firms failed early in the 20th century as the Gothic movement was superseded by newer styles. At the same time there were also some interesting developments where stained glass artists took studios in shared facilities. Examples include the Glass House in London set up by Mary Lowndes and Alfred J. Drury and An Túr Gloine in Dublin, which was run by Sarah Purser and included artists such as Harry Clarke.

A revival occurred in the middle of the century because of a desire to restore thousands of church windows throughout Europe destroyed as a result of World War II bombing. German artists led the way. Much work of the period is mundane and often was not made by its designers, but industrially produced.

Other artists sought to transform an ancient art form into a contemporary one, sometimes using traditional techniques while exploiting the medium of glass in innovative ways and in combination with different materials. The use of slab glass set in concrete was a 20th-century innovation. Gemmail, a technique developed by the French artist Jean Crotti in 1936 and perfected in the 1950s, is a type of stained glass where adjacent pieces of glass are overlapped without using lead cames to join the pieces, allowing for greater diversity and subtlety of colour. Many famous works by late 19th- and early 20th-century painters, notably Picasso, have been reproduced in gemmail. A major exponent of this technique is the German artist Walter Womacka.

Among the early well-known 20th-century artists who experimented with stained glass as an Abstract art form were Theo van Doesburg and Piet Mondrian. In the 1960s and 1970s the Expressionist painter Marc Chagall produced designs for many stained glass windows that are intensely coloured and crammed with symbolic details. Important 20th-century stained glass artists include John Hayward, Douglas Strachan, Ervin Bossanyi, Louis Davis, Wilhelmina Geddes, Karl Parsons, John Piper, Patrick Reyntiens, Johannes Shreiter, Judith Schaechter, Paul Woodroffe, Jean René Bazaine at Saint Séverin, Sergio de Castro at Couvrechef- La Folie (Caen), Hamburg-Dulsberg and Romont (Switzerland), and the Loire Studio of Gabriel Loire at Chartres. The west windows of England's Manchester Cathedral, by Tony Hollaway, are some of the most notable examples of symbolic work.

In Germany, stained glass development continued with the inter-war work of Johan Thorn Prikker and Josef Albers, and the postwar achievements of Joachim Klos, Johannes Schreiter and Ludwig Shaffrath. Trends included the abandonment of figurative designs and of painting on glass in favour of a mix of biomorphic and rigorously geometric abstraction and the calligraphic non-functional use of leads. The works of Ludwig Schaffrath demonstrate the late 20th-century trends in the use of stained glass for architectural purposes, filling entire walls with coloured and textured glass. In the 1970s young British stained-glass artists such as Brian Clarke were influenced by the large scale and abstraction in German twentieth-century glass.

In the United States, there is a 100-year-old trade organization, The Stained Glass Association of America, whose purpose is to function as a publicly recognized organization to assure survival of the craft by offering guidelines, instruction and training to craftspersons. The SGAA also sees its role as defending and protecting its craft against regulations that might restrict its freedom as an architectural art form. The current president is B. Gunar Gruenke of the Conrad Schmitt Studios. Today there are academic establishments that teach the traditional skills. One of these is Florida State University's Master Craftsman Program, which recently completed a 30 ft (9.1 m) high stained-glass windows installed in Bobby Bowden Field at Doak Campbell Stadium.

De Stijl abstraction by Theo van Doesburg, Netherlands (1917)

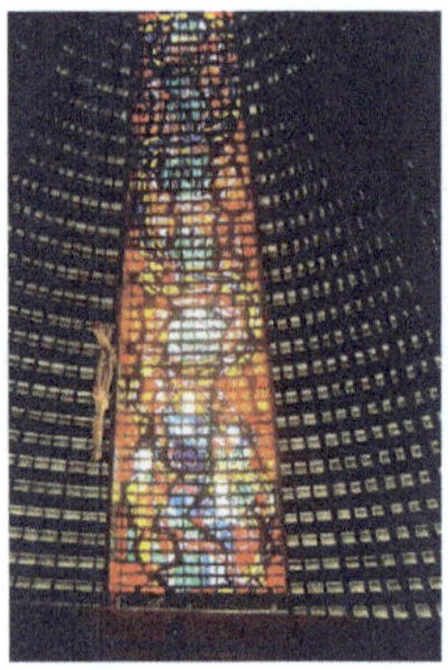

One of four 64-metre (210 ft)-high stained glass panels, Rio de Janeiro Cathedral, Brazil

The *Bald Eagle*, from commercial studios working with traditional techniques, Dryden High School, USA

Thin slices of agate set into lead and glass, Grossmünster, Zürich, Switzerland, by Sigmar Polke (2009)

Sergio de Castro, *7th Day of Creation*, Church of the Benedictines, Caen (France).

Sergio de Castro, detail of Jonah window for the Collegiate of Romont (Switzerland).

Combining Ancient and Modern Traditions

Mid-20th-century window showing a continuation of ancient and 19th-century methods applied to a modern historical subject. Florence Nightingale window at St Peters, Derby, made for the Derbyshire Royal Infirmary

Figurative design using the lead lines and minimal glass paint in the 13th-century manner combined with the texture of Cathedral glass, Ins, Switzerland

- *St Michael and the Devil* at the church of St Michael Paternoster Row, by English artist John Hayward combines traditional methods with a distinctive use of shard-like sections of glass.

Postmodern design of etched glass text and cathedral glass books set as a trompe l'oeil, LGBT pride Libraries, University of Western Ontario, artist: Lynette Richards

Buildings Incorporating Stained Glass Windows

Churches

Stained glass windows were commonly used in churches for decorative and informative purposes. Many windows are donated to churches by members of the congregation as memorials of loved ones. For more information on the use of stained glass to depict religious subjects.

- Important examples
    - o Cathedral of Chartres, in France, 11th- to 13th-century glass

- Canterbury Cathedral, in England, 12th to 15th century plus 19th- and 20th-century glass
- York Minster, in England, 11th- to 15th-century glass
- Sainte-Chapelle, in Paris, 13th- and 14th-century glass
- Florence Cathedral, Italy, 15th-century glass designed by Uccello, Donatello and Ghiberti
- St. Andrew's Cathedral, Sydney, Australia, early complete cycle of 19th-century glass, Hardman of Birmingham.
- Coventry Cathedral, England, mid-20th-century glass by various designers, the large baptistry window being by John Piper
- Brown Memorial Presbyterian Church, extensive collection of windows by Louis Comfort Tiffany

Synagogues In addition to Christian churches, stained glass windows have been incorporated into Jewish temple architecture for centuries. Jewish communities in the United States saw this emergence in mid-19th century, with such notable examples as the sanctuary depiction of the Ten Commandments in New York's Congregation Anshi Chesed. From the mid-20th century to the present, stained glass windows have been a ubiquitous feature of American synagogue architecture. Styles and themes for synagogue stained glass artwork are as diverse as their church counterparts. As with churches, synagogue stained glass windows are often dedicated by member families in exchange for major financial contributions to the institution.

Places of Worship

The dazzling display of medieval glass at Sainte-Chapelle, Paris

The chancel windows of the Church of Our Lady, Koblenz, Germany

Late 20th-century stained glass from Temple Ohev Sholom, Harrisburg, Pennsylvania by Ascalon Studios.

Houses

Stained glass windows in houses were particularly popular in the Victorian era and many domestic examples survive. In their simplest form they typically depict birds and flowers in small panels, often surrounded with machine-made *cathedral glass* which, despite what the name suggests, is pale-coloured and textured. Some large homes have splendid examples of secular pictorial glass. Many small houses of the 19th and early 20th centuries have leadlight windows.

- Prairie style homes
- The houses of Frank Lloyd Wright

Shabaka (stained glass) at the Palace of Shaki Khans

Public and commercial use of stained glass Stained glass has often been used as a decorative element in public buildings, initially in places of learning, government or justice but increasingly in other public and commercial places such as banks, retailers and railway stations. Public houses in some countries make extensive use of stained glass and leaded lights to create a comfortable atmosphere and retain privacy.

Stained glass in Dowlat Abad Garden at Yazd, Iran

Stained glass in the Town Hall, Liberec, Czech Republic

Windows of the Hungarian Room, University of Pittsburgh

The Federal Palace, Switzerland

Abstract design by Marcelle Ferron at a Metro station in Montreal, Canada

Windows by Mordecai Ardon at the Jewish National and University Library, Jerusalem

Sculpture

The Four Seasons (1978) by Leonard French at La Trobe University Sculpture Park in Melbourne. Australia

Fused glass sculpture (2012) by Carlo Roccella Glass Sculpture in Paris. France

Crown Glass (Window)

Crown Glass

Crown glass was an early type of window glass. In this process, glass was blown into a "crown" or hollow globe. This was then transferred from the blowpipe to a punty and then flattened by reheating and spinning out the bowl-shaped piece of glass (bullion) into a flat disk by centrifugal force, up to 5 or 6 feet (1.5 to 1.8 metres) in diameter. The glass was then cut to the size required.

The thinnest glass was in a band at the edge of the disk, with the glass becoming thicker and more opaque toward the center. Known as a bullseye, the thicker center area around the punty mark was used for less expensive windows. In order to fill large window spaces with the best glass, many small diamond shapes would be cut from the edge of the disk and these would be mounted in a lead lattice work and fitted into the window frame.

Crown glass was one of the two most common processes for making window glass until the 19th century. The other was blown plate. The process was first perfected by French glassmakers in the 1320s, notably around Rouen, and was a trade secret. As a result, crown glass was not made in London until 1678.

Crown glass is one of many types of hand-blown glass. Other methods include: broad sheet, blown plate, polished plate and cylinder blown sheet. These methods of manufacture lasted at least until the end of the 19th century. The early 20th century marks the move away from hand-blown to machine manufactured glass such as rolled plate, machine drawn cylinder sheet, flat drawn sheet, single and twin ground polished plate and float glass.

Chemically Strengthened Glass

Chemically strengthened glass is a type of glass that has increased strength as a result of a post-production chemical process. When broken, it still shatters in long pointed splinters similar to float glass. For this reason, it is not considered a safety glass and must be laminated if a safety glass is required. However, chemically strengthened glass is typically six to eight times the strength of float glass.

The glass is chemically strengthened by a surface finishing process. Glass is submersed in a bath containing a potassium salt (typically potassium nitrate) at 300°C (572°F). This causes sodium ions in the glass surface to be replaced by potassium ions from the bath solution.

These potassium ions are larger than the sodium ions and therefore wedge into the gaps left by the smaller sodium ions when they migrate to the potassium nitrate solution. This replacement of ions causes the surface of the glass to be in a state of compression and the core in compensating tension. The surface compression of chemically strengthened glass may reach up to 690 MPa.

There also exists a more advanced two-stage process for making chemically strengthened glass, in which the glass article is first immersed in a sodium nitrate bath at 450°C (842°F), which enriches the surface with sodium ions. This leaves more sodium ions on the glass for the immersion in potassium nitrate to replace with potassium ions. In this way, the use of a sodium nitrate bath increases the potential for surface compression in the finished article.

Chemical strengthening results in a strengthening similar to toughened glass. However, the process does not use extreme variations of temperature and therefore chemically strengthened glass has little or no bow or warp, optical distortion or strain pattern. This differs from toughened glass, in which slender pieces can be significantly bowed.

Also unlike toughened glass, chemically strengthened glass may be cut after strengthening, but loses its added strength within the region of approximately 20 mm of the cut. Similarly, when the surface of chemically strengthened glass is deeply scratched, this area loses its additional strength.

Chemically strengthened glass was used for the aircraft canopy of some fighter aircraft.

References

- Streeter, Victor Lyle; Wylie, E. Benjamin; Bedford, Keith W. (1998). Fluid Mechanics. McGraw-Hill. ISBN 0-07-062537-9.
- Incropera, Frank P.; DeWitt, David P. (2007). Fundamentals of Heat and Mass Transfer. Wiley. ISBN 0-471-45728-0.
- Serway, Raymond A. (1996). Physics for Scientists & Engineers (4th ed.). Saunders College Publishing. ISBN 0-03-005932-1.
- Smits, Alexander J.; Dussauge, Jean-Paul (2006). Turbulent shear layers in supersonic flow. Birkhäuser. p. 46. ISBN 0-387-26140-0
- Hirshfelder, J. O.; Curtis, C. F.; Bird, R. B. (1964). Molecular theory of gases and liquids (1st ed.). Wiley. ISBN 0-471-40065-3.
- Maples, Robert E. (2000). Petroleum Refinery Process Economics (2nd ed.). Pennwell Books. ISBN 0-87814-779-9.
- Lide, D. R., ed. (2005). CRC Handbook of Chemistry and Physics (86th ed.). Boca Raton (FL): CRC Press. ISBN 0-8493-0486-5.
- Hill, Russ (1999). Coated Glass Applications and Markets. Fairfield, CA: BOC Coating Technology. pp. 1–4. ISBN 0-914289-01-2.
- Carmody, John , Stephen Selkowitz, Lisa Heschong (1996). Residential windows : a guide to new technologies and energy performance (1st. ed.). New York: Norton. ISBN 0-393-73004-2.
- [Fundamentals of corrosion - Mechanisms, Causes and Preventative Methods]. Philip A. Schweitzer, Taylor and Francis Group, LLC (2010) ISBN 978-1-4200-6770-5, p. 25.
- M.I. Ojovan, W.E. Lee. New Developments in Glassy Nuclear Wasteforms. Nova Science Publishers, New York (2007) ISBN 1600217834 pp. 100 ff.
- Corrosion of Glass, Ceramics and Ceramic Superconductors. D.E. Clark, B.K. Zoitos (eds.), William Andrew Publishing/Noyes (1992) ISBN 081551283X.
- Moynihan, C. et al. in The Glass Transition and the Nature of the Glassy State, Eds. M. Goldstein and R. Simha, Ann. N.Y. Acad. Sci., Vol. 279 (1976) ISBN 0890720533

3

Architectural Glass: An Integrated Study

The glass that is used as building material is known as architectural glass. Glass is usually used for internal partitions within a building. Glass bricks, annealing, laminated glass, heatable glass and self-cleaning glass are some of the examples given in this chapter. This section will provide an integrated understanding of architectural glass.

Architectural Glass

Architectural glass is glass that is used as a building material. It is most typically used as transparent glazing material in the building envelope, including windows in the external walls. Glass is also used for internal partitions and as an architectural feature. When used in buildings, glass is often of a safety type, which include reinforced, toughened and laminated glasses.

A building in Canterbury, England, which displays its long history in different building styles and glazing of every century from the 16th to the 20th included.

Cast Glass

Cast glass windows, albeit with poor optical qualities, began to appear in the most important buildings in Rome and the most luxurious villas of Herculaneum and Pompeii.

Crown Glass

One of the earliest methods of glass window manufacture was the *crown glass* method. Hot blown glass was cut open opposite the pipe, then rapidly spun on a table before it could cool. Centrifugal force shaped the hot globe of glass into a round, flat sheet. The sheet would then be broken off the pipe and trimmed to form a rectangular window to fit into a frame.

The concentric arcs that distort some of these panes indicate they are crown glass, possibly of the 16th century.

At the center of a piece of crown glass, a thick remnant of the original blown bottle neck would remain, hence the name "bullseye." Optical distortions produced by the bullseye could be reduced by grinding the glass. The development of diaper latticed windows was in part because three regular diamond-shaped panes could be conveniently cut from a piece of Crown glass, with minimum waste and with minimum distortion.

This method for manufacturing flat glass panels was very expensive and could not be used to make large panes. It was replaced in the 19th century by the cylinder, sheet, and rolled plate processes, but it is still used in traditional construction and restoration.

Cylinder Glass

In this manufacturing process, glass is blown into a cylindrical iron mould. The ends are cut off and a cut is made down the side of the cylinder. The cut cylinder is then placed in an oven where the cylinder unrolls into a flat glass sheet.

Drawn Sheet Glass (Fourcault Process)

The uneven surface of old glass is visible in the reflection on this window pane.

Drawn Sheet glass was made by dipping a leader into a vat of molten glass then pulling that leader straight up while a film of glass hardened just out of the vat – this is known as the Fourcault process. This film or ribbon was pulled up continuously held by tractors on both edges while it cooled. After 12 metres or so it was cut off the vertical ribbon and tipped down to be further cut. This glass is clear but has thickness variations due to small temperature changes just out of the vat as it was hardening. These variations cause lines of slight distortions. This glass may still be seen in older houses. Float glass replaced this process.

Cast Plate Glass

Developed by James Hartley in 1848. The glass is taken from the furnace in large iron ladles, which are carried upon slings running on overhead rails; from the ladle the glass is thrown upon the cast-iron bed of a rolling-table; and is rolled into sheet by an iron roller, the process being similar to that employed in making plate-glass, but on a smaller scale. The sheet thus rolled is roughly trimmed while hot and soft, so as to remove those portions of glass which have been spoiled by immediate contact with the ladle, and the sheet, still soft, is pushed into the open mouth of an annealing tunnel or temperature-controlled oven called a lehr, down which it is carried by a system of rollers.

Polished Plate Glass

The polished plate glass process starts with sheet or rolled plate glass. This glass is dimensionally inaccurate and often created visual distortions. These rough panes were ground flat and then polished clear. This was a fairly expensive process.

Before the float process, mirrors were plate glass as sheet glass had visual distortions that were akin to those seen in amusement park or funfair mirrors.

Rolled Plate (Figured) Glass

Figure rolled glass

The elaborate patterns found on figured (or 'Cathedral') rolled-plate glass are produced in a similar fashion to the rolled plate glass process except that the plate is cast between two rollers, one of which carries a pattern. On occasion, both rollers can carry a pattern. The pattern is impressed upon the sheet by a printing roller which is brought down upon the glass as it leaves the main rolls while still soft. This glass shows a pattern in high relief. The glass is then annealed in a lehr.

The glass used for this purpose is typically whiter in colour than the clear glasses used for other applications.

This glass can be laminated or toughened depending on the depth of the pattern to produce a safety glass.

Float Glass

Ninety percent of the world's flat glass is produced by the float glass process invented in the 1950s by Sir Alastair Pilkington of Pilkington Glass, in which molten glass is poured onto one end of a molten tin bath. The glass floats on the tin, and levels out as it spreads along the bath, giving a smooth face to both sides. The glass cools and slowly solidifies as it travels over the molten tin and leaves the tin bath in a continuous ribbon. The glass is then annealed by cooling in an oven called a lehr. The finished product has near-perfect parallel surfaces.

The side of the glass that has been in contact with the tin has a very small amount of the tin embedded in its surface. This quality makes that side of the glass easier to be coated in order to turn it into a mirror, however that side is also softer and easier to scratch.

Glass is produced in standard metric thicknesses of 2, 3, 4, 5, 6, 8, 10, 12, 15, 19 and 22 mm. Molten glass floating on tin in a nitrogen/hydrogen atmosphere will spread out to a thickness of about 6 mm and stop due to surface tension. Thinner glass is made by stretching the glass while it floats on the tin and cools. Similarly, thicker glass is pushed back and not permitted to expand as it cools on the tin.

Prism Glass

Prism glass is architectural glass which was frequently used around the turn of the 20th century to provide lighting to underground spaces and areas that would otherwise be too difficult to light. Prism glass uses a unique convex lens design that illuminates more than ordinary glass. Prism glass can sometimes be found on sidewalks and in this form is known as vault lighting. It is similar to deck prisms that were used to light spaces below deck on sailing ships.

Glass Block

Glass bricks used in a wall

Sidewalk skylight (also named 'pavement light') made of Glass brick in Burlington House, London

Glass block, also known as glass brick, is an architectural element made from glass used in areas where privacy or visual obscuration is desired while admitting light, such as underground parking garages, washrooms, and municipal swimming baths. Glass block was originally developed in the early 1900s to provide natural light in industrial factories.

Annealed Glass

Annealed glass is glass without internal stresses caused by heat treatment, i.e., rapid cooling, or by toughening or heat strengthening. Glass becomes annealed if it is heated above a transition point then allowed to cool slowly, without being quenched. Float glass is annealed during the process of manufacture. However, most toughened glass is made from float glass that has been specially heat-treated.

Annealed glass breaks into large, jagged shards that can cause serious injury and is considered a hazard in architectural applications. Building codes in many parts of the world restrict the use of annealed glass in areas where there is a high risk of breakage and injury, for example in bathrooms, door panels, fire exits and at low heights in schools or domestic houses.

Laminated Glass

Broken tempered laminated glass "wet blanket effect"

Laminated glass is manufactured by bonding two or more layers of glass together with an interlayer, such as PVB, under heat and pressure, to create a single sheet of glass. When broken, the interlayer keeps the layers of glass bonded and prevents it from breaking apart. The interlayer can also give the glass a higher sound insulation rating.

There are several types of laminated glasses manufactured using different types of glass and interlayers which produce different results when broken.

Laminated glass that is made up of annealed glass is normally used when safety is a concern, but tempering is not an option. Windshields are typically laminated glasses. When broken, the PVB layer prevents the glass from breaking apart, creating a "spider web" cracking pattern.

Tempered laminated glass is designed to shatter into small pieces, preventing possible injury. When both pieces of glass are broken it produces a "wet blanket" effect and it will fall out of its opening.

Heat strengthened laminated glass is stronger than annealed, but not as strong as tempered. It is often used where security is a concern. It has a larger break pattern than tempered, but because it holds its shape (unlike the "wet blanket" effect of tempered laminated glass) it remains in the opening and can withstand more force for a longer period of time, making it much more difficult to get through.

Toughened Glass (Tempered Glass)

Toughened glass in a vandalized phone booth in Britain

Toughened (or tempered) glass is a type of safety glass that has increased strength and will usually shatter in small, square pieces when broken. It is used when strength, thermal resistance and safety are important considerations. Using toughened glass on automobile windshields would be a problem when a small stone hits the windshield at speed, as it would shatter into small squares endangering the driver and passengers. In commercial structures it is used in unframed assemblies such as frameless doors, structurally loaded applications and door lites and vision lites adjacent to doors. Toughened glass is typically four to six times the strength of annealed glass.

Heat-strengthened Glass

Heat-strengthened glass is glass that has been heat treated to induce surface compression, but not to the extent of causing it to "dice" on breaking in the manner of tempered glass. On breaking, heat-strengthened glass breaks into sharp pieces that are typically somewhat smaller than those found on breaking annealed glass, and is intermediate in strength between annealed and toughened glasses.

Chemically Strengthened Glass

Chemically strengthened glass is a type of glass that has increased strength. When broken it still shatters in long pointed splinters similar to float (annealed) glass. For this reason, it is not considered a safety glass and must be laminated if a safety glass is required. Chemically strengthened glass is typically six to eight times the strength of annealed glass.

The glass is chemically strengthened by submerging the glass in a bath containing a potassium salt (typically potassium nitrate) at 450 °C (842 °F). This causes sodium ions in the glass surface to be replaced by potassium ions from the bath solution.

Unlike toughened glass, chemically strengthened glass may be cut after strengthening, but loses its added strength within the region of approximately 20 mm of the cut. Similarly, when the surface of chemically strengthened glass is deeply scratched, this area loses its additional strength.

Chemically strengthened glass was used on some fighter aircraft canopies.

Low-emissivity Glass

Glass coated with a low-emissivity substance can reflect radiant infrared energy, encouraging radiant heat to remain on the same side of the glass from which it originated, while letting visible light pass. This often results in more efficient windows because radiant heat originating from indoors in winter is reflected back inside, while infrared heat radiation from the sun during summer is reflected away, keeping it cooler inside.

Heatable Glass

Electrically heatable glass is a relatively new product, which helps to find solutions while designing buildings and vehicles. The idea of heating glass is based on usage of energy-efficient low-emissive glass that is generally simple silicate glass with special metallic oxides coating. Heatable glass can be used in all kinds of standard glazing systems, made of wood, plastic, aluminum or steel.

Self-cleaning Glass

A recent (2001 Pilkington Glass) innovation is so-called self-cleaning glass, aimed at building, automotive and other technical applications. A nanometre-scale coating of titanium dioxide on the outer surface of glass introduces two mechanisms which lead to the self-cleaning property. The first is a photo-catalytic effect, in which ultra-violet rays catalyse the breakdown of organic compounds on the window surface; the second is a hydrophilic effect in which water is attracted to the surface of the glass, forming a thin sheet which washes away the broken-down organic compounds.

Insulated Glazing

Insulated glazing, or double glazing, consists of a window or glazing element of two or more layers of glazing separated by a spacer along the edge and sealed to create a dead air space between the layers. This type of glazing has functions of thermal insulation and noise reduction. When the space is filled with an inert gas it is part of energy conservation sustainable architecture design for low energy buildings.

Evacuated Glazing

A 1994 innovation for insulated glazing is evacuated glass, which as yet is produced commercially only in Japan and China. The extreme thinness of evacuated glazing offers many new architectural possibilities, particularly in building conservation and historicist architecture, where evacuated glazing can replace traditional single glazing, which is much less energy-efficient.

An evacuated glazing unit is made by sealing the edges of two glass sheets, typically by using a solder glass, and evacuating the space inside with a vacuum pump. The evacuated space between the two sheets can be very shallow and yet be a good insulator, yielding insulative window glass with nominal thicknesses as low as 6 mm overall. The reasons for this low thickness are deceptively complex, but the potential insulation is good essentially because there can be no convection or gaseous conduction in a vacuum.

Unfortunately, evacuated glazing does have some disadvantages; its manufacture is complicated and difficult. For example, a necessary stage in the manufacture of evacuated glazing is outgassing; that is, heating it to liberate any gases adsorbed on the inner surfaces, which could otherwise later escape and destroy the vacuum. This heating process currently means that evacuated glazing cannot be toughened or heat-strengthened. If an evacuated safety glass is required, the glass must be laminated. The high temperatures necessary for outgassing also tend to destroy the highly effective "soft" low-emissivity coatings that are often applied to one or both of the internal surfaces (i.e. the ones facing the air gap) of other forms of modern insulative glazing, in order to prevent loss of heat through infrared radiation. Slightly less effective "hard" coatings are still suitable for evacuated glazing, however.

Furthermore, because of the atmospheric pressure present on the outside of an evacuated glazing unit, its two glass sheets must somehow be held apart in order to prevent them flexing together and touching each other, which would defeat the object of evacuating the unit. The task of holding the panes apart is performed by a grid of spacers, which typically consist of small stainless steel discs that are placed around 20 mm apart. The spacers are small enough that they are visible only at very close distances, typically up to 1 m. However, the fact that the spacers will conduct some heat often leads in cold weather to the formation of temporary, grid-shaped patterns on the surface of an evacuated window, consisting either of small circles of interior condensation centred around the spacers, where the glass is slightly colder than average, or, when there is dew outside, small circles on the exterior face of the glass, in which the dew is absent because the spacers make the glass near them slightly warmer.

The conduction of heat between the panes, caused by the spacers, tends to limit evacuated glazing's overall insulative effectiveness. Nevertheless, evacuated glazing is still as insulative as much thicker conventional double glazing and tends to be stronger, since the two constituent glass sheets are pressed together by the atmosphere, and hence react practically as one thick sheet to bending forces. Evacuated glazing also offers very good sound insulation in comparison with other popular types of window glazing.

Building Code Seismic Requirements

The most current building code enforced in most jurisdictions in the United States is the 2006 International Building Code (IBC, 2006). The 2006 IBC references for the 2005 edition of the standard Minimum Design Loads for buildings and other Structures prepared by the American Society of Civil Engineers (ASCE, 2005) for its seismic provisions. ASCE 7-05 contains specific requirements for nonstructural components including requirements for architectural glass.

Glass Brick

Glass brick, also known as glass block, is an architectural element made from glass. Glass bricks provide visual obscuration while admitting light. The glass block was originally developed in the early 1900s to provide natural light in manufacturing plants.

Glass bricks are produced for both wall and floor applications. Glass blocks for use in floors are normally manufactured as a single solid piece, or as a hollow glass block with thicker side walls than the standard wall blocks. These blocks are normally cast into a reinforced concrete gridwork or set into a metal frame, allowing multiple units to be combined to span over openings in basements and roofs. Glass wall blocks should not be used in flooring applications. Hollow glass wall blocks are manufactured as two separate halves and, whilst the glass is still molten, the two pieces are pressed together and annealed. The resulting glass blocks will have a partial vacuum at the hollow centre.

Glass block wall in Chicago.

Standards and Grading

Glass blocks in Europe are manufactured in accordance with the European Standard EN1052-2. The International Standard is ISO TC 160/SG1. The Standards allow for variation in sizes and production irregularity. Blocks fall within three classifications; Class1, Class 2 and Class 3 with Class 1 being the highest and best rating with a maximum permissible deviation from designed size and rectangularity of 1 mm.

Construction Methods

Glass wall blocks are fixed together to form complete walls by several methods – the most common method of construction is to bed the blocks together in a Portland cement-based mortar with reinforcing rods of steel placed within the mortar as recommended by the project architect or block manufacturer.

Other methods of construction include several proprietary systems whereby the mortar is replaced by timber or PVC extrusions.

R Value

Glass brick has an r value between 1.75 and 1.96, close to that of thermopane windows. There are newer Glass Blocks injected with Argon gas and having a layer of low-emissivity glass between the halves, which increases the insulative (U) value to 1.5 $W/m^2{\cdot}K$, which is between triple glazed windows(1.8 $W/m^2{\cdot}K$) and specialty double glazed windows with advanced frame and coatings(1.2 $W/m^2{\cdot}K$).

Specialty Types

Specialist glass blocks are produced for various applications including:

Bullet and Vandal Resistance

Bullet and vandal resistant blocks are generally solid glass or have very thick side walls similar to pavement blocks.

Fire Resistance

Fire resistance of varying degrees can be achieved by several methods. Standard production hollow wall block will offer little fire resistance; however, resistance is improved by utilising specially produced hollow blocks with thicker sidewalls, or the inclusion of a special layer of fire resisting material between the two halves of the block during manufacture. Some manufacturers of glass blocks have developed a method of bonding two glass blocks together with adhesive, producing blocks of up to 160 mm (6½") thick with enhanced fire resistance. It is important that the block manufacturer's recommendations are followed with regards to the installation of fire resisting glass block walls, as without special construction techniques, the wall will not achieve the desired fire resistance.

Gas Insulated

A recent innovation in the manufacture of glass blocks is the inclusion of argon gas within the hollow centre of glass wall blocks. This advancement in production technique has resulted in a glass block which is able to offer significantly improved thermal insulation properties.

Colored

Some hollow glass wall blocks are available in coloured variants. These coloured variants fall into two categories; those that are manufactured with coloured glass which are UV stable and can be used in the same locations as standard clear glass blocks. The other method by which coloured glass blocks are achieved is to inject a coloured material, dye or transparent paint into the hollow centre of the blocks to form a permanent coating. This method of producing coloured blocks enables vibrant colours to be achieved which are not possible with coloured glass. The downside of this production method is that the coloured coating may not be UV stable and can fade in bright sunshine over time and may therefore, not be suitable for all locations.

Examples of Architectural Use

- Crown Fountain in Chicago, United States.
- Maison de Verre (for House of Glass) in Paris, France
- Michigan State Capitol in Lansing, Michigan
- Hermes Store in Ginza, Tokyo, Japan by Renzo Piano
- Streamline Moderne

Crown Fountain is composed of glass brick towers.

Glass brick labyrinths, used to obscure clear visual line of sight into women's washroom on the left, and men's washroom on the right.

Glass block floor in the Michigan State Capitol

Annealing (Glass)

Annealing of glass is a process of slowly cooling hot glass objects after they have been formed, to relieve residual internal stresses introduced during manufacture. Especially for smaller, simpler objects, annealing may be incidental to the process of manufacture, but in larger or more complex products it commonly demands a special process of annealing in a temperature-controlled kiln known as a lehr. Annealing of glass is critical to its durability. Glass that has not been properly annealed retains thermal stresses caused by quenching, which indefinitely decrease the strength and reliability of the product. Inadequately annealed glass is likely to crack or shatter when subjected to relatively small temperature changes or to mechanical shock or stress. It even may fail spontaneously.

To anneal glass it is necessary to heat it to its annealing temperature, at which its viscosity, η, drops to 10^{13} Poise ("Poise" is a measure of absolute viscosity; 1 poise = 1 dyne-second/cm^2). For most kinds of glass this annealing temperature is in the range of 850–900 °F (454 - 482 °C), and is the so-called stress-relief point or annealing point of the glass. At such a viscosity the glass still is too hard for significant external deformation without breaking, but it is soft enough to relax internal strains by microscopic flow in response to the intense stresses they introduce internally. The piece then heat-soaks until its temperature is even throughout and the stress relaxation is adequate. The time necessary for this step varies depending on the type of glass and its maximum thickness. The glass then is permitted to cool at a predetermined rate until its temperature passes the strain point ($\eta = 10^{14.5}$ Poise), below which even microscopic internal flow effectively stops and annealing stops with it. It then is safe to cool the product to room temperature at a rate limited by the heat capacity, thickness, thermal conductivity, and thermal expansion coefficient of the glass. After annealing is complete the material can be cut to size, drilled, or polished without risk of its internal stresses shattering it.

At the *annealing point* ($\eta = 10^{13}$ Poise) stresses relax within several minutes, while at the *strain point* ($\eta = 10^{14.5}$ Poise) stresses relax within several hours. Stresses acquired at temperatures above the strain point, and not relaxed by annealing, remain in the glass indefinitely and may cause either short-term or much delayed failure. Stresses resulting from cooling too rapidly below the strain point largely are considered temporary, although they may be adequate to promote short-term failure.

Laminated Glass

Automobile windshield with "spider web" cracking typical of laminated safety glass.

Laminated glass is a type of safety glass that holds together when shattered. In the event of breaking, it is held in place by an interlayer, typically of polyvinyl butyral (PVB) or ethylene-vinyl acetate (EVA), between its two or more layers of glass. The interlayer keeps the layers of glass bonded even when broken, and its high strength prevents the glass from breaking up into large sharp pieces. This produces a characteristic "spider web" cracking pattern when the impact is not enough to completely pierce the glass. In the case of the EVA, the thermoset EVA, offers a complete bounding (cross-linking) with the material whether it is glass, polycarbonate, P.E.T. or other types products.

Laminated glass is normally used when there is a possibility of human impact or where the glass could fall if shattered and also for architectural applications. Skylight glazing and automobile windshields typically use laminated glass. In geographical areas requiring hurricane-resistant construction, laminated glass is often used in exterior storefronts, curtain walls and windows.

Laminated glass is also used to increase the sound insulation rating of a window, where it significantly improves sound attenuation compared to unlaminated glass panes of the same thickness. For this purpose a special "acoustic PVB" compound is used for the interlayer. In the case of the EVA material, no additional acoustic material is required, since the EVA provides sound insulation. An additional property of laminated glass for windows is that a PVB and EVA interlayer can block essentially most ultraviolet radiation. A thermoset EVA could block up to 99.9% of the UV rays.

History

In 1902, the French *Le Carbone* corporation obtained a patent for coating glass objects with celluloid in order to render them less susceptible to cracking or breaking.

Laminated glass was invented in 1903 by the French chemist Édouard Bénédictus (1878-1930), inspired by a laboratory accident. A glass flask had become coated with the plastic cellulose nitrate and when dropped shattered but did not break into pieces. However, it was not until 1909 that Benedictus filed a patent. In 1911, he formed the *Société du Verre Triplex*, which fabricated a glass-plastic composite to reduce injuries in car accidents. Production of Triplex glass was slow and painstaking, making it expensive. It was not immediately widely adopted by automobile manufacturers, but laminated glass was widely used in the eyepieces of gas masks during World War I. In 1912, the process was licensed to *The English Triplex Safety Glass Company*. Subsequently, in the United States, both Libbey Owens-Ford and Du Pont de Nemours with Pittsburg Plate Glass produced Triplex.

Firefighters breaking through a laminated windshield

Meanwhile, in 1905, John Crewe Wood, a solicitor in Swindon, Wiltshire, England, patented a laminated glass for use as windscreens / windshields. The layers of glass were bonded together by Canada balsam. In 1906 he founded the Safety Motor Screen Co. to produce and sell his product.

In 1927, the Canadian chemists Howard W. Matheson and Frederick W. Skirrow invented the plastic polyvinyl butyral (PVB). By 1936, United States companies had discovered that laminated "safety glass" consisting of a layer of polyvinyl butyral between two layers of glass would not discolor and was not easily penetrated during accidents. Within five years, the new safety glass had virtually replaced its predecessor.

In the Road Traffic Act of 1930, the British parliament required new cars to use windscreens of "safety glass".

By 1939 some 600,000 square feet (56,000 m^2) of "Indestructo" safety glass was being used every year in vehicles produced at the Ford Motor Company works in Dagenham, England. "Indestructo" safety glass was manufactured by British Indestructo Glass, Ltd. of London. This was the laminated glass used by the Ford Motor Company in 1939, chosen because "it gives the most complete protection. In addition to being splinter-proof, it is crystal clear and permanently non-discolourable." This quote hints at some of the technical issues, problems and concerns that stopped laminated glass from being widely used in automobiles immediately after it was invented.

Modern laminated glass is produced by bonding two or more layers of ordinary annealed glass (or tempered glass) together with a plastic interlayer, usually polyvinyl butyral or ethylene-vinyl acetate (EVA). The PVB or the EVA are sandwiched by the glass, which is either passed through a series of rollers, or vacuum bagging systems, and ovens, or autoclaves, to expel any air pockets. Then it is heated to form the initial melting. These constructions are then heated under pressure in an autoclave or oven, to achieve the final bounded product(fully crosslinked in the case of the thermoset EVA). The tint at the top of some car windshields is in the PVB. Also, colored P.E.T. films can be combined with the thermoset EVA material, during the laminating process, in order to obtain a colored glass.

Once a thermoset EVA is properly laminated during the process, the glass could be exposed frameless and there will be no water/moisture infiltration, the yellowing index is very low and it shouldn't delaminate, due to the high level of bounding (crosslinking).

Specifications

A typical laminated makeup is 2.5 mm glass / 0.38 mm interlayer / 2.5 mm glass. This gives a final product that would be referred to as 5.38 laminated glass.

Multiple laminates and thicker glass increases the strength. Bullet-resistant glass is usually constructed using polycarbonate, thermoplastic materials, thermoset EVA, and layers of laminated glass. A similar glass is often used in airliners on the front windows, often three sheets of 6 mm toughened glass with thick PVB between them.

Newer developments have increased the thermoplastic family for the lamination of glass. Beside PVB, important thermoplastic glass lamination materials today are ethylene-vinyl acetate (EVA), thermoset EVA ethylene-vinyl acetate (EVA) and thermoplastic polyurethane (TPU). The adhesion of PVB/TPU and EVA is not only high to glass, but also to polyester (PE) interlayers. Since 2004 metallised and electroconductive polyethylene terephthalate (PET) interlayers are used as substrate for light emitting diodes and laminated to or between glass.

- Top layer: Glass
- Interlayer: Transparent thermoplastic materials (TPU or PVB, EVA) or transparent thermoset material (EVA)
- Interlayer: LED (light emitting diodes)on transparent conductive Polymer

- Interlayer: Transparent thermoplastic materials (TPU or PVB, EVA) or transparent thermoset material (EVA)
- Bottom layer: Glass

Laminated glass is also sometimes used in glass sculptures.

Manufacture

There are several laminated glass manufacturing processes:

1. using two or more pieces of glass bonded between one or more pieces of adhesives; such as PVB or EVA, using heat and pressure.
2. using two or more pieces of glass and polycarbonate, bonded together with aliphatic urethane or EVA interlayer under heat and pressure.
3. interlaid with a cured resin or EVA.

Each manufacturing process may include glass lites of equal or unequal thickness.

Cutting

Plastic interlayers in laminated glass make its cutting difficult. There is an unsafe practice of cutting both sides separately, pouring a flammable liquid such as denatured alcohol into the crack, and igniting it to melt the interlayer to separate the pieces. The following safer methods were recommended by the UK Government's Health and Safety Executive in 2005:

- Special purpose laminated cutting tables
- Vertically-inclined saw frames
- A blowlamp or hot air blower.
- High pressure abrasive waterjet.

Repair

According to the United States National Windshield Repair Association, laminated glass repair is possible for minor impact damage using a process that involves drilling into the fractured glass to reach the lamination layer. Special clear adhesive resin is injected under pressure and then cured with ultraviolet light. When done properly, the strength and clarity is sufficiently restored for most safety related purposes. The process is widely used to repair large industrial automotive windshields where the damage does not interfere with the view of the driver.

Disposal

Waste disposal of laminated glass is no longer permitted in landfill in most European countries as the End of Life Vehicles Directive (ELV) is implemented. A study by University of Surrey and Pilkington Glass proposes that waste laminated glass be placed into a separating device such as a rolling mill where the glass is fragmented and the larger cullet is mechanically detached from the inner film. The application of heat then melts the laminating plastic, usually polyvinyl butyral (PVB), enabling both the glass and the interior film to be recycled. The PVB recycling process is a simple procedure of melting and reshaping it.

Toughened Glass

Toughened or tempered glass is a type of safety glass processed by controlled thermal or chemical treatments to increase its strength compared with normal glass. Tempering puts the outer surfaces into compression and the inner surfaces into tension. Such stresses cause the glass, when broken, to crumble into small granular chunks instead of splintering into jagged shards as plate glass (aka: annealed glass) creates. The granular chunks are less likely to cause injury.

As a result of its safety and strength, toughened glass is used in a variety of demanding applications, including passenger vehicle windows, shower doors, architectural glass doors and tables, refrigerator trays, mobile screen protectors, as a component of bulletproof glass, for diving masks, and various types of plates and cookware.

Properties

Toughened glass is physically and thermally stronger than regular glass. The greater contraction of the inner layer during manufacturing induces compressive stresses in the surface of the glass balanced by tensile stresses in the body of the glass. For glass to be considered toughened, this compressive stress on the surface of the glass should be a minimum of 69 megapascals (10,000 psi). For it to be considered safety glass, the surface compressive stress should exceed 100 megapascals (15,000 psi). The greater the surface stress, the smaller the glass particles will be when broken.

Toughened glass of car rear window. Variations in glass stress are clearly seen when photographed through a polarizing filter (bottom picture).

It is this compressive stress that gives the toughened glass increased strength. This is because any surface flaws tend to be pressed closed by the retained compressive forces, while the core layer remains relatively free of the defects which could cause a crack to begin.

Any cutting or grinding must be done prior to tempering. Cutting, grinding, and sharp impacts after tempering will cause the glass to fracture.

The strain pattern resulting from tempering can be observed with polarized light or by using a pair of polarizing sun glasses.

Uses

Safety approval markings on an automobile vent window made for a Chrysler car by PPG.

Toughened glass is used when strength, thermal resistance, and safety are important considerations.

Automobiles

The most commonly encountered tempered glass is that used for side and rear windows in automobiles.

A shattered driver's side window reveals small fragments that still cling together.

It is used for its characteristic of shattering into small cubes rather than large shards and is sometimes referred to as safety glass in this context. (The windscreen or windshield is instead made of laminated glass, which will not shatter when broken.)

Buildings and Structures

Toughened glass is also used in buildings for unframed assemblies (such as frameless glass doors), structurally loaded applications, and any other application that would become dangerous in the event of human impact. Tempered and heat strengthened glass can be three to seven times stronger than annealed glass. Building codes in the United States require tempered or laminated glass in several situations including some skylights, near doorways and stairways, large windows, windows which extend close to floor level, sliding doors, elevators, fire department access panels, and near swimming pools.

Food Service

"Rim-tempered" indicates that a limited area, such as the rim of the glass or plate, is tempered and is popular in food service. However, there are also specialist manufacturers that offer a fully tempered/toughened drinkware solution that can bring increased benefits in the form of strength and thermal shock resistance. In some countries these products are specified in venues that require increased performance levels or have a requirement for a safer glass due to intense usage.

Tempered glass has also seen increased usage in bars and pubs, particularly in the United Kingdom and Australia, to prevent broken glass being used as a weapon.

Cooking and Baking

Some forms of tempered glass are used for cooking and baking. Manufacturers and brands include Glasslock, Pyrex, Corelle, and Arc International.

Mobile Devices

Most touchscreen mobile devices use some form of toughened glass (such as Corning's Gorilla Glass), as do some aftermarket screen protectors for these devices.

Manufacturing

Toughened glass can be made from annealed glass via a thermal tempering process. The glass is placed onto a roller table, taking it through a furnace that heats it well above its transition temperature of 564 °C (1,047 °F) to around 620 °C (1,148 °F). The glass is then rapidly cooled with forced air drafts while the inner portion remains free to flow for a short time.

An alternative chemical toughening process involves forcing a surface layer of glass at least 0.1 mm thick into compression by ion exchange of the sodium ions in the glass surface with potassium ions (which are 30% larger), by immersion of the glass into a bath of molten potassium nitrate. Chemical toughening results in increased toughness compared with thermal toughening and can be applied to glass objects of complex shapes.

Advantages

Tempered safety glass, which has been laminated, often does not fall out of its frame when it breaks – usually due to the anti-splinter film applied on the glass, as seen in this grocery store meat case.

The term "toughened glass" is generally used to describe fully tempered glass but is sometimes used to describe heat-strengthened glass as both types undergo a thermal "toughening" process.

There are two main types of heat-treated glass: heat-strengthened and fully tempered. Heat-strengthened glass is twice as strong as annealed glass while fully tempered glass has typically four to six times the strength of annealed glass and withstands heating in microwave ovens. The difference is the residual stress in the edge and glass surface. Fully tempered glass in the US is generally rated above 65 megapascals (9,400 psi) in pressure-resistance, while heat-strengthened glass is between 40 and 55 megapascals (5,800 and 8,000 psi).

The tempering process does not change the stiffness of the glass. Annealed glass undergoes a similar deflection compared to tempered glass under the same load, but tempered glass can take a higher load and, therefore, deflects further before breaking.

Disadvantages

Toughened glass must be cut to size or pressed to shape before toughening, and cannot be re-worked once toughened. Polishing the edges or drilling holes in the glass is carried out before the toughening process starts. Because of the balanced stresses in the glass, damage to any portion will eventually result in the glass shattering into thumbnail-sized pieces. The glass is most susceptible to breakage due to damage to the edge of the glass, where the tensile stress is the greatest, but shattering can also occur in the event of a hard impact in the middle of the glass pane or if the impact is concentrated (for example, striking the glass with a hardened point).

Using toughened glass can pose a security risk in some situations because of the tendency of the glass to shatter completely upon hard impact rather than leaving shards in the window frame.

The surface of tempered glass does exhibit surface waves caused by contact with flattening rollers, if it has been formed using this process. This waviness is a significant problem in manufacturing of thin film solar cells. The float glass process can be used to provide low-distortion sheets with very flat and parallel surfaces.

History

Francois Barthelemy Alfred Royer de la Bastie (1830–1901) of Paris, France is credited with first developing a method of tempering glass by quenching almost molten glass in a heated bath of oil or grease in 1874, the method patented in England on August 12, 1874, patent number 2783. Tempered glass is sometimes known as *Bastie glass* after de la Bastie. In 1877 the German Frederick Siemens developed a different process, sometimes called compressed glass or Siemens glass, producing a tempered glass stronger than the Bastie process by pressing the glass in cool molds. The first patent on a whole process to make tempered glass was held by chemist Rudolph A. Seiden who was born in 1900 in Austria and immigrated to the United States in 1935.

Though the underlying mechanism was not known at the time, the effects of "tempering" glass have been known for centuries. In about 1660, Prince Rupert of the Rhine brought the discovery of what are now known as "Prince Rupert's Drops" to the attention of King Charles II. These are teardrop-shaped bits of glass which are produced by allowing a molten drop of glass to fall into a bucket of water, thereby rapidly cooling it. They can withstand a blow from a hammer on the bulbous end without breaking, but the drops will disintegrate explosively into powder if the tail end is even slightly damaged.

Self-cleaning Glass

Self-cleaning glass is a specific type of glass with a surface that keeps itself free of dirt and grime.

Introduction

The field of self-cleaning coatings on glass is divided into two categories: hydrophobic and hydrophilic. These two types of coating both clean themselves through the action of water, the former by rolling droplets and the latter by sheeting water that carries away dirt. Hydrophilic coatings based on titania, however, have an additional property: they can chemically break down absorbed dirt in sunlight.

The requirements for a self-cleaning hydrophobic surface are a very high static water contact angle θ, the condition often quoted is $\theta > 160°$, and a very low roll-off angle, i.e. the minimum inclination angle necessary for a droplet to roll off the surface.

Self-cleaning Surfaces

Several techniques are known for the patterning of hydrophobic surfaces through the use of moulded polymers and waxes, by physical processing methods such as ion etching and compression of polymer beads, and by chemical methods such as plasma-chemical roughening, which can all result in ultra-hydrophobic coatings. While these surfaces are effective self-cleaners, they suffer from a number of drawbacks which have so far prevented widespread application. Batch processing a hydrophobic material is a costly and time-consuming technique, and the coatings produced are usually hazy, precluding applications on lenses and windows, and fragile materials. The second class of self-cleaning surfaces are hydrophilic surfaces which do not rely solely on the flow of water to wash away dirt. These coatings chemically break down dirt when exposed to light, a process known as photocatalysis. Despite the commercialization of a hydrophilic self-cleaning coating in a number of products, the field is far from mature; investigations into the fundamental mechanisms of self-cleaning and characterizations of new coatings are regularly published in the primary literature.

The Discovery of Self-cleaning Behavior

The first self-cleaning glass was based on a thin film titania coating. The film can be applied by spin coating of organo-titanate chelated precursor (for example titanium iso-tetrapropoxide chelated by acetylacetone), followed by heat treatment at elevated temperatures to burn the organic residues and to form the anatase phase. In that case, sodium might diffuse from the glass into the nascent titanium dioxide, causing a degradation in the hydrophilic/catalytic effect unless preventive measures are taken. The glass cleans itself in two stages. The "photocatalytic" stage of the process breaks down the organic dirt on the glass using ultraviolet light and makes the glass superhydrophilic (normally glass is hydrophobic). During the following "superhydrophilic" stage, rain washes away the dirt, leaving almost no streaks, because water spreads evenly on superhydrophilic surfaces.

The First Commercial Product

In 2001, Pilkington Glass announced the development of the first self-cleaning windows, Pilkington Activ™, and in the following months several other major glass companies released similar products. As a result, glazing is perhaps the largest commercial application of self-cleaning coatings to date. All of these windows are coated with a thin transparent layer of titanium dioxide. This coating acts to clean the window in two stages, using two distinct properties: photocatalysis and hydrophilicity. In sunlight, photocatalysis causes the coating to chemically break down organic dirt adsorbed onto the window. When the glass is wet by rain or other water, hydrophilicity reduces contact angles to very low values, causing the water to form a thin layer rather than droplets, and this layer washes dirt away.

The use of Titanium Dioxide in Self-cleaning Applications

Titanium dioxide has become the material of choice for self-cleaning windows, and hydrophilic self-cleaning surfaces in general, because of its favorable physical and chemical properties. Not only is titanium dioxide highly efficient at photocatalysing dirt in sunlight and reaching the superhydrophilic state, it is also non-toxic, chemically inert in the absence of light, inexpensive, relatively easy to handle and deposit into thin films and is an established household chemical that is used as a pigment in cosmetics and paint and as a food additive.

The Mechanism

The anatase phase is the most photocatalytic among its polymorphic structures. Moreover, ultraviolet irradiation creates surface oxygen vacancies at bridging sites, resulting in the conversion of relevant Ti^{4+} sites to Ti^{3+} sites which are favourable for dissociative water adsorption. These defects presumably influence the affinity to chemisorbed water of their surrounding sites, forming hydrophilic domains, whereas the rest of the surface remains oleophilic. Hydrophilic domains are areas where dissociative water is adsorbed, associated with oxygen vacancies that are preferentially photogenerated along the direction of the (110) plane; the same direction in which oxygen bridging sites align.

Other Applications

It is speculated that other possible application areas are computer monitors and PDA screens, where fingerprints are undesirable.

Titanium dioxide–based glass cannot decompose thick non-transparent deposits, such as paint or silicone, waterstop fingerprints or bleeding after weathering, or stucco dust produced during construction.

Since 2001 the TC24 "Coatings on Glass" committee International Commission on Glass has been trying to set up test methods for evaluation of photocatalytic self-cleaning coatings on glass.

Brands

- The Pilkington Activ brand by Pilkington is claimed by the company to be the first self-cleaning glass. Pilkington Activ™ consists of a 20–30 nm layer of nanocrystalline anatase titanium dioxide deposited by an atmospheric pressure chemical vapor deposition technique onto soda-lime silicate float glass. The result is a product with extremely favorable visible transmission and reflectance properties; Activ™ has a visible reflectance of around 7% and a visible haze of less than 1%, but absorbs 20% of incident solar UV light which is used in the self-cleaning process. The coating is also robust and cannot be damaged by application of Scotch tape or moderate mechanical abrasion; Pilkington claim that the coating will last the lifetime of the window frame.
- The SunClean brand by PPG Industries also uses a coating of titanium dioxide, applied by a patented process.
- Neat Glass by Cardinal Glass Industries has a titanium dioxide layer less than 10 nm thick applied by magnetron sputtering
- SGG Aquaclean (1st generation, hydrophilic only, 2002) and Bioclean (2nd generation, both photoactive and hydrophilic, 2003) by Saint-Gobain. The Bioclean coating is applied by chemical vapor deposition.
- REIBORG HIKARI by Nippon Sheet Glass. Hyoumenkagaku Vol. 26, No. 11, pp. 700–703, 2005, T.Anzaki, movie

Insulated Glazing

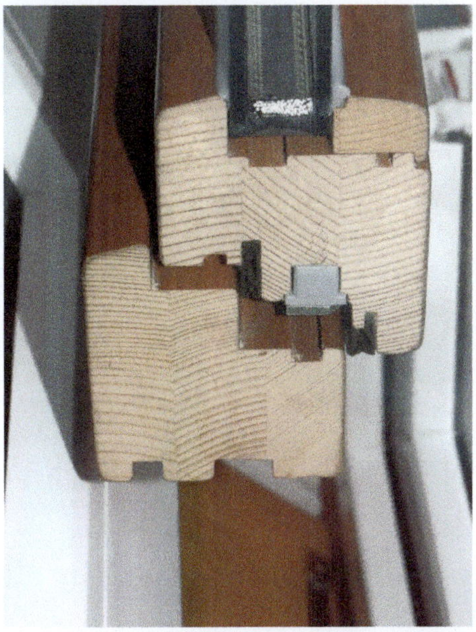

EURO 68 wooden window profile with insulated glazing

Insulated glazing (IG), more commonly known as double glazing (or double-pane, and increasingly triple glazing/pane), consists of two or three glass window panes separated by a vacuum or gas filled space to reduce heat transfer across a part of the building envelope.

Insulated glass units (IGUs) are manufactured with glass in range of thickness from 3 to 10 mm (1/8" to 3/8") or more in special applications. Laminated or tempered glass may also be used as part of the construction. Most units are produced with the same thickness of glass used on both panes but special applications such as acoustic attenuation or security may require wide ranges of thicknesses to be incorporated in the same unit.

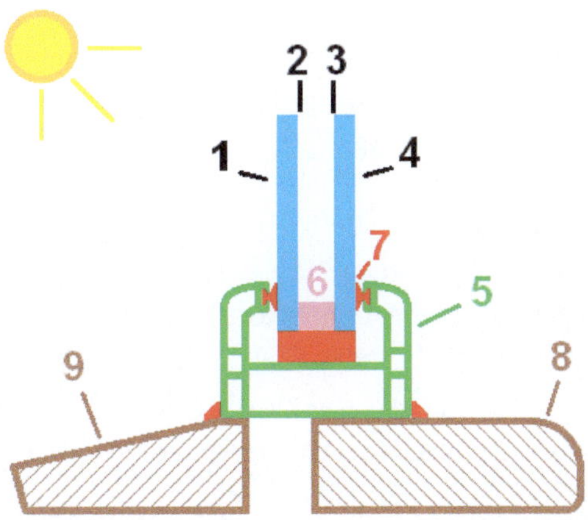

A sectioned diagram of a fixed insulated glass unit (IGU), indicating the numbering convention used in this article. Surface #1 is facing outside, Surface #2 is the inside surface of the exterior pane, Surface #3 is the outside surface of the interior pane, and Surface #4 is the inside surface of interior pane. The window frame is labelled #5, a spacer is indicated as #6, seals are shown in red (#7), the internal reveal is on the right hand side (#8) and the exterior windowsill on the left (#9)

Double-hung and Storm Windows

A typical installation of insulated glass windows with uPVC frames

Insulated glazing is an evolution from older technologies known as double-hung windows and storm windows. Traditional double-hung windows used a single pane of glass to separate the interior and exterior spaces.

- In the summer, a window screen would be installed on the exterior over the double-hung window to keep out animals and insects.
- In the winter, the screen was removed and replaced with a storm window, which created a two-layer separation between the interior and exterior spaces, increasing window insulation in cold winter months. To permit ventilation the storm window may be hung from removable hinge loops and swung open using folding metal arms. No screening was usually possible with open storm windows, though in the winter, insects typically are not active.

Traditional storm windows and screens are time consuming and very labor-intensive, requiring removal and storage of the storm windows in the spring, and reinstallation in the fall and storage of the screens. The weight of the large storm window frame and glass makes replacement on upper-stories of tall buildings a difficult task requiring repeatedly climbing a ladder with each window and trying to hold the window in place while securing retaining clips around the edges. However, current reproductions of these old-style storm windows can be made with detachable glass in the bottom pane that can be replaced with a detachable screen when desired. This eliminates the need for changing the entire storm window according to the seasons.

Insulated glazing forms a very compact multi-layer sandwich of air and glass, which eliminates the need for storm windows. Screens may also be left installed year-round with insulated glazing, and can be installed in a manner that permits installation and removal from inside the building, eliminating the requirement to climb up the exterior of the house to service the windows. It is possible to retrofit insulated glazing into traditional double-hung frames, though this would require significant modification to the wood frame due to the increased thickness of the IG assembly. .

Modern window units with IG typically completely replace the older double-hung unit, and include other improvements such as better sealing between the upper and lower windows, and spring-operated weight balancing that removes the need for large hanging weights inside the wall next to the windows, allowing for more insulation around the window and reducing air leakage, provides robust protection against the sun and will keep the house cool in the hot summer and warm in winter. Stop up to 70% in heat loss and 77% in heat gain with double-glazed windows and thermally broken frames These spring-operated balancing mechanisms also typically permit the top of the windows to swing inward, permitting cleaning of the exterior of the IG window from inside the building.

Spacer

Spacer: 10/14/20 mm

The glass panes are separated by a "spacer". A spacer is the piece that separates the two panes of glass in an insulating glass system, and seals the gas space between them. Historically, spacers were made primarily of metal and fiber, which manufacturers thought provided more durability.

However, metal spacers conduct heat (unless the metal is thermally improved), undermining the ability of the insulated glass unit (IGU) to reduce heat flow. It may also result in water or ice forming at the bottom of the sealed unit because of the sharp temperature difference between the window and surrounding air. To reduce heat transfer through the spacer and increase overall thermal performance, manufacturers may make the spacer out of a less-conductive material such as structural foam. A spacer made of aluminum that also contains a highly structural thermal barrier reduces condensation on the glass surface and improves insulation, as measured by the overall U-value.

- A spacer that reduces heat flow in glazing configurations may also have characteristics for sound dampening where external noise is an issue.
- Typically, spacers are filled with or contain desiccant to remove moisture trapped in the gas space during manufacturing, thereby lowering the dew point of the gas in that space, and preventing condensation from forming on surface #2 when the outside glass pane temperature falls.
- New technology has emerged to combat the heat loss from traditional spacer bars, including improvements to the structural performance and long-term-durability of improved metal (aluminum with a thermal barrier) and foam spacers.

Construction

IGUs are often manufactured on a made to order basis on factory production lines, but standard units are also available. The width and height dimensions, the thickness of the glass panes and the type of glass for each pane as well as the overall thickness of the unit must be supplied to the manufacturer. On the assembly line, spacers of specific thicknesses are cut and assembled into the required overall width and height dimensions and filled with desiccant. On a parallel line, glass panes are cut to size and washed to be optically clear.

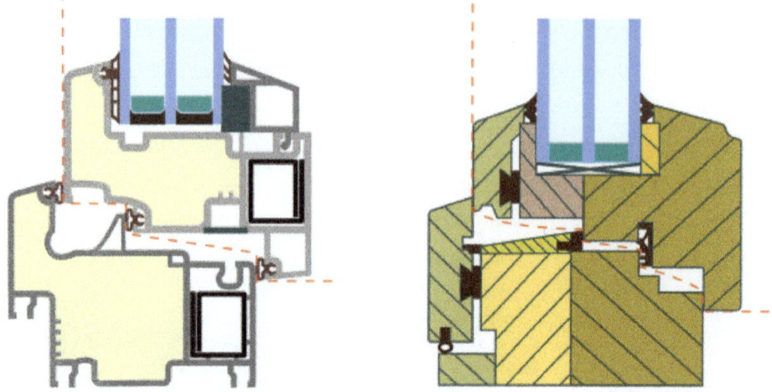

Examples of modern plastic and wooden window profiles with insulated glazing

An adhesive sealant (polyisobutylene) is applied to the face of the spacer on each side and the panes pressed against the spacer. If the unit is gas filled, two holes are drilled into the spacer of the assembled unit, lines are attached to draw out the air out of the space and replacing it (or leaving just vacuum) with the desired gas. The lines are then removed and holes sealed to contain the gas. The more modern technique is to use an online gas filler, which eliminates the need to drill holes in the spacer. The units are then sealed on the edge side using either polysulfide or silicone sealant or similar material to prevent humid outside air from entering the unit. The desiccant will remove traces of humidity from the air space so that no water appears on the inside faces (no condensation) of the glass panes facing the air space during cold weather. Some manufacturers have developed specific processes that combine the spacer and desiccant into a single step application system.

The insulating glazing unit, consisting of two glass panes bound together into a single unit with a seal between the edges of the panes, was patented in the United States by Thomas Stetson in 1865. It was developed into a commercial product in the 1930s, when several patents were filed, and a product was announced by the Libbey-Owens-Ford Glass Company in 1944. Their product was sold under the Thermopane brand name, which had been registered as a trademark in 1941. The Thermopane technology differs significantly from contemporary IGUs. The two panes of glass were welded together by a glass seal, and the two panes were separated by less than the 0.5 inches (1.3 cm) typical of modern units. The brand name Thermopane has entered the vocabulary of the glazing industry as the genericized trademark for any IGU.

Thermal Performance

The maximum insulating efficiency of a standard IGU is determined by the thickness of the space. Typically, most sealed units achieve maximum insulating values using a space of 16–19 mm (0.63–0.75 in) when measured at the centre of the IGU.

IGU thickness is a compromise between maximizing insulating value and the ability of the framing system used to carry the unit. Some residential and most commercial glazing systems can accommodate the ideal thickness of a double-paned unit. Issues arise with the use of triple glazing to further reduce heat loss in an IGU. The combination of thickness and weight results in units that are too unwieldy for most residential or commercial glazing systems, particularly if these panes are contained in moving frames or sashes.

This trade-off does not apply to vacuum insulated glass (VIG), or evacuated glazing, as heat loss due to convection is eliminated, leaving radiation losses and conduction through the edge seal. These VIG units have most of the air removed from the space between the panes, leaving a nearly-complete vacuum. VIG units which are currently on the market are hermetically sealed along their perimeter with solder glass, that is, a glass frit having a reduced melting point. Such a glass seal is rigid, and will experience increasing stress with increasing temperature differential across the unit. This stress may prevent vacuum glazing from being used when the temperature differential is too great. One manufacturer provides a recommendation of 35 °C.

Vacuum technology is also used in some non-transparent insulation products called vacuum insulated panels.

An older-established way to improve insulation performance is to replace air in the space with a lower thermal conductivity gas. Gas convective heat transfer is a function of viscosity and specific heat. Monatomic gases such as argon, krypton and xenon are often used since (at normal temperatures) they do not carry heat in rotational modes, resulting in a lower heat capacity than poly-atomic gases. Argon has a thermal conductivity 67% that of air, krypton has about half the conductivity of argon. Krypton and xenon are very expensive. These gases are used because they are non-toxic, clear, odorless, chemically inert, and commercially available because of their widespread application in industry. Some manufacturers also offer sulfur hexafluoride as an insulating gas, especially to insulate sound. It has only 2/3 the conductivity of argon, but it is stable, inexpensive and dense. However, sulfur hexafluoride is an extremely potent greenhouse gas that contributes to global warming. In Europe, SF 6 falls under the F-Gas directive which ban or control its usage for several applications. Since 1 January 2006, SF 6 is banned as a tracer gas and in all applications except high-voltage switchgear.

In general, the more effective a fill gas is at its optimum thickness, the thinner the optimum thickness is. For example, the optimum thickness for krypton is lower than for argon, and lower for argon than for air. However, since it is difficult to determine whether the gas in an IGU has become mixed with air at time of manufacture (or becomes mixed with air once installed), many designers prefer to use thicker gaps than would be optimum for the fill gas if it were pure. Argon is commonly used in insulated glazing as it is the most affordable. Krypton, which is considerably more expensive, is not generally used except to produce very thin double glazing units or extremely high performance triple-glazed units. Xenon has found very little application in IGUs because of cost.

Heat Insulating Properties

The effectiveness of insulated glass can be expressed as an R-value. The higher the R-value, the greater is its resistance to heat transfer. A standard IGU consisting of clear uncoated panes of glass (or lites) with air in the cavity between the lites typically has an R-value of 0.35 $K{\cdot}m^2/W$.

Using US customary units, a rule of thumb in standard IGU construction is that each change in the component of the IGU results in an increase of 1 R-value to the efficiency of the unit. Adding argon gas increases the efficiency to about R-3. Using low emissivity glass on surface #2 will add another R-value. Properly designed triple-glazed IGUs with low emissivity coatings on surfaces #2 and #4 and filled with argon gas in the cavities result in IG units with R-values as high as R-5. Certain vacuum insulated glass units (VIGU) or multi-chambered IG units using coated plastic films result in R-values as high as R-12.5

Additional layers of glazing provide the opportunity for improved insulation. While the standard double glazing is most widely used, triple glazing is not uncommon, and quadruple glazing is produced for very cold environments such as Alaska. Even quintuple glazing (four cavities, five panes) is available - with mid-pane insulation factors equivalent to walls.

Acoustic Insulating Properties

In some situations the insulation is in reference to noise mitigation. In these circumstances a large air space improves the noise insulation quality or sound transmission class. Asymmetric double glazing, using different thicknesses of glass rather than the conventional symmetrical systems (equal glass thicknesses used for both lites) will improve the acoustic attenuation properties of the IGU. If standard air spaces are used, sulfur hexafluoride can be used to replace or augment an inert gas and improve acoustical attenuation performance.

Other glazing material variations affect acoustics. The most widely used glazing configurations for sound dampening include laminated glass with varied thickness of the interlayer and thickness of the glass. Including a structural, thermally improved aluminum thermal barrier air spacer in the insulating glass can improve acoustical performance by reducing the transmission of exterior noise sources in the fenestration system.

Reviewing the glazing system components, including the air space material used in the insulating glass, can ensure overall sound transmission improvement.

Longevity

The life of an IGU varies depending on the quality of materials used, size of gap between inner and outer pane, temperature differences, workmanship and location of installation both in terms of facing direction and geographic location, as well as the treatment the unit receives. IG units typically last from 10 to 25 years, with windows facing the equator often lasting less than 12 years. IGUs typically carry a warranty for 10 to 20 years depending upon the manufacturer. If IGUs are altered (such as installation of a solar control film) the warranty may be voided by the manufacturer.

The Insulating Glass Manufacturers Alliance (IGMA) undertook an extensive study to characterize the failures of commercial insulating glass units over a 25-year period.

For a standard construction IG unit, condensation collects between the layers of glass when the perimeter seal has failed and when the desiccant has become saturated, and can generally only be eliminated by replacing the IGU. Seal failure and subsequent replacement results in a significant factor in the overall cost of owning IGUs.

Large temperature differences between the inner and outer panes stress the spacer adhesives, which can eventually fail. Units with a small gap between the panes are more prone to failure because of the increased stress.

Atmospheric pressure changes combined with wet weather can, in rare cases, eventually lead to the gap filling with water.

The flexible sealing surfaces preventing infiltration around the window unit can also degrade or be torn or damaged. Replacement of these seals can be difficult to impossible, due to IG windows commonly using extruded channel frames without seal retention screws or plates. Instead, the edge seals are installed by pushing an arrow-shaped indented one-way flexible lip into a slot on the extruded channel, and often cannot be easily extracted from the extruded slot to be replaced.

In Canada, since the beginning of 1990, there are some companies offering servicing of failed IG units. They provide open ventilation to the atmosphere by drilling hole(s) in the glass and/or spacer. This solution often reverses the visible condensation, but cannot clean the interior surface of the glass and staining that may have occurred after long term exposure to moisture. They may offer a warranty from 5 to 20 years. This solution lowers the insulating value of the window, but it can be a "green" solution when the window is still in good condition. If the IG unit had a gas fill (e.g. argon or krypton or a mixture) the gas is naturally dissipated and the R-value suffers.

Since 2004, there are also some companies offering the same restoration process for failed double-glazed units in the UK, and there is one company offering restoration of failed IG units in Ireland since 2010.

Thermal Stress Cracking

Temperature differences across the surface of glass panes can lead to cracking of the glass. This typically occurs where the glass is partially shaded and one section is heated in sunlight, but can also occur due to thermal differences along the edge of the glass where it is secured to the frame, which acts as a heat sink.

Thermal expansion creates a pressure differential between the warm and cool sections, and a crack may form which relieves the stress. In situations where thermal stress cracking has occurred, it can be prevented by using a thicker standard window glass which is structurally stronger and more resistant to cracking, or by using thin toughened glass to increase strength.

Estimating Heat Loss from Double-glazed Windows

Given the thermal properties of the sash, frame, and sill, and the dimensions of the glazing and thermal properties of the glass, the heat transfer rate for a given window and set of conditions can be calculated. This can be calculated in kW (kilowatts), but more usefully for cost benefit calculations can be stated as kWH pa (kilowatt hours per annum), based on the typical conditions over a year for a given location.

The glass panels in double-glazed windows transmit heat in both directions by radiation, across the panes by convection, and by conduction around the perimeter seals. The actual rates will vary with the conditions throughout the year, and while solar gain is much welcomed in the winter, it may result in increased air conditioning costs in the summer. The unwanted heat transfer can be mitigated by for example using curtains in the winter and using sun shades in the summer. In an attempt to provide a useful comparison between alternative window constructions the British

Fenestration Rating Council have defined a "Window Energy Rating" WER, ranging from A for the best down through B and C etc. This takes into account a combination of the heat loss through the window (U value, the reciprocal of R-value), the solar gain (g value), and loss through air leakage around the frame (L value). For example, an A Rated window will in a typical year gain as much heat from solar gain as it loses in other ways (however the majority of this gain will occur during the summer months, when the heat may not be needed by the building occupant). This provides better thermal performance than a typical wall.

Flat Glass

Plate glass is often used in windows, such as at this cafe in Tel Aviv, Israel overlooking the Mediterranean Sea.

Fragment of a Roman window glass plate dated to 1st to 4th century A.D.

Flat glass, sheet glass or plate glass is a type of glass, initially produced in plane form, commonly used for windows, glass doors, transparent walls, and windscreens. For modern architectural and automotive applications, the flat glass is sometimes bent after production of the plane sheet. Flat glass stands in contrast to *container glass* (used for bottles, jars, cups) and *glass fibre* (used for thermal insulation, in fiberglass composites, and optical communication).

Glass for flat glass has a higher magnesium oxide and sodium oxide content than container glass, and a lower silica, calcium oxide, and aluminium oxide content. (From the lower soluble oxide content comes the better chemical durability of container glass against water, which is required especially for storage of beverages and food).

Most flat glass is soda-lime glass, produced by the float glass process. Other processes for making flat glass include:

- Rolling (Rolled plate glass, Figure rolled glass)
- Overflow downdraw method
- Blown plate method
- Broad sheet method
- Window crown glass technique
- Cylinder blown sheet method
- Fourcault process
- Machine drawn cylinder sheet method
- Plate polishing

References

- Behr, R A (2009). Architectural glass to resist seismic and extreme climatic events. Woodhead Publishing Limited. ISBN 978-1-84569-369-5.
- Werner Vogel: "Glass Chemistry"; Springer-Verlag Berlin and Heidelberg GmbH & Co. K; 2nd revised edition (November 1994), ISBN 3-540-57572-3
- Jester, Thomas C., ed. (2014). Twentieth-Century Building Materials: History and Conservation. Getty Publications. p. 273. ISBN 9781606063255.
- Headley, Megan (2014). "EVA Finds Popularity Among Decorative Fabricators". NewsAnalysis:Trends US Glass, Volume 49, Issue 4 - April 2014.
- "Repair of Laminated Auto Glass Standard 02.13.2007 (Second Draft of Proposed Standard)" (PDF). Retrieved February 2014. Check date values in: |access-date= (help)
- "Development and quality control of vacuum glazing by N. Ng and L. So; University of Sydney". Glassfiles.com. Retrieved 2011-04-05.

4

Natural Occurrence of Glass

Libyan desert glass is a substance that is mostly found in the Libyan Desert. Darwin glass, just like Libyan Desert glass is found naturally. Darwin glass is found in the south of Queenstown. The other glasses mentioned in the following section are tektite, volcanic glass, tachylite, obsidian, Pele's tears, pumice etc. This chapter strategically incorporates the natural occurrences of glass.

Libyan Desert Glass

Libyan desert glass

Libyan desert glass (LDG), or great sand sea glass is a substance found in areas in the Libyan Desert. Fragments of desert glass can be found over areas of tens of square kilometers.

Geologic Origin

The origin of Libyan glass is uncertain. Meteoritic origins have long been considered possible, and recent research links the glass to impact features, such as zircon-breakdown, vaporized quartz and meteoritic metals, and to an impact crater. Some geologists associate the glass with radiative melting from meteoric large aerial bursts, making it analogous to trinitite created from sand exposed to the thermal radiation of a nuclear explosion. Libyan glass has been dated as having formed about 26 million years ago. It was knapped and used to make tools during the Pleistocene.

Tutankhamun's pectoral features a scarab carved from desert glass.

Darwin Glass

Assorted fragments of Darwin glass (scale in mm)

Darwin glass is a natural glass found south of Queenstown in West Coast, Tasmania. It takes its name from Mount Darwin in the West Coast Range, where it was first reported, and later gave its name to Darwin Crater, a probable impact crater, and the inferred source of the glass.

Occurrence

Fragments of Darwin glass are found scattered over a 410 km^2 area. Such an area is called a strewn field. On slopes and flat ground between 250 and 500 m elevation, the glass occurs with quartzite fragments buried under peat and soil. The peat is normally around 20 cm thick, and the quartzite fragment horizon is typically 30 cm thick. On mountain peaks higher than 500 m, the bedrock is directly exposed to the air, and Darwin glass occurs occasionally on the surface. In valleys below

220 m the Darwin glass is buried below peat and sediments. The glass occurs north, west and south from the crater. Its distribution extends to Kelly Basin and the lower northeast shore of Macquarie Harbour. Northwards it extends almost to the Lyell Highway and Crotty Dam. Darwin glass is rare in the crater itself.

In controlled excavations of gravel deposits the abundance of Darwin glass was found to vary from 0.3 to 47 kg/m^3. The highest abundance was found about 2 km from the crater, with the average abundance estimated at 3.4 kg/m^3 of gravel over a 50 km^2 study area near the crater. From this it can be estimated that about 25000 tons of Darwin glass, or about 10000 m^3, occurs in this 50 km^2 area. The amount of glass is large compared with the size of the crater. Preservation is helped by acid ground water which does not dissolve the glass, but this alone cannot explain the glass abundance. There is so much glass present that the glass must have been more copiously produced than in other meteorite impacts of similar size.

Nature

The glass is light to dark green, white or black. The glass takes the form of twisted masses, fragments or chunks up to 10 cm. Internally it has a flowing texture defined by lines of elliptical bubbles. There are two kinds of Darwin glass when composition is measured. Type 1 is normally white or green whereas type 2 is normally black to dark green. The dark glass contains less silica and more magnesium and iron than the light green glass. The dark glass is also enriched in chromium, nickel and cobalt. A possible explanation for the chemical differences is that, in addition to being mainly composed of melted local metamorphic rocks, the type 2 glass also contains a component of extraterrestrial material from the meteorite. Darwin glass has been dated at about 816,000 years old using argon dating methods.

Crater

The glass is an impactite resulting from the melting of local rocks due to the impact of a large meteorite. The assumed source is a 1.2-kilometer-wide topographic depression known as Darwin Crater. The crater is filled with 230 m of sediments and breccia. A crater of that size would be created by a meteorite 20 to 50 m in diameter and its impact with Earth would release 20 megatons of energy.

Tektite

Tektites are gravel-size bodies composed of black, green, brown or gray natural glass formed from terrestrial debris ejected during meteorite impacts. They generally range in size from millimeters to centimeters. Millimeters-size tektites are known as microtektites.

Tektites are characterized by:

1. a fairly homogeneous composition;
2. an extremely low content of water and other volatiles;

3. an abundance of lechatelierite;
4. a general lack of microscopic crystals known as microlites and chemical relation to the local bedrock or local sediments;
5. their distribution within geographically extensive strewnfields

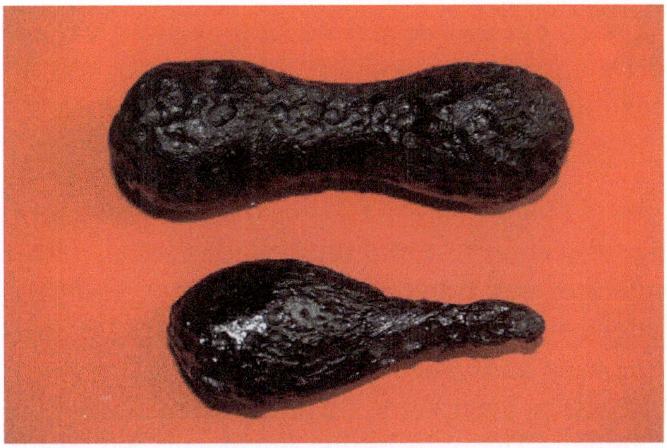

Two splash-form tektites, molten terrestrial ejecta from a meteorite impact

Characteristics

Although tektites are superficially similar to some terrestrial volcanic glasses (obsidians), they have unusual distinctive physical characteristics that distinguish them from such glasses. First, they are completely glassy and lack any microlites or phenocrysts, unlike terrestrial volcanic glasses. Second, although high in silica (>65 wt%), the bulk chemical and isotopic composition of tektites is closer to those of shales and similar sedimentary rocks and quite different from the bulk chemical and isotopic composition of terrestrial volcanic glasses. Third, tektites contain virtually no water (<0.02 wt%), unlike terrestrial volcanic glasses. Fourth, the flow-banding within tektites often contains particles and bands of lechatelierite, which are not found in terrestrial volcanic glasses. Finally, a few tektites contain partly melted inclusions of shocked and unshocked mineral grains, i.e. quartz, apatite, and zircon, as well as coesite.

The difference in water content can be used to distinguish tektites from terrestrial volcanic glasses. When heated to their melting point, terrestrial volcanic glasses will turn into a foamy glass because of their content of water and other volatiles. Unlike terrestrial volcanic glass, a tektite will produce only a few bubbles at most when heated to its melting point, because of its much lower water and other volatiles content.

Classification

On the basis of morphology and physical characteristics, tektites have traditionally been divided into four groups. The tektites which have been found on land have traditionally been subdivided into three groups: (1) splash-form (normal) tektites, (2) aerodynamically shaped tektites, and (3) Muong Nong-type (layered) tektites. Splash-form and aerodynamically shaped tektites are only differentiated on the basis of their appearance and some of their physical characteristics. Splash-form tektites are centimeter-sized tektites that are shaped like spheres, ellipsoids, teardrops, dumbbells, and other forms characteristic of isolated molten bodies. They are regarded as having formed from the solidification of rotating liquids, and not atmospheric ablation. Aerodynamically shaped tektites, which are mainly part of the Australasian strewn field, are splash-form tektites (buttons) which display a secondary ring or flange. The secondary ring or flange is argued as having been produced during the high-speed reentry and ablation of a solidified splash-form tektite into the atmosphere. Muong Nong tektites are typically larger, greater than 10 cm in size and 24 kg in weight, irregular, and layered tektites. They have a chunky, blocky appearance, exhibit a layered structure with abundant vesicles, and contain mineral inclusions, such as zircon, baddeleyite, chromite, rutile, corundum, cristobalite and coesite.

Microtektites, the fourth group of tektites, are tektites that are less than 1 mm in size. They exhibit a variety of shapes ranging from spherical to dumbbell, disc, oval, and teardrop. The color of microtektites ranges from colorless and transparent to yellowish and pale brown. They frequently contain bubbles and lechatelierite inclusions. Microtektites are typically found in deep-sea sediments that are of the same ages as those of the four known strewn fields. Microtektites of the Australasian strewn field have also been found on land within Chinese loess deposits, and in sediment-filled joints and decimeter-sized weathering pits developed within glacially eroded granite outcrops of the Victoria Land Transantarctic Mountains, Antarctica.

A very rare aerodynamically shaped Australite - Shallow Bowl

Occurrence

Since 1963, it has been known that the majority of known tektites occur only within four geographically extensive strewn fields: the Australasian, Central European, Ivory Coast, and North American strewn fields. As summarized by Koeberl, the tektites within each strewn field are related to each other with respect to the criteria of petrological, physical, and chemical properties as well as their age. In addition, three of the four strewn fields have been clearly linked with impact craters using those same criteria. Recognized types of tektites, grouped according to their known strewn fields, their associated craters, and ages are:

- Australasian strewnfield (no associated crater identified, age: 0.77–0.78 million years):

    - Australites (Australia, dark, mostly black);
    - Indochinites (South East Asia, dark, mostly black);
    - Rizalite (Philippines, black).
- Central European strewnfield (Nördlinger Ries impact crater (24 km), Germany, age: 15 million years):
    - Moldavites (Czech Republic, green).
- Ivory Coast strewnfield (Lake Bosumtwi impact crater (10 km), Ghana, age: 1 million years):
    - Ivorites (Ivory Coast, black).
- North American paras (Chesapeake Bay impact crater (40 km), United States – age: 34 million years):
    - Bediasites (Texas – black to dark brown, some with metallic finish);
    - Georgiaites (Georgia – green).

Comparing the number of known impact craters versus the number of known strewn fields, Artemieva considered essential factors such as the crater must exceed a certain diameter to produce distal ejecta, and that the event must be relatively recent. Limiting to diameters 10 km or more and younger than 50 Ma, the study yielded a list of 13 candidate craters, of which the youngest eight are given below,

Name	Location	Age (million years)	Diameter (km)	Strewn field
?	Indochina?	0.78	32-114?	Australasian strewn field
Zhamanshin	Kazakhstan	0.9 ± 0.1	14	?
Bosumtwi	Ghana	1.07	10	Ivory Coast strewn field
Elgygytgyn	Siberia	3.5 ± 0.5	18	?
Karakul	Tajikistan	<5	52	?
Karla	Russia	5 ± 1	10	?
Ries	Germany	15.1 ± 0.1	24	Central European strewn field
Chesapeake Bay	USA	35.5 ± 0.3	40	North American strewn field
Popigai	Siberia	35.7 ± 0.2	100	?

Preliminary papers in the late 1970s suggested either Zhamanshin or Elgygytgyn as the source of the Australasian strewnfield.

Povenmire and others have proposed the existence of an additional tektite strewn field, the Central American strewn field. Evidence for this reported tektite strewn field consists of tektites recovered from western Belize in the area of the villages of Bullet Tree Falls, Santa Familia and Billy White. This area lies about 55 km east-southeast of Tikal where 13 tektites, two of which were dated as being 820,000 years old, of unknown origin were found. A limited amount of evidence is interpreted as indicating that the proposed Central American strewn field likely covers Belize, Honduras, Guatemala, Nicaragua and possibly parts of southern Mexico. It is speculated that the hypothesized Pantasma Impact Crater in northern Nicaragua might be the source of these tektites.

Age

The ages of tektites from the four strewnfields have been determined using radiometric dating methods. The age of moldavites, a type of tektite found in Czech Republic, was determined to be 14 million years, which agrees well with the age determined for the Nördlinger Ries crater (a few hundred kilometers away in Germany) by radiometric dating of Suevite (an impact breccia found at the crater). Similar agreements exist between tektites from the North American strewnfield and the Chesapeake Bay impact crater and between tektites from the Ivory Coast strewnfield and the Lake Bosumtwi-Crater. Ages of tektites have usually been determined by either the K-Ar method, fission-track dating, the Ar-Ar technique, or combination of these techniques.

Origins

Terrestrial Source Theory

A simple, spherical splash-form Indochinite tektite

The overwhelming consensus of Earth and planetary scientists is that tektites consist of terrestrial debris that was ejected during the formation of an impact crater. During the extreme conditions created by an hypervelocity meteorite impact, near-surface terrestrial sediments and rocks were either melted, vaporized, or some combination of these and ejected from an impact crater. After ejection from the impact crater, the material formed millimeter- to centimeter-sized bodies of molten material, which as they re-entered the atmosphere, rapidly cooled to form tektites that fell to Earth to create a layer of distal ejecta hundreds or thousands of kilometers away from the impact site.

The terrestrial source for tektites is supported by well-documented evidence. The chemical and isotopic composition of tektites indicates that they are derived from the melting of silica-rich crustal and sedimentary rocks, which are not found on the Moon. In addition, some tektites contain relict mineral inclusions (quartz, zircon, rutile, chromite, and monazite) that are characteristic of terrestrial sediments and crustal and sedimentary source rocks. Also, three of the four tektite strewnfields have been linked by their age and chemical and isotopic composition to known impact craters. A number of different geochemical studies of tektites from the Australasian strewnfield concluded that these tektites consist of melted Jurassic sediments or sedimentary rocks that were weathered and deposited about 167 Ma ago. Their geochemistry suggests that the source of Australasian tektites is a single sedimentary formation with a narrow range of stratigraphic ages close to 170 Ma more or less. This effectively refutes multiple impact hypotheses.

A moldavite tektite

Although it is widely accepted that the formation of and widespread distribution of tektites requires the intense (superheated) melting of near-surface sediments and rocks at the impact site and the following high-velocity ejection of this material from the impact crater, the exact processes involved remain poorly understood. One possible mechanism for the formation of tektites is by the jetting of highly shocked and superheated melt during the initial contact/compression stage of impact crater formation. Alternatively, various mechanisms involving the dispersal of shock-melted material by an expanding vapor plume, which is created by a hypervelocity impact, have been used to explain the formation of tektites. Any mechanism by which tektites are created must explain chemical data that suggest that parent material from which tektites were created came from near-surface rocks and sediments at an impact site. In addition, the scarcity of known strewn fields relative to the number of identified Impact craters indicate that very special and rarely met circumstances are required in order for tektites to be created by a meteorite impact.

Nonterrestrial Source Theories

Though the meteorite impact theory of tektite formation is widely accepted, there has been considerable controversy about their origin in the past. In contrast to a terrestrial impact source for tektites, it was argued that tektites consist of material that was ejected from the Moon by major hydrogen-driven lunar volcanic eruptions and then drifted through space to later fall to Earth as tektites. The major proponents of the lunar origin of tektites include

NASA scientist John A. O'Keefe, NASA aerodynamicist Dean R. Chapman, meteorite and tektite collector Darryl Futrell, and long-time tektite researcher Hal Povenmire. From the 1950s to the 1990s, O'Keefe argued for the lunar origin of tektites based upon their chemical, i.e. rare-earth, isotopic, and bulk, composition and physical properties. Chapman used complex orbital computer models and extensive wind tunnel tests to argue that the so-called Australasian tektites originated from the Rosse ejecta ray of the large crater Tycho on the Moon's nearside. O'Keefe, Povenmire, and Futrell claimed on the basis of behavior of glass melts that the homogenization, which is called "fining", of silica melts that characterize tektites could not be explained by the terrestrial-impact theory. They also argued that the terrestrial-impact theory could not explain the vesicules and extremely low water and other volatile content of tektites. Futrell also reported the presence of microscopic internal features within tektites, which argued for a volcanic origin.

Aerodynamically shaped Australite; the button shape caused by ablation of molten glass in the atmosphere.

At one time, theories advocating the lunar origin of tektites enjoyed considerable support as part of a spirited controversy about the origin of tektites that occurred during the 1960s. Starting with the publication of research concerning lunar samples returned from the Moon, the consensus of Earth and planetary scientists shifted in favor of theories advocating a terrestrial impact versus lunar volcanic origin. For example, one problem with the lunar origin theory is that the arguments for it that are based upon the behavior of glass melts use data from pressures and temperatures that are vastly uncharacteristic of and unrelated to the extreme conditions of hypervelocity impacts. In addition, various studies have shown that hypervelocity impacts are likely quite capable of producing low volatile melts with extremely low water content. The consensus of Earth and planetary scientists regards the chemical, i.e. rare-earth, isotopic, and bulk composition evidence as decisively demonstrating that tektites are derived from terrestrial crustal rock, i.e. sedimentary rocks, that are unlike any known lunar crust.

In 1960, another non-terrestrial hypothesis for the origin of tektites was proposed by the Russian-born mathematician Matest M. Agrest, who suggested that tektites were formed as a result of nuclear blasts produced by extraterrestrial beings. He used this as an argument to support his paleocontact hypothesis.

Volcanic Glass

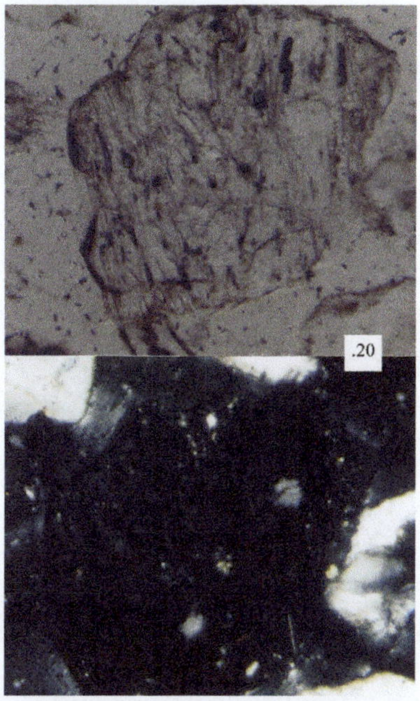

A sand grain of volcanic glass under the petrographic microscope. Its amorphous nature makes it go extinct in cross-polarized light (bottom frame). Scale box in millimeters.

Volcanic glass is the amorphous (uncrystallized) product of rapidly cooling magma. Like all types of glass, it is a state of matter intermediate between the close-packed, highly ordered array of a crystal and the highly disordered array of gas. Volcanic glass can refer to the interstitial, or matrix, material in an aphanitic (fine grained) volcanic rock or can refer to any of several types of vitreous igneous rocks. Most commonly, it refers to obsidian, a rhyolitic glass with high silica content.

Other types of volcanic glass include:

- Pumice, which is considered a glass because it has no crystal structure.
- Apache tears, a kind of nodular obsidian.
- Tachylite (also spelled tachylyte), a basaltic glass with relatively low silica content.
- Sideromelane, a less common form tachylyte.
- Palagonite, a basaltic glass with relatively low silica content.
- Hyaloclastite, a hydrated tuff-like breccia of sideromelane and palagonite.
- Pele's hair, threads or fibers of volcanic glass, usually basaltic.
- Pele's tears, tear-like drops of volcanic glass, usually basaltic.
- Limu o Pele (Pele's seaweed), thin sheets and flakes of brownish-green to near-clear volcanic glass, usually basaltic.

Tachylite

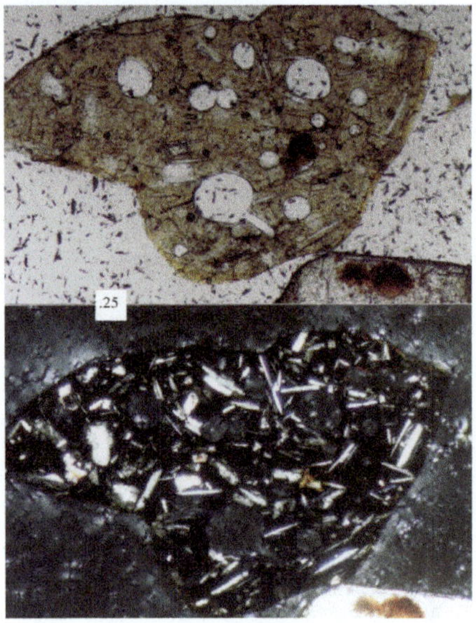

Basaltic glass in thin section, plane light top; cross-polarized light below

Tachylite (also spelled tachylyte) is a vitreous form of basaltic volcanic glass. This "glass" is formed naturally by the rapid cooling of molten basalt. It is a basic type of igneous rock that is decomposable by acids and readily fusible. The color is a black or dark-brown, and it has a greasy-looking, resinous luster. It is very brittle and occurs in dikes, veins and intrusive masses.

Tachylites have the appearance of pitch and are often more or less vesicular and sometimes spherulitic. They are very brittle and break down readily under a hammer. Small crystals of feldspar or olivine are sometimes visible in them with the unaided eye. All tachylites weather rather easily and by oxidation of their iron become dark brown or red. Three modes of occurrence characterize this rock. In all cases they are found under conditions which imply rapid cooling, but they are much less common than acid volcanic glasses (or obsidians), the reason being apparently that the basic rocks have a stronger tendency to crystallize, partly because they are more liquid and the molecules have more freedom to arrange themselves in crystalline order.

Geographic Occurrences

Scoria Sources

The fine scoria ashes or "cinders" thrown out by basaltic volcanoes are often spongy masses of tachylite with only a few larger crystals or phenocrysts imbedded in black glass. Such tachylite bombs and scoria are frequent in Iceland, Auvergne, Stromboli and Etna, and are very common also in the ash beds or tuffs of older date, such as occur in Skye, Midlothian and Fife, Derbyshire, and elsewhere. Basic pumices of this kind are exceedingly widespread on the bottom of the sea, either dispersed in the pelagic red clay and other deposits or forming layers coated with oxides of manganese precipitated on them from the sea water. These tachylite fragments, which are usually much decomposed by the oxidation and hydration of their ferrous compounds, have taken on a dark red color. This altered basic glass is known as "palagonite "; concentric bands of it often surround kernels of unaltered tachylite, and are so soft that they are easily cut with a knife. In the palagonite the minerals are also decomposed and are represented only by pseudomorphs. The fresh tachylite glass, however, often contains lozenge-shaped crystals of plagioclase feldspar and small prisms of augite and olivine, but all these minerals very frequently occur mainly as microlites or as beautiful skeletal growths with sharply-pointed corners or ramifying processes. Palagonite tuffs are found also among the older volcanic rocks. In Iceland a broad stretch of these rocks, described as "the palagonite formation," is said to cross the island from south-west to north-east. Some of these tuffs are fossiliferous; others are intercalated with glacial deposits. The lavas with which they occur are mostly olivine-basalts. Palagonite tuffs are found in Sicily, the Eifel, Hungary, Canary Islands, and other places.

Lava Flow Sources

Flaked stone artefacts from Australia, made of tachylite

A second mode of occurrence of tachylite is in the form of lava flows. Basaltic rocks often contain a small amount of glassy ground-mass, and in the limburgites this becomes more important and conspicuous, but vitreous types are far less common in these than in the acid lavas. In the Hawaiian Islands, however, the volcanoes have poured out vast floods of black basalt, containing feldspar, augite, olivine, and iron ores in a black glassy base. They are highly liquid when discharged, and the rapid cooling that ensues on their emergence to the air prevents crystallization taking place completely. Many of them are spongy or vesicular, and their upper surfaces are often exceedingly rough and jagged, while at other times they assume rounded wave-like forms on solidification. Great caves are found where the crust has solidified and the liquid interior has subsequently flowed away, and stalactites and stalagmites of black tachylite adorn the roofs and floors. On section these growths show usually a central cavity enclosed by walls of dark brown glass in which skeletons and microliths of augite, olivine and feldspar lie imbedded. From the crater of Mt. Kilauea, thin clouds of steam rise constantly, and as the bubbles of vapor are liberated from the molten rock they carry into the air with them thin fibers of basalt that solidify at once and assume the form of tachylite threads. Under the microscope they prove to be nearly completely glassy with small circular air vesicles sometimes drawn out to long tubes. Only in the Hawaiian Islands are glassy basaltic lavas of this kind at all common. A small outcrop at Spring Hill in Victoria, Australiahas tachylite which has been exploited as a material for making Aboriginal flaked stone implements.

Dike and Sill Sources

A third mode of occurrence of tachylite is as the margins and thin offshoots of dikes or sills of basalt and diabase. They are sometimes only a fraction of an inch in thickness, resembling a thin layer of pitch or tar on the edge of a crystalline diabase dike, but veins several inches thick are sometimes found. In these situations tachylite is rarely vesicular, but it often shows very pronounced fluxion banding accentuated by the presence of rows of spherulites that are visible as dark brown rounded spots. The spherulites have a distinct radiate structure and sometimes exhibit zones of varying color. The non-spherulitic glassy portion is sometimes perlitic, and these rocks are always brittle. Common crystals are olivine, augite and feldspar, with swarms of minute dusty black grains of magnetite. At the extreme edges the glass is often perfectly free from crystalline products, but it merges rapidly into the ordinary crystalline diabase, which in a very short distance may contain no vitreous base whatever. The spherulites may form the greater part of the mass, they may be a quarter of an inch in diameter and are occasionally much larger than this. These coarsely spherulitic rocks pass over into the variolites by increasing coarseness in the fibers of their spherulites, which soon become recognizable as needles of feldspar or feathery growths of augite. The ultimate product of decomposition in this case also is a red palagonitic substance, but owing to the absence of steam cavities the tachylite selvages of dikes are more often found in a fresh state than the basic lapilli in ash-beds. Many occurrences of basaltic pitchstones have been reported from Skye, Mull, and the western part of Scotland; they are found also in connection with the intrusive diabase sills in the north of England and the center of Scotland. In the Saar district of Germany similar rocks occur, some of which have been described as weisselbergites (from Weisselberg).

Other localities for tachylites of this group are New Providence, Silesia and Sweden.

Obsidian

Obsidian is a naturally occurring volcanic glass formed as an extrusive igneous rock.

It is produced when felsic lava extruded from a volcano cools rapidly with minimal crystal growth. Obsidian is commonly found within the margins of rhyolitic lava flows known as obsidian flows, where the chemical composition (high silica content) induces a high viscosity and polymerization degree of the lava. The inhibition of atomic diffusion through this highly viscous and polymerized lava explains the lack of crystal growth. Obsidian is hard and brittle; it therefore fractures with very sharp edges, which were used in the past in cutting and piercing tools, and it has been used experimentally as surgical scalpel blades.

Origin and Properties

Obsidian talus at Obsidian Dome, California

Polished snowflake obsidian, formed through the inclusion of cristobalite crystals

... among the various forms of glass we may reckon Obsian glass, a substance very similar to the stone found by Obsius in Ethiopia.

The translation into English of *Natural History* written by Pliny the Elder of Rome shows a few sentences on the subject of a volcanic glass called Obsian, so named from its resemblance to a stone (*obsiānus lapis*) found in Ethiopia by Obsius, a Roman explorer.

Obsidian is the rock formed as a result of quickly cooled lava, which is the parent material. Tektites were once thought by many to be obsidian produced by lunar volcanic eruptions, though few scientists now adhere to this hypothesis.

Obsidian is mineral-like, but not a true mineral because as a glass it is not crystalline; in addition, its composition is too complex to comprise a single mineral. It is sometimes classified as a mineraloid. Though obsidian is usually dark in color similar to mafic rocks such as basalt, obsidian's composition is extremely felsic. Obsidian consists mainly of SiO_2 (silicon dioxide), usually 70% or more. Crystalline rocks with obsidian's composition include granite and rhyolite. Because obsidian is metastable at the Earth's surface (over time the glass becomes fine-grained mineral crystals), no obsidian has been found that is older than Cretaceous age. This breakdown of obsidian is accelerated by the presence of water. Having a low water content when newly formed, typically less than 1% water by weight, obsidian becomes progressively hydrated when exposed to groundwater, forming perlite.

Pure obsidian is usually dark in appearance, though the color varies depending on the presence of impurities. Iron and other transition elements may give the obsidian a dark brown to black color. Very few samples are nearly colorless. In some stones, the inclusion of small, white, radially clustered crystals of cristobalite in the black glass produce a blotchy or snowflake pattern (*snowflake obsidian*). Obsidian may contain patterns of gas bubbles remaining from the lava flow, aligned along layers created as the molten rock was flowing before being cooled. These bubbles can produce interesting effects such as a golden sheen (*sheen obsidian*). An iridescent, rainbow-like sheen (*rainbow obsidian*) is caused by inclusions of magnetite nanoparticles.

Occurrence

Glass Mountain, a large obsidian flow at Medicine Lake Volcano

Obsidian can be found in locations which have experienced rhyolitic eruptions. It can be found in Argentina, Armenia, Azerbaijan, Australia, Canada, Chile, Georgia, Greece, El Salvador, Guatemala, Iceland, Italy, Japan, Kenya, Mexico, New Zealand, Papua New Guinea, Peru, Scotland, Turkey and the United States. Obsidian flows which may be hiked on are found within the calderas of Newberry Volcano and Medicine Lake Volcano in the Cascade Range of western North America, and at Inyo Craters east of the Sierra Nevada in California. Yellowstone National Park has a mountainside containing obsidian located between Mammoth Hot Springs and the Norris Geyser Basin, and deposits can be found in many other western U.S. states including Arizona, Colorado, New Mexico, Texas, Utah, Washington, Oregon and Idaho. Obsidian can also be found in the eastern U.S. states of Virginia, as well as Pennsylvania and North Carolina.

There are only four major deposit areas in the central Mediterranean: Lipari, Pantelleria, Palmarola and Monte Arci.

Ancient sources in the Aegean were Melos and Giali.

Acigöl town and the Göllü Dağ volcano were the most important sources in central Anatolia, one of the more important source areas in the prehistoric Near East.

Historical Use

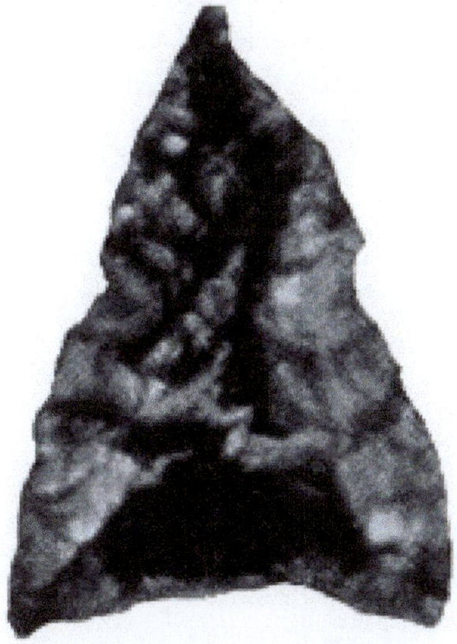

Obsidian arrowhead

The first known archaeological evidence of usage was in Kariandusi and other sites of the Acheulian age (beginning 1.5 million years BP) dated 700,000 BC, although the number of objects found at these sites were very low relative to the Neolithic. Use of obsidian in pottery of the Neolithic in the area around Lipari was found to be significantly less at a distance representing two weeks journeying. Anatolian sources of obsidian are known to have been the material used in the Levant and modern-day Iraqi Kurdistan from a time beginning sometime about 12,500 BC. The first attested civilized use is from excavations at Tell Brak dated the late fifth millennia. Obsidian was valued in Stone Age cultures because, like flint, it could be fractured to produce sharp blades or arrowheads. Like all glass and some other types of naturally occurring rocks, obsidian breaks with a characteristic conchoidal fracture. It was also polished to create early mirrors. Modern archaeologists have developed a relative dating system, obsidian hydration dating, to calculate the age of obsidian artifacts.

Middle East

Obsidian tools from Tilkitepe, Turkey, 5th millennium BC. Museum of Anatolian Civilizations

In the Ubaid in the 5th millennium BC, blades were manufactured from obsidian extracted from outcrops located in modern-day Turkey. Ancient Egyptians used obsidian imported from the eastern Mediterranean and southern Red Sea regions. Obsidian was also used in ritual circumcisions because of its deftness and sharpness. In the eastern Mediterranean area the material was used to make tools, mirrors and decorative objects.

Obsidian has also been found in Gilat, a site in the western Negev in Israel. Eight obsidian artifacts dating to the Chalcolithic Age found at this site were traced to obsidian sources in Anatolia. Neutron activation analysis (NAA) on the obsidian found at this site helped to reveal trade routes and exchange networks previously unknown.

Americas

Obsidian worked into plates and other wares by Victor Lopez Pelcastre of Nopalillo, Epazoyucan, Hidalgo. On display at the Museo de Arte Popular, Mexico City.

Lithic analysis can be instrumental in understanding prehispanic groups in Mesoamerica. A careful analysis of obsidian in a culture or place can be of considerable use to reconstruct commerce, production, distribution and thereby understand economic, social and political aspects of a civilization. This is the case in Yaxchilán, a Maya city where even warfare implications have been studied linked with obsidian use and its debris. Another example is the archeological recovery at coastal Chumash sites in California indicating considerable trade with the distant site of Casa Diablo, California in the Sierra Nevada Mountains.

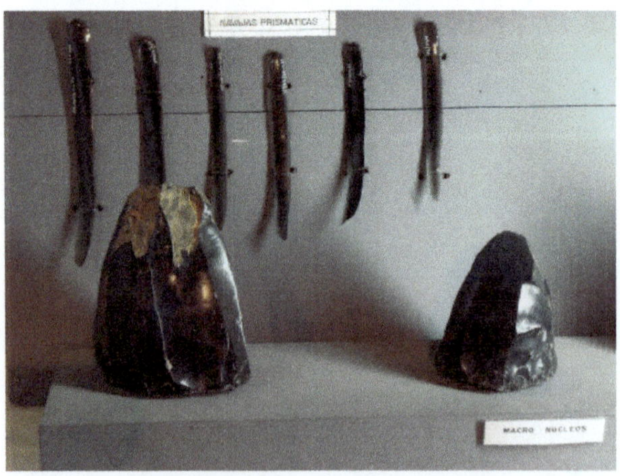

Raw obsidian and obsidian blades from the Mayan site of Takalik Abaj

Pre-Columbian Mesoamericans' use of obsidian was extensive and sophisticated; including carved and worked obsidian for tools and decorative objects. Mesoamericans also made a type of sword with obsidian blades mounted in a wooden body. Called a *macuahuitl*, the weapon was capable of inflicting terrible injuries, combining the sharp cutting edge of an obsidian blade with the ragged cut of a serrated weapon.

Native American people traded obsidian throughout the Americas. Each volcano and in some cases each volcanic eruption produces a distinguishable type of obsidian, making it possible for archaeologists to trace the origins of a particular artifact. Similar tracing techniques have allowed obsidian to be identified in Greece also as coming from Melos, Nisyros or Yiali, islands in the Aegean Sea. Obsidian cores and blades were traded great distances inland from the coast.

In Chile obsidian tools from Chaitén Volcano have been found as far away as in Chan-Chan 400 km (250 mi) north of the volcano and also in sites 400 km south of it.

Easter Island

Obsidian was also used on Rapa Nui (Easter Island) for edged tools such as *Mataia* and the pupils of the eyes of their Moai (statues).

Current Use

Obsidian can be used to make extremely sharp knives, and obsidian blades are a type of glass knife made using naturally occurring obsidian instead of manufactured glass. Obsidian is used by some surgeons for scalpel blades, although this is not approved by the US Food and Drug Administration (FDA) for use on humans. Well-crafted obsidian blades, as with any glass knife, can have a cutting edge many times sharper than high-quality steel surgical scalpels, the cutting edge of the blade being only about 3 nanometers thick. Even the sharpest metal knife has a jagged, irregular blade when viewed under a strong enough microscope; when examined even under an electron microscope an obsidian blade is still smooth and even. One study found that obsidian incisions produced fewer inflammatory cells and less granulation tissue at seven days, in a group of rats, although no differences were found after 21 days. Don Crabtree produced obsidian blades for surgery and other purposes, and has written articles on the subject. Obsidian scalpels may currently be purchased for surgical use on research animals.

Pig carved in snowflake obsidian, 10 centimeters (4 in) long. The markings are spherulites.

Obsidian is also used for ornamental purposes and as a gemstone. It presents a different appearance depending on how it is cut: in one direction it is jet black, while in another it is glistening gray. "Apache tears" are small rounded obsidian nuggets often embedded within a grayish-white perlite matrix.

Plinths for audio turntables have been made of obsidian since the 1970s; e.g. the grayish-black SH-10B3 plinth by Technics.

Perlite

Expanded perlite

Perlite is an amorphous volcanic glass that has a relatively high water content, typically formed by the hydration of obsidian. It occurs naturally and has the unusual property of greatly expanding when heated sufficiently. It is an industrial mineral and a commercial product useful for its low density after processing.

Properties

Perlite boulders with fireweed in foreground

Perlite softens when it reaches temperatures of 850–900 °C (1,560–1,650 °F). Water trapped in the structure of the material vaporises and escapes, and this causes the expansion of the material to 7–16 times its original volume. The expanded material is a brilliant white, due to the reflectivity of the trapped bubbles. Unexpanded ("raw") perlite has a bulk density around 1100 kg/m³ (1.1 g/cm³), while typical expanded perlite has a bulk density of about 30–150 kg/m³ (0.03-0.150 g/cm³).

Typical Analysis of Perlite

- 70–75% silicon dioxide: SiO_2
- 12–15% aluminium oxide: Al_2O_3
- 3–4% sodium oxide: Na_2O
- 3–5% potassium oxide: K_2O
- 0.5-2% iron oxide: Fe_2O_3
- 0.2–0.7% magnesium oxide: MgO
- 0.5–1.5% calcium oxide: CaO
- 3–5% loss on ignition (chemical / combined water)

Production and Uses

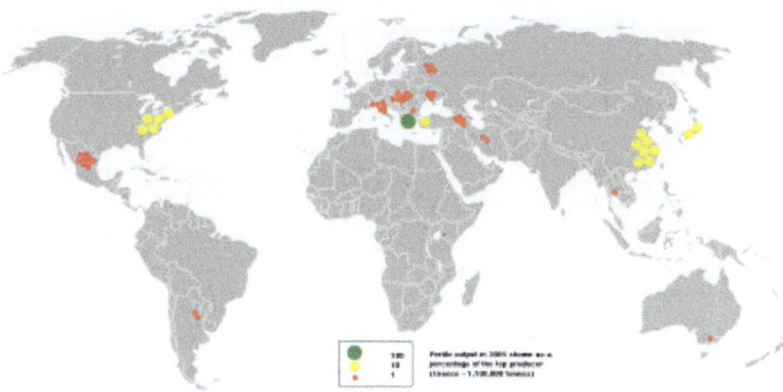

Perlite output in 2005

Perlite mine in Owens Valley, California.

Perlite is a non-renewable resource. The world reserves of perlite are estimated at 700 million tonnes. In 2011, 1.7 million tonnes were produced, mostly by Greece (500,000 t), United States (375,000 t) and Turkey (220,000 t). However, no information for China – a leading producer – was available. As of 2003, Greece was a leader in processed perlite production; however, estimates of perlite production from the report *USGS Mineral Commodity Summaries 2013* indicate that the U.S. may have overtaken Greece.

Because of its low density and relatively low price (about US$50 per tonne of unexpanded perlite), many commercial applications for perlite have developed. In the construction and manufacturing fields, it is used in lightweight plasters, concrete and mortar (masonry), insulation and ceiling tiles. In horticulture, perlite can be used as a soil amendment or alone as a medium for hydroponics or for starting cuttings. When used as an amendment it has high permeability / low water retention and helps prevent soil compaction. Perlite is an excellent filtration aid and is used extensively as an alternative to diatomaceous earth. The popularity of perlite usage as a filter medium is growing considerably worldwide. Perlite filters are fairly commonplace in filtering beer before it is bottled.

Small quantities of perlite are also used in foundries, cryogenic insulation, and in ceramics as a clay additive. It is also used by the explosives industry. Due to thermal and mechanical stability, non-toxicity, and high resistance against microbial attacks and organic solvents, perlite is widely used in biotechnological applications. Perlite was found to be an excellent support for immobilization of biocatalysts such as enzymes for bioremediation and sensing applications.

Estimated perlite consumption in the U.S. by application	
Fraction	**Use**
53%	building construction products
14%	horticultural aggregate
14%	fillers
8%	filter aid
11%	other

Cost of unexpanded perlite	
End of year	**Price in the U.S. $ per metric ton**
2001	$36.30
2002	$36.50
2003	$38.20
2004	$41.80
2005	$40.50
2006	$42.90
2007	$45.30
2008	$48.00
2009	$49.00

Substitutes

Perlite can be substituted for all of its uses. Substitutes include:

- Diatomite
- Expanded clay
- Shale
- Pumice

- Slag
- Vermiculite

Occupational Safety

People can be exposed to perlite in the workplace by breathing in dust, skin contact, and eye contact.

United States

The Occupational Safety and Health Administration (OSHA) has set the legal limit (permissible exposure limit) for perlite exposure in the workplace as 15 mg/m^3 total exposure and 5 mg/m^3 respiratory exposure over an 8-hour workday. The National Institute for Occupational Safety and Health (NIOSH) has set a recommended exposure limit (REL) of 10 mg/m^3 total exposure and 5 mg/m^3 respiratory exposure over an 8-hour workday.

Pele's Tears

Assorted shapes of Pele's tears from Mauna Ulu. U.S. dime for scale in lower right.

Pele's tears is a geological term for small pieces of solidified lava drops formed when airborne particles of molten material fuse into tearlike drops of volcanic glass. Pele's tears are jet black in color and are often found on one end of a strand of Pele's hair. Pele's tears is primarily a scientific term used by volcanologists.

Pele's tears, like Pele's hair, is named after Pele, the Hawaiian fire goddess of volcanoes.

Introduction

Lava fountains produce spectacular displays of volcanic activity with lava spurting out of the vent in a showering cascade of hot molten droplets. Many of the clots aggregate to form the familiar types of lava flows that we see in Hawaiian and Strombolian eruptions. However, there is also another interesting process that is occurring at the same time that is relatively unknown. This is the formation of glass spherules or droplets of solidified lava known as Pele's tears. These are black in color and, as you can see from the photograph, they vary in shape and size.

Formation

The formation of these tears is a complex process depending on a number of different factors. Imagine a tiny droplet of lava being erupted from a lava fountain. Whilst it is travelling through the air two things are happening: It is cooling down very rapidly (a process known as quenching) and it is being deformed. The deformation of a droplet depends on the speed at which it is erupted from the volcano, its surface tension, the viscosity (thickness) of the magma and the resistance it experiences as it travels through the air.

Pele's tears are also found entangled within fine strands of volcanic glass known as Pele's hair and it was considered that they formed together under similar conditions. Shimozura (1994) investigated this further and found that the velocity of the erupting lava was the main factor in determining whether Pele's tears or Pele's hair were formed. If the velocity of the erupting magma is high then Pele's hair is formed and if the velocity is low the formation of Pele's tears is favoured.

Under the microscope very tiny Pele's tears (less than 1μm) can be found within the cavities of Pele's hair. This would suggest they formed prior to being trapped within the strand and supports a different mechanism for formation. It has been considered that they became trapped during transport in the eruptive plume.

Scientific Importance

Pele's tears are interesting to volcanologists because trapped within the glass droplet are bubbles of gas and particles called vesicles. When these are analysed they can provide a great deal of information about the mechanisms of an eruption. For instance, the shape of a vesicle can provide an indication of the velocity of the eruption. When vesicles form within the lava they are spherical in shape. If the eruption is turbulent the vesicles will deform and become elongated in shape. If the velocity is low they will retain their spherical shape as in the case of Pele's tears.

The composition of a magma chamber includes lava, crystals and gases known as volatiles. At high pressures within the magma chamber, gases are dissolved in the melt and are soluble. As the magma rises, the pressure decreases and the gases come out of solution. Some of these gases escape and some become trapped in the melt. When the gases trapped within the vesicles of Pele's tears are analysed they can provide a great deal of information about the chemical composition of the magma chamber. This information can be used to determine the explosive nature of an eruption and help to understand the complex processes occurring within volcanos. This is important in the management of natural hazards.

Palagonite

Palagonite is an alteration product from the interaction of water with volcanic glass of chemical composition similar to basalt. Palagonite can also result from the interaction between water and basalt melt. The water flashes to steam on contact with the hot lava and the small fragments of lava react with the steam to form the light colored palagonite tuff cones common in areas of basaltic eruptions in contact with water. An example is found in the pyroclastic cones of the Galapagos Islands. Charles Darwin recognized the origin of these cones during his visit to the islands. Palagonite can also be formed by a slower weathering of lava into palagonite, resulting in a thin, yellow-orange rind on the surface of the rock. The process of conversion of lava to palagonite is called *palagonitization*.

Palagonite layers at Moya Beach, Mayotte.

Palagonite soil is a light yellow-orange dust, comprising a mixture of particles ranging down to sub-micrometer sizes, usually found mixed with larger fragments of lava. The color is indicative of the presence of iron in the +3 oxidation state, embedded in an amorphous matrix.

Palagonite tuff is a tuff composed of sideromelane fragments and coarser pieces of basaltic rock, embedded in a palagonite matrix. A composite of sideromelane aggregate in palagonite matrix is called hyaloclastite.

Palagonite on Mars

Based on infrared spectroscopy, the fine-grained component of Mauna-Kea palagonite is the terrestrial material with the best match to the spectral properties of Martian dust, and is believed to be similar in composition and in origin to dusty component of the surface regolith of Mars. The palagonitic tephra from a cinder cone in Hawaii has been used to create Martian regolith simulant for researchers. The spectroscopic signature of palagonitic alteration on Mars is used as evidence for the existence of water on Mars.

Pumice

Pumice, called pumicite in its powdered or dust form, is a volcanic rock that consists of highly vesicular rough textured volcanic glass, which may or may not contain crystals. It is typically light colored. Scoria is another vesicular volcanic rock that differs from pumice in having larger vesicles, thicker vesicle walls and being dark colored and denser.

Pumice is created when super-heated, highly pressurized rock is violently ejected from a volcano. The unusual foamy configuration of pumice happens because of simultaneous rapid cooling and rapid depressurization. The depressurization creates bubbles by lowering the solubility of gases (including water and CO_2) that are dissolved in the lava, causing the gases to rapidly exsolve (like the bubbles of CO_2 that appear when a carbonated drink is opened). The simultaneous cooling and depressurization freezes the bubbles in a matrix. Eruptions under water are rapidly cooled and the large volume of pumice created can be a shipping hazard for cargo ships.

Specimen of highly porous pumice from Teide volcano on Tenerife, Canary Islands. Density of specimen approximately 0.25 g/cm³ (0.14 oz/cu in); scale in centimeters.

Properties

Pumice is composed of highly microvesicular glass pyroclastic with very thin, translucent bubble walls of extrusive igneous rock. It is commonly, but not exclusively of silicic or felsic to intermediate in composition (e.g., rhyolitic, dacitic, andesite, pantellerite, phonolite, trachyte), but basaltic and other compositions are known. Pumice is commonly pale in color, ranging from white, cream, blue or grey, to green-brown or black. It forms when volcanic gases exsolving from viscous magma form bubbles that remain within the viscous magma as it cools to glass. Pumice is a common product of explosive eruptions (plinian and ignimbrite-forming) and commonly forms zones in upper parts of silicic lavas. Pumice has an average porosity of 90%, and initially floats on water.

Scoria differs from pumice in being denser. With larger vesicles and thicker vesicle walls, it sinks rapidly. The difference is the result of the lower viscosity of the magma that forms scoria. When larger amounts of gas are present, the result is a finer-grained variety of pumice known as pumicite. Pumice is considered a glass because it has no crystal structure. Pumice varies in density according to the thickness of the solid material between the bubbles; many samples float in water. After the explosion of Krakatoa, rafts of pumice drifted through the Pacific Ocean for up to 20 years, with tree trunks floating among them. In fact, pumice rafts disperse and support several marine species. In 1979, 1984 and 2006, underwater volcanic eruptions near Tonga created large pumice rafts, some as large as 30 kilometers (19 mi) that floated hundreds of kilometres to Fiji.

There are two main forms of vesicles. Most pumice contains tubular microvesicles that can impart a silky or fibrous fabric. The elongation of the microvesicles occurs due to ductile elongation in the volcanic conduit or, in the case of pumiceous lavas, during flow. The other form of vesicles are subspherical to spherical and result from high vapor pressure during eruption.

Illustrates the porous nature in detail.

Rocks from the Bishop tuff, uncompressed with pumice on left; compressed with fiamme on right.

A 15-centimeter (5.9 in) piece of pumice supported by a rolled-up U.S. 20-dollar bill demonstrates its very low density.

Uses

Pumice is widely used to make lightweight concrete or insulative low-density cinder blocks. When used as an additive for cement, a fine-grained version of pumice called pozzolan is mixed with lime to form a light-weight, smooth, plaster-like concrete. This form of concrete was used as far back as Roman times. Roman engineers used it to build the huge dome of the Pantheon and as construction material for many aqueducts.

It is also used as an abrasive, especially in polishes, pencil erasers, cosmetic exfoliants, and the production of stone-washed jeans. "Pumice stones" are often used in beauty salons during the pedicure process to remove dry and excess skin from the bottom of the foot as well as calluses. It was also used in ancient Greek and Roman times to remove excess hair. Finely ground pumice is added to some toothpastes and heavy-duty hand cleaners (such as Lava soap) as a mild abrasive. Pumice is also used as a growing substrate for growing horticultural crops. Some brands of chinchilla dust bath are made of powdered pumice.

Owing to its high demand particularly for water filtration, chemical spill containment, cement manufacturing, horticulture and increasingly for the pet industry, the mining of pumice in environmentally sensitive areas has been under more scrutiny after such an operation was stopped in the U.S. state of Oregon, at Rock Mesa in the southern part of the Three Sisters Wilderness.

Gardening

Pumice is mixed with soil to provide good aeration for the plants. It is also used for hydroponics as a soilless growing medium.

Diamond

Diamond is a metastable allotrope of carbon, where the carbon atoms are arranged in a variation of the face-centered cubic crystal structure called a diamond lattice. Diamond is less stable than graphite, but the conversion rate from diamond to graphite is negligible at standard conditions.

Diamond is renowned as a material with superlative physical qualities, most of which originate from the strong covalent bonding between its atoms. In particular, diamond has the highest hardness and thermal conductivity of any bulk material. Those properties determine the major industrial application of diamond in cutting and polishing tools and the scientific applications in diamond knives and diamond anvil cells.

Because of its extremely rigid lattice, it can be contaminated by very few types of impurities, such as boron and nitrogen. Small amounts of defects or impurities (about one per million of lattice atoms) color diamond blue (boron), yellow (nitrogen), brown (lattice defects), green (radiation exposure), purple, pink, orange or red. Diamond also has relatively high optical dispersion (ability to disperse light of different colors).

Most natural diamonds are formed at high temperature and pressure at depths of 140 to 190 kilometers (87 to 118 mi) in the Earth's mantle. Carbon-containing minerals provide the carbon source, and the growth occurs over periods from 1 billion to 3.3 billion years (25% to 75% of the age of the Earth). Diamonds are brought close to the Earth's surface through deep volcanic eruptions by magma, which cools into igneous rocks known as kimberlites and lamproites. Diamonds can also be produced synthetically in a HPHT method which approximately simulates the conditions in the Earth's mantle. An alternative, and completely different growth technique is chemical vapor deposition (CVD). Several non-diamond materials, which include cubic zirconia and silicon carbide and are often called diamond simulants, resemble diamond in appearance and many properties. Special gemological techniques have been developed to distinguish natural diamonds, synthetic diamonds, and diamond simulants.

History

Diamonds are thought to have been first recognized and mined in India, where significant alluvial deposits of the stone could be found many centuries ago along the rivers Penner, Krishna and Godavari. Diamonds have been known in India for at least 3,000 years but most likely 6,000 years.

Diamonds have been treasured as gemstones since their use as religious icons in ancient India. Their usage in engraving tools also dates to early human history. The popularity of diamonds has risen since the 19th century because of increased supply, improved cutting and polishing techniques, growth in the world economy, and innovative and successful advertising campaigns.

In 1772, Antoine Lavoisier used a lens to concentrate the rays of the sun on a diamond in an atmosphere of oxygen, and showed that the only product of the combustion was carbon dioxide, proving that diamond is composed of carbon. Later in 1797, Smithson Tennant repeated and expanded that experiment. By demonstrating that burning diamond and graphite releases the same amount of gas, he established the chemical equivalence of these substances.

The most familiar uses of diamonds today are as gemstones used for adornment, a use which dates back into antiquity, and as industrial abrasives for cutting hard materials. The dispersion of white light into spectral colors is the primary gemological characteristic of gem diamonds. In the 20th century, experts in gemology developed methods of grading diamonds and other gemstones based on the characteristics most important to their value as a gem. Four characteristics, known informally as the *four Cs*, are now commonly used as the basic descriptors of diamonds: these are *carat* (its weight), *cut* (quality of the cut is graded according to proportions, symmetry and polish), *color* (how close to white or colorless; for fancy diamonds how intense is its hue), and *clarity* (how free is it from inclusions). A large, flawless diamond is known as a paragon.

Natural History

The formation of natural diamond requires very specific conditions—exposure of carbon-bearing materials to high pressure, ranging approximately between 45 and 60 kilobars (4.5 and 6 GPa), but at a comparatively low temperature range between approximately 900 and 1,300 °C (1,650 and 2,370 °F). These conditions are met in two places on Earth; in the lithospheric mantle below relatively stable continental plates, and at the site of a meteorite strike.

Formation in Cratons

The conditions for diamond formation to happen in the lithospheric mantle occur at considerable depth corresponding to the requirements of temperature and pressure. These depths are estimated between 140 and 190 kilometers (87 and 118 mi) though occasionally diamonds have crystallized at depths about 300 km (190 mi). The rate at which temperature changes with increasing depth into the Earth varies greatly in different parts of the Earth. In particular, under oceanic plates the temperature rises more quickly with depth, beyond the range required for diamond formation at the depth required. The correct combination of temperature and pressure is only found in the thick, ancient, and stable parts of continental plates where regions of lithosphere known as *cratons* exist. Long residence in the cratonic lithosphere allows diamond crystals to grow larger.

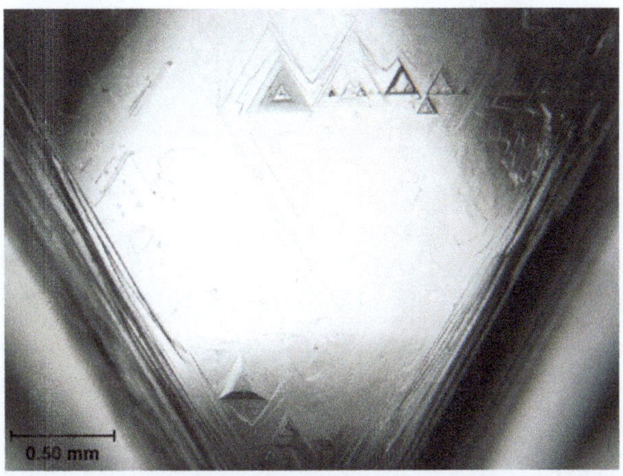

One face of an uncut octahedral diamond, showing trigons (of positive and negative relief) formed by natural chemical etching

Through studies of carbon isotope ratios (similar to the methodology used in carbon dating, except with the stable isotopes C-12 and C-13), it has been shown that the carbon found in diamonds comes from both inorganic and organic sources. Some diamonds, known as *harzburgitic*, are formed from inorganic carbon originally found deep in the Earth's mantle. In contrast, *eclogitic* diamonds contain organic carbon from organic detritus that has been pushed down from the surface of the Earth's crust through subduction before transforming into diamond. These two different source of carbon have measurably different ^{13}C:^{12}C ratios. Diamonds that have come to the Earth's surface are generally quite old, ranging from under 1 billion to 3.3 billion years old. This is 22% to 73% of the age of the Earth.

Diamonds occur most often as euhedral or rounded octahedra and twinned octahedra known as *macles*. As diamond's crystal structure has a cubic arrangement of the atoms, they have many facets that belong to a cube, octahedron, rhombicosidodecahedron, tetrakis hexahedron or disdyakis dodecahedron. The crystals can have rounded off and unexpressive edges and can be elongated. Sometimes they are found grown together or form double "twinned" crystals at the surfaces of the octahedron. These different shapes and habits of some diamonds result from differing external circumstances. Diamonds (especially those with rounded crystal faces) are commonly found coated in *nyf*, an opaque gum-like skin.

Transport from Mantle

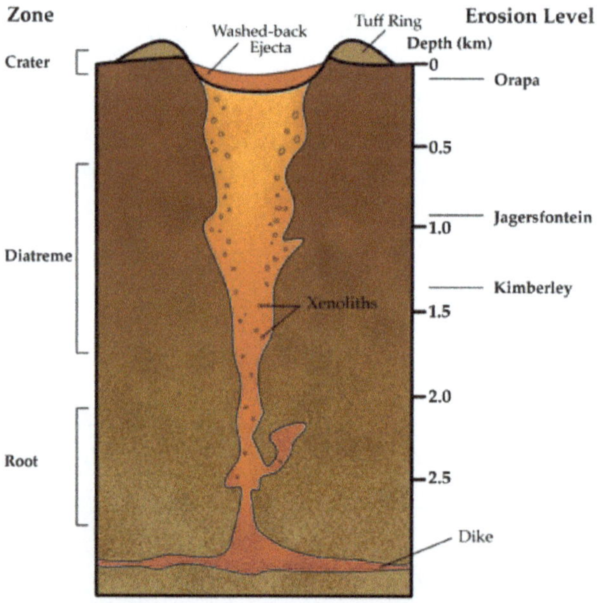

Schematic diagram of a volcanic pipe

Diamond-bearing rock is carried from the mantle to the Earth's surface by deep-origin volcanic eruptions. The magma for such a volcano must originate at a depth where diamonds can be formed—150 km (93 mi) or more (three times or more the depth of source magma for most volcanoes). This is a relatively rare occurrence. These typically small surface volcanic craters extend downward in formations known as volcanic pipes. The pipes contain material that was transported toward the surface by volcanic action, but was not ejected before the volcanic activity ceased. During eruption these pipes are open to the surface, resulting in open circulation; many xenoliths of surface rock and even wood and fossils are found in volcanic pipes. Diamond-bearing volcanic pipes are closely related to the oldest, coolest regions of continental crust (cratons). This is because cratons are very thick, and their lithospheric mantle extends to great enough depth that diamonds are stable. Not all pipes contain diamonds, and even fewer contain enough diamonds to make mining economically viable.

The magma in volcanic pipes is usually one of two characteristic types, which cool into igneous rock known as either kimberlite or lamproite. The magma itself does not contain diamond; instead, it acts as an elevator that carries deep-formed rocks (xenoliths), minerals (xenocrysts), and fluids upward. These rocks are characteristically rich in magnesium-bearing olivine, pyroxene, and amphibole minerals which are often altered to serpentine by heat and fluids during and after eruption. Certain *indicator minerals* typically occur within diamantiferous kimberlites and are used as mineralogical tracers by prospectors, who follow the indicator trail back to the volcanic pipe which may contain diamonds. These minerals are rich in chromium (Cr) or titanium (Ti), elements which impart bright colors to the minerals. The most common indicator minerals are chromium garnets (usually bright red chromium-pyrope, and occasionally green ugrandite-series garnets), eclogitic garnets, orange titanium-pyrope, red high-chromium spinels, dark chromite, bright green chromium-diopside, glassy green olivine, black picroilmenite, and magnetite. Kimberlite deposits are known as *blue ground* for the deeper serpentinized part of the deposits, or as *yellow ground* for the near surface smectite clay and carbonate weathered and oxidized portion.

Once diamonds have been transported to the surface by magma in a volcanic pipe, they may erode out and be distributed over a large area. A volcanic pipe containing diamonds is known as a *primary source* of diamonds. *Secondary sources* of diamonds include all areas where a significant number of diamonds have been eroded out of their kimberlite or lamproite matrix, and accumulated because of water or wind action. These include alluvial deposits and deposits along existing and ancient shorelines, where loose diamonds tend to accumulate because of their size and density. Diamonds have also rarely been found in deposits left behind by glaciers (notably in Wisconsin and Indiana); in contrast to alluvial deposits, glacial deposits are minor and are therefore not viable commercial sources of diamond.

Space Diamonds

Not all diamonds found on Earth originated on Earth. Primitive interstellar meteorites were found to contain carbon possibly in the form of diamond. A type of diamond called carbonado that is found in South America and Africa may have been deposited there via an asteroid impact (not formed from the impact) about 3 billion years ago. These diamonds may have formed in the intrastellar environment, but as of 2008, there was no scientific consensus on how carbonado diamonds originated.

Diamonds can also form under other naturally occurring high-pressure conditions. Very small diamonds of micrometer and nanometer sizes, known as *microdiamonds* or *nanodiamonds* respectively, have been found in meteorite impact craters. Such impact events create shock zones of high pressure and temperature suitable for diamond formation. Impact-type microdiamonds can be used as an indicator of ancient impact craters. Popigai crater in Russia may have the world's largest diamond deposit, estimated at trillions of carats, and formed by an asteroid impact.

Scientific evidence indicates that white dwarf stars have a core of crystallized carbon and oxygen nuclei. The largest of these found in the universe so far, BPM 37093, is located 50 light-years (4.7×10^{14} km) away in the constellation Centaurus. A news release from the Harvard-Smithsonian Center for Astrophysics described the 2,500-mile (4,000 km)-wide stellar core as a *diamond*.

Material Properties

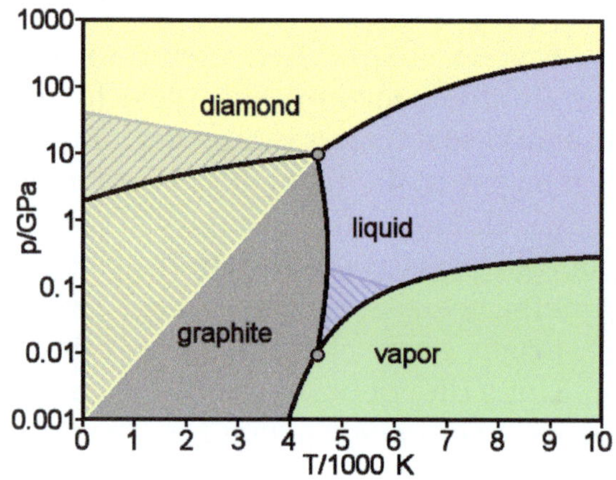

Theoretically predicted phase diagram of carbon

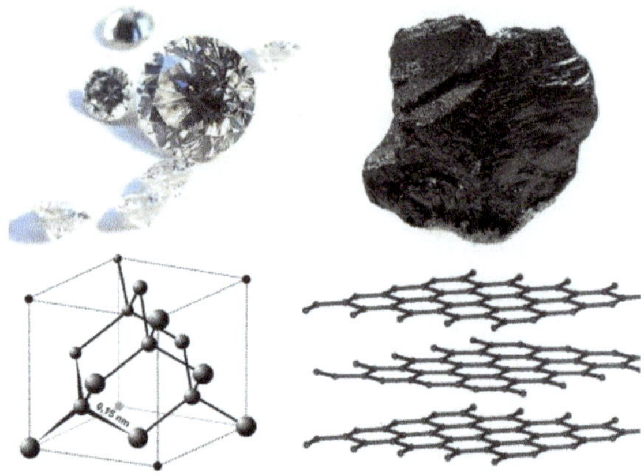

Diamond and graphite are two allotropes of carbon: pure forms of the same element that differ in structure.

A diamond is a transparent crystal of tetrahedrally bonded carbon atoms in a covalent network lattice (sp^3) that crystallizes into the diamond lattice which is a variation of the face centered cubic structure. Diamonds have been adapted for many uses because of the material's exceptional physical characteristics. Most notable are its extreme hardness and thermal conductivity (900–2320 $W\cdot m^{-1}\cdot K^{-1}$), as well as wide bandgap and high optical dispersion. Above 1700 °C (1973 K / 3583 °F) in vacuum or oxygen-free atmosphere, diamond converts to graphite; in air, transformation starts at ~700 °C. Diamond's ignition point is 720 – 800 °C in oxygen and 850 – 1000 °C in air. Naturally occurring diamonds have a density ranging from 3.15–3.53 g/cm³, with pure diamond close to 3.52 g/cm³. The chemical bonds that hold the carbon atoms in diamonds together are weaker than those in graphite. In diamonds, the bonds form an inflexible three-dimensional lattice, whereas in graphite, the atoms are tightly bonded into sheets, which can slide easily over one another, making the overall structure weaker. In a diamond, each carbon atom is surrounded by neighboring four carbon atoms forming a tetrahedral shaped unit.

Hardness

Diamond is the hardest known natural material on both the Vickers and the Mohs scale. Diamond's hardness has been known since antiquity, and is the source of its name.

Diamond hardness depends on its purity, crystalline perfection and orientation: hardness is higher for flawless, pure crystals oriented to the <111> direction (along the longest diagonal of the cubic diamond lattice). Therefore, whereas it might be possible to scratch some diamonds with other materials, such as boron nitride, the hardest diamonds can only be scratched by other diamonds and nanocrystalline diamond aggregates.

The hardness of diamond contributes to its suitability as a gemstone. Because it can only be scratched by other diamonds, it maintains its polish extremely well. Unlike many other gems, it is well-suited to daily wear because of its resistance to scratching—perhaps contributing to its popularity as the preferred gem in engagement or wedding rings, which are often worn every day.

The extreme hardness of diamond in certain orientations makes it useful in materials science, as in this pyramidal diamond embedded in the working surface of a Vickers hardness tester.

The hardest natural diamonds mostly originate from the Copeton and Bingara fields located in the New England area in New South Wales, Australia. These diamonds are generally small, perfect to semiperfect octahedra, and are used to polish other diamonds. Their hardness is associated with the crystal growth form, which is single-stage crystal growth. Most other diamonds show more evidence of multiple growth stages, which produce inclusions, flaws, and defect planes in the crystal lattice, all of which affect their hardness. It is possible to treat regular diamonds under a combination of high pressure and high temperature to produce diamonds that are harder than the diamonds used in hardness gauges.

Somewhat related to hardness is another mechanical property *toughness*, which is a material's ability to resist breakage from forceful impact. The toughness of natural diamond has been measured as 7.5–10 $MPa \cdot m^{1/2}$. This value is good compared to other ceramic materials, but poor compared to most engineering materials such as engineering alloys, which typically exhibit toughnesses over 100 $MPa \cdot m^{1/2}$. As with any material, the macroscopic geometry of a diamond contributes to its resistance to breakage. Diamond has a cleavage plane and is therefore more fragile in some orientations than others. Diamond cutters use this attribute to cleave some stones, prior to faceting. "Impact toughness" is one of the main indexes to measure the quality of synthetic industrial diamonds.

Pressure Resistance

Used in so-called diamond anvil experiments to create high-pressure environments, diamonds are able to withstand crushing pressures in excess of 600 gigapascals (6 million atmospheres).

Electrical Conductivity

Other specialized applications also exist or are being developed, including use as semiconductors: some blue diamonds are natural semiconductors, in contrast to most diamonds, which are excellent electrical insulators. The conductivity and blue color originate from boron impurity. Boron substitutes for carbon atoms in the diamond lattice, donating a hole into the valence band.

Substantial conductivity is commonly observed in nominally undoped diamond grown by chemical vapor deposition. This conductivity is associated with hydrogen-related species adsorbed at the surface, and it can be removed by annealing or other surface treatments.

Surface Property

Diamonds are naturally lipophilic and hydrophobic, which means the diamonds' surface cannot be wet by water but can be easily wet and stuck by oil. This property can be utilized to extract diamonds using oil when making synthetic diamonds. However, when diamond surfaces are chemically modified with certain ions, they are expected to become so hydrophilic that they can stabilize multiple layers of water ice at human body temperature.

The surface of diamonds is partially oxidized. The oxidized surface can be reduced by heat treatment under hydrogen flow. That is to say, this heat treatment partially removes oxygen-containing functional groups. But diamonds (sp^3C) are unstable against high temperature (above about 400 °C (752 °F)) under atmospheric pressure. The structure gradually changes into sp^2C above this temperature. Thus, diamonds should be reduced under this temperature.

Chemical Stability

Diamonds are not very reactive. Under room temperature diamonds do not react with any chemical reagents including strong acids and bases. A diamond's surface can only be oxidized at temperatures above about 850 °C (1,560 °F) in air. Diamond also reacts with fluorine gas above about 700 °C (1,292 °F).

Color

Diamond has a wide bandgap of 5.5 eV corresponding to the deep ultraviolet wavelength of 225 nanometers. This means pure diamond should transmit visible light and appear as a clear colorless crystal. Colors in diamond originate from lattice defects and impurities. The diamond crystal lattice is exceptionally strong and only atoms of nitrogen, boron and hydrogen can be introduced into diamond during the growth at significant concentrations (up to atomic percents). Transition metals nickel and cobalt, which are commonly used for growth of synthetic diamond by high-pressure high-temperature techniques, have been detected in diamond as individual atoms; the maximum concentration is 0.01% for nickel and even less for cobalt. Virtually any element can be introduced to diamond by ion implantation.

Brown diamonds at the National Museum of Natural History in Washington, D.C.

The most famous colored diamond, Hope Diamond in 1974.

Nitrogen is by far the most common impurity found in gem diamonds and is responsible for the yellow and brown color in diamonds. Boron is responsible for the blue color. Color in diamond has two additional sources: irradiation (usually by alpha particles), that causes the color in green diamonds; and plastic deformation of the diamond crystal lattice. Plastic deformation is the cause of color in some brown and perhaps pink and red diamonds. In order of rarity, yellow diamond is followed by brown, colorless, then by blue, green, black, pink, orange, purple, and red. "Black", or Carbonado, diamonds are not truly black, but rather contain numerous dark inclusions that give the gems their dark appearance. Colored diamonds contain impurities or structural defects that cause the coloration, while pure or nearly pure diamonds are transparent and colorless. Most diamond impurities replace a carbon atom in the crystal lattice, known as a carbon flaw. The most common impurity, nitrogen, causes a slight to intense yellow coloration depending upon the type and concentration of nitrogen present. The Gemological Institute of America (GIA) classifies low saturation yellow and brown diamonds as diamonds in the *normal color range*, and applies a grading scale from "D" (colorless) to "Z" (light yellow). Diamonds of a different color, such as blue, are called *fancy colored* diamonds, and fall under a different grading scale.

In 2008, the Wittelsbach Diamond, a 35.56-carat (7.112 g) blue diamond once belonging to the King of Spain, fetched over US$24 million at a Christie's auction. In May 2009, a 7.03-carat (1.406 g) blue diamond fetched the highest price per carat ever paid for a diamond when it was sold at auction for 10.5 million Swiss francs (6.97 million euro or US$9.5 million at the time). That record was however beaten the same year: a 5-carat (1.0 g) vivid pink diamond was sold for $10.8 million in Hong Kong on December 1, 2009.

Identification

Diamonds can be identified by their high thermal conductivity. Their high refractive index is also indicative, but other materials have similar refractivity. Diamonds cut glass, but this does not positively identify a diamond because other materials, such as quartz, also lie above glass on the Mohs scale and can also cut it. Diamonds can scratch other diamonds, but this can result in damage to one or both stones. Hardness tests are infrequently used in practical gemology because of their potentially destructive nature. The extreme hardness and high value of diamond means that gems are typically polished slowly using painstaking traditional techniques and greater attention to detail than is the case with most other gemstones; these tend to result in extremely flat, highly polished facets with exceptionally sharp facet edges. Diamonds also possess an extremely high refractive index and fairly high dispersion. Taken together, these factors affect the overall appearance of a polished diamond and most diamantaires still rely upon skilled use of a loupe (magnifying glass) to identify diamonds 'by eye'.

Industry

A round brilliant cut diamond set in a ring

The diamond industry can be separated into two distinct categories: one dealing with gem-grade diamonds and another for industrial-grade diamonds. Both markets value diamonds differently.

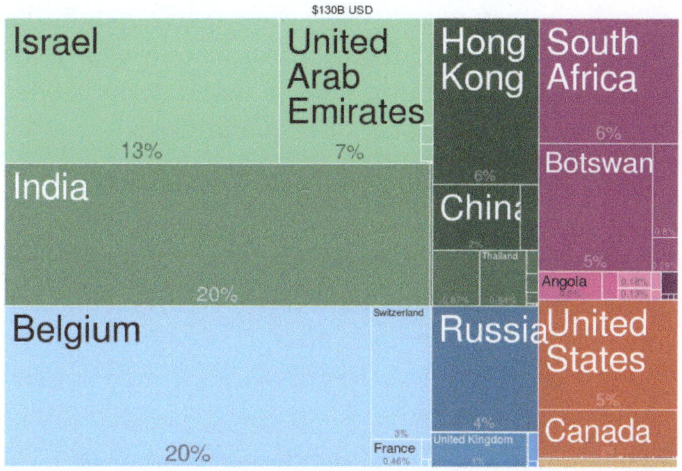

Diamond Exports by Country (2014) from Harvard Atlas of Economic Complexity

Gem-grade Diamonds

A large trade in gem-grade diamonds exists. Although most gem-grade diamonds are sold newly polished, there is a well-established market for resale of polished diamonds (e.g. pawnbroking, auctions, second-hand jewelry stores, diamantaires, bourses, etc.). One hallmark of the trade in gem-quality diamonds is its remarkable concentration: wholesale trade and diamond cutting is limited to just a few locations; in 2003, 92% of the world's diamonds were cut and polished in Surat, India. Other important centers of diamond cutting and trading are the Antwerp diamond district in Belgium, where the International Gemological Institute is based, London, the Diamond District in New York City, the Diamond Exchange District in Tel Aviv, and Amsterdam. One contributory factor is the geological nature of diamond deposits: several large primary kimberlite-pipe mines each account for significant portions of market share (such as the Jwaneng mine in Botswana, which is a single large-pit mine that can produce between 12,500,000 carats (2,500 kg) to 15,000,000 carats (3,000 kg) of diamonds per year). Secondary alluvial diamond deposits, on the other hand, tend to be fragmented amongst many different operators because they can be dispersed over many hundreds of square kilometers (e.g., alluvial deposits in Brazil).

The production and distribution of diamonds is largely consolidated in the hands of a few key players, and concentrated in traditional diamond trading centers, the most important being Antwerp, where 80% of all rough diamonds, 50% of all cut diamonds and more than 50% of all rough, cut and industrial diamonds combined are handled. This makes Antwerp a de facto "world diamond capital". The city of Antwerp also hosts the Antwerpsche Diamantkring, created in 1929 to become the first and biggest diamond bourse dedicated to rough diamonds. Another important diamond center is New York City, where almost 80% of the world's diamonds are sold, including auction sales.

The De Beers company, as the world's largest diamond mining company, holds a dominant position in the industry, and has done so since soon after its founding in 1888 by the British imperialist Cecil Rhodes. De Beers is currently the world's largest operator of diamond production facilities (mines) and distribution channels for gem-quality diamonds. The Diamond Trading Company (DTC) is a subsidiary of De Beers and markets rough diamonds from De Beers-operated mines. De Beers and its subsidiaries own mines that produce some 40% of annual world diamond production. For most of the 20th century over 80% of the world's rough diamonds passed through De Beers, but by 2001–2009 the figure had decreased to around 45%, and by 2013 the company's market share had further decreased to around 38% in value terms and even less by volume. De Beers sold off the vast majority of its diamond stockpile in the late 1990s – early 2000s and the remainder largely represents working stock (diamonds that are being sorted before sale). This was well documented in the press but remains little known to the general public.

As a part of reducing its influence, De Beers withdrew from purchasing diamonds on the open market in 1999 and ceased, at the end of 2008, purchasing Russian diamonds mined by the largest Russian diamond company Alrosa. As of January 2011, De Beers states that it only sells diamonds from the following four countries: Botswana, Namibia, South Africa and Canada. Alrosa had to suspend their sales in October 2008 due to the global energy crisis, but the company reported that it had resumed selling rough diamonds on the open market by October 2009. Apart from Alrosa, other important diamond mining companies include BHP Billiton, which is the world's largest mining company; Rio Tinto Group, the owner of Argyle (100%), Diavik (60%), and Murowa (78%) diamond mines; and Petra Diamonds, the owner of several major diamond mines in Africa.

Diamond polisher in Amsterdam

Further down the supply chain, members of The World Federation of Diamond Bourses (WFDB) act as a medium for wholesale diamond exchange, trading both polished and rough diamonds. The WFDB consists of independent diamond bourses in major cutting centers such as Tel Aviv, Antwerp, Johannesburg and other cities across the USA, Europe and Asia. In 2000, the WFDB and The International Diamond Manufacturers Association established the World Diamond Council to prevent the trading of diamonds used to fund war and inhumane acts. WFDB's additional activities include sponsoring the World Diamond Congress every two years, as well as the establishment of the *International Diamond Council* (IDC) to oversee diamond grading.

Once purchased by Sightholders (which is a trademark term referring to the companies that have a three-year supply contract with DTC), diamonds are cut and polished in preparation for sale as gemstones ('industrial' stones are regarded as a by-product of the gemstone market; they are used for abrasives). The cutting and polishing of rough diamonds is a specialized skill that is concentrated in a limited number of locations worldwide. Traditional diamond cutting centers are Antwerp, Amsterdam, Johannesburg, New York City, and Tel Aviv. Recently, diamond cutting centers have been established in China, India, Thailand, Namibia and Botswana. Cutting centers with lower cost of labor, notably Surat in Gujarat, India, handle a larger number of smaller carat diamonds, while smaller quantities of larger or more valuable diamonds are more likely to be handled in Europe or North America. The recent expansion of this industry in India, employing low cost labor, has allowed smaller diamonds to be prepared as gems in greater quantities than was previously economically feasible.

Diamonds which have been prepared as gemstones are sold on diamond exchanges called *bourses*. There are 28 registered diamond bourses in the world. Bourses are the final tightly controlled step in the diamond supply chain; wholesalers and even retailers are able to buy relatively small lots of diamonds at the bourses, after which they are prepared for final sale to the consumer. Diamonds can be sold already set in jewelry, or sold unset ("loose"). According to the Rio Tinto Group, in 2002 the diamonds produced and released to the market were valued at US$9 billion as rough diamonds, US$14 billion after being cut and polished, US$28 billion in wholesale diamond jewelry, and US$57 billion in retail sales.

Cutting

The Darya-I-Nur Diamond—an example of unusual diamond cut and jewelry arrangement

Mined rough diamonds are converted into gems through a multi-step process called "cutting". Diamonds are extremely hard, but also brittle and can be split up by a single blow. Therefore, diamond cutting is traditionally considered as a delicate procedure requiring skills, scientific knowledge, tools and experience. Its final goal is to produce a faceted jewel where the specific angles between the facets would optimize the diamond luster, that is dispersion of white light, whereas the number and area of facets would determine the weight of the final product. The weight reduction upon cutting is significant and can be of the order of 50%. Several possible shapes are considered, but the final decision is often determined not only by scientific, but also practical considerations. For example, the diamond might be intended for display or for wear, in a ring or a necklace, singled or surrounded by other gems of certain color and shape. Some of them may be considered as classical, such as round, pear, marquise, oval, hearts and arrows diamonds, etc. Some of them are special, produced by certain companies, for example, Phoenix, Cushion, Sole Mio diamonds, etc.

The most time-consuming part of the cutting is the preliminary analysis of the rough stone. It needs to address a large number of issues, bears much responsibility, and therefore can last years in case of unique diamonds. The following issues are considered:

- The hardness of diamond and its ability to cleave strongly depend on the crystal orientation. Therefore, the crystallographic structure of the diamond to be cut is analyzed using X-ray diffraction to choose the optimal cutting directions.
- Most diamonds contain visible non-diamond inclusions and crystal flaws. The cutter has to decide which flaws are to be removed by the cutting and which could be kept.
- The diamond can be split by a single, well calculated blow of a hammer to a pointed tool, which is quick, but risky. Alternatively, it can be cut with a diamond saw, which is a more reliable but tedious procedure.

After initial cutting, the diamond is shaped in numerous stages of polishing. Unlike cutting, which is a responsible but quick operation, polishing removes material by gradual erosion and is extremely time consuming. The associated technique is well developed; it is considered as a routine and can be performed by technicians. After polishing, the diamond is reexamined for possible flaws, either remaining or induced by the process. Those flaws are concealed through various diamond enhancement techniques, such as repolishing, crack filling, or clever arrangement of the stone in the jewelry. Remaining non-diamond inclusions are removed through laser drilling and filling of the voids produced.

Marketing

Marketing has significantly affected the image of diamond as a valuable commodity.

N. W. Ayer & Son, the advertising firm retained by De Beers in the mid-20th century, succeeded in reviving the American diamond market. And the firm created new markets in countries where no diamond tradition had existed before. N. W. Ayer's marketing included product placement, advertising focused on the diamond product itself rather than the De Beers brand, and associations with celebrities and royalty. Without advertising the De Beers brand, De Beers was advertising its competitors' diamond products as well, but this was not a concern as De Beers dominated the diamond market throughout the 20th century. De Beers' market share dipped temporarily to 2nd place in the global market below Alrosa in the aftermath of the global economic crisis of 2008, down to less than 29% in terms of carats mined, rather than sold. The campaign lasted for decades but was effectively discontinued by early 2011. De Beers still advertises diamonds, but the advertising now mostly promotes its own brands, or licensed product lines, rather than completely "generic" diamond products. The campaign was perhaps best captured by the slogan "a diamond is forever". This slogan is now being used by De Beers Diamond Jewelers, a jewelry firm which is a 50%/50% joint venture between the De Beers mining company and LVMH, the luxury goods conglomerate.

Brown-colored diamonds constituted a significant part of the diamond production, and were predominantly used for industrial purposes. They were seen as worthless for jewelry (not even being assessed on the diamond color scale). After the development of Argyle diamond mine in Australia in 1986, and marketing, brown diamonds have become acceptable gems. The change was mostly due to the numbers: the Argyle mine, with its 35,000,000 carats (7,000 kg) of diamonds per year, makes about one-third of global production of natural diamonds; 80% of Argyle diamonds are brown.

Industrial-grade Diamonds

A scalpel with synthetic diamond blade

Close-up photograph of an angle grinder blade with tiny diamonds shown embedded in the metal

A diamond knife blade used for cutting ultrathin sections (typically 70 to 350 nm) for transmission electron microscopy.

Industrial diamonds are valued mostly for their hardness and thermal conductivity, making many of the gemological characteristics of diamonds, such as the 4 Cs, irrelevant for most applications. 80% of mined diamonds (equal to about 135,000,000 carats (27,000 kg) annually), are unsuitable for use as gemstones, and used industrially. In addition to mined diamonds, synthetic diamonds found industrial applications almost immediately after their invention in the 1950s; another 570,000,000 carats (114,000 kg) of synthetic diamond is produced annually for industrial use (in 2004; in 2014 it's 4,500,000,000 carats (900,000 kg), 90% of which is produced in China). Approximately 90% of diamond grinding grit is currently of synthetic origin.

The boundary between gem-quality diamonds and industrial diamonds is poorly defined and partly depends on market conditions (for example, if demand for polished diamonds is high, some lower-grade stones will be polished into low-quality or small gemstones rather than being sold for industrial use). Within the category of industrial diamonds, there is a sub-category comprising the lowest-quality, mostly opaque stones, which are known as bort.

Industrial use of diamonds has historically been associated with their hardness, which makes diamond the ideal material for cutting and grinding tools. As the hardest known naturally occurring material, diamond can be used to polish, cut, or wear away any material, including other diamonds. Common industrial applications of this property include diamond-tipped drill bits and saws, and the use of diamond powder as an abrasive. Less expensive industrial-grade diamonds, known as bort, with more flaws and poorer color than gems, are used for such purposes. Diamond is not suitable for machining ferrous alloys at high speeds, as carbon is soluble in iron at the high temperatures created by high-speed machining, leading to greatly increased wear on diamond tools compared to alternatives.

Specialized applications include use in laboratories as containment for high pressure experiments, high-performance bearings, and limited use in specialized windows. With the continuing advances being made in the production of synthetic diamonds, future applications are becoming feasible. The high thermal conductivity of diamond makes it suitable as a heat sink for integrated circuits in electronics.

Mining

Siberia's Udachnaya diamond mine

Approximately 130,000,000 carats (26,000 kg) of diamonds are mined annually, with a total value of nearly US$9 billion, and about 100,000 kg (220,000 lb) are synthesized annually.

Roughly 49% of diamonds originate from Central and Southern Africa, although significant sources of the mineral have been discovered in Canada, India, Russia, Brazil, and Australia. They are mined from kimberlite and lamproite volcanic pipes, which can bring diamond crystals, originating from deep within the Earth where high pressures and temperatures enable them to form, to the surface. The mining and distribution of natural diamonds are subjects of frequent controversy such as concerns over the sale of *blood diamonds* or *conflict diamonds* by African paramilitary groups. The diamond supply chain is controlled by a limited number of powerful businesses, and is also highly concentrated in a small number of locations around the world.

Only a very small fraction of the diamond ore consists of actual diamonds. The ore is crushed, during which care is required not to destroy larger diamonds, and then sorted by density. Today, diamonds are located in the diamond-rich density fraction with the help of X-ray fluorescence, after which the final sorting steps are done by hand. Before the use of X-rays became commonplace, the separation was done with grease belts; diamonds have a stronger tendency to stick to grease than the other minerals in the ore.

Historically, diamonds were found only in alluvial deposits in Guntur and Krishna district of the Krishna River delta in Southern India. India led the world in diamond production from the time of their discovery in approximately the 9th century BC to the mid-18th century AD, but the commercial potential of these sources had been exhausted by the late 18th century and at that time India was eclipsed by Brazil where the first non-Indian diamonds were found in 1725. Currently, one of the most prominent Indian mines is located at Panna.

Diamond extraction from primary deposits (kimberlites and lamproites) started in the 1870s after the discovery of the Diamond Fields in South Africa. Production has increased over time and now an accumulated total of 4,500,000,000 carats (900,000 kg) have been mined since that date. Twenty percent of that amount has been mined in the last five years, and during the last 10 years, nine new mines have started production; four more are waiting to be opened soon. Most of these mines are located in Canada, Zimbabwe, Angola, and one in Russia.

In the U.S., diamonds have been found in Arkansas, Colorado, Wyoming, and Montana. In 2004, the discovery of a microscopic diamond in the U.S. led to the January 2008 bulk-sampling of kimberlite pipes in a remote part of Montana. The Crater of Diamonds State Park in Arkansas is open to the public, and is the only mine in the world where members of the public can dig for diamonds.

Today, most commercially viable diamond deposits are in Russia (mostly in Sakha Republic, for example Mir pipe and Udachnaya pipe), Botswana, Australia (Northern and Western Australia) and the Democratic Republic of the Congo. In 2005, Russia produced almost one-fifth of the global diamond output, according to the British Geological Survey. Australia boasts the richest diamantiferous pipe, with production from the Argyle diamond mine reaching peak levels of 42 metric tons per year in the 1990s. There are also commercial deposits being actively mined in the Northwest Territories of Canada and Brazil. Diamond prospectors continue to search the globe for diamond-bearing kimberlite and lamproite pipes.

Political Issues

In some of the more politically unstable central African and west African countries, revolutionary groups have taken control of diamond mines, using proceeds from diamond sales to finance their operations. Diamonds sold through this process are known as *conflict diamonds* or *blood diamonds.* Major diamond trading corporations continue to fund and fuel these conflicts by doing business with armed groups.

In response to public concerns that their diamond purchases were contributing to war and human rights abuses in central and western Africa, the United Nations, the diamond industry and diamond-trading nations introduced the Kimberley Process in 2002. The Kimberley Process aims to ensure that conflict diamonds do not become intermixed with the diamonds not controlled by such rebel groups. This is done by requiring diamond-producing countries to provide proof that the money they make from selling the diamonds is not used to fund criminal or revolutionary activities. Although the Kimberley Process has been moderately successful in limiting the number of conflict diamonds entering the market, some still find their way in. According to the International Diamond Manufacturers Association, conflict diamonds constitute 2–3% of all diamonds traded. Two major flaws still hinder the effectiveness of the Kimberley Process: (1) the relative ease of smuggling diamonds across African borders, and (2) the violent nature of diamond mining in nations that are not in a technical state of war and whose diamonds are therefore considered "clean".

Unsustainable diamond mining in Sierra Leone

The Canadian Government has set up a body known as the Canadian Diamond Code of Conduct to help authenticate Canadian diamonds. This is a stringent tracking system of diamonds and helps protect the "conflict free" label of Canadian diamonds.

Synthetics, Simulants, and Enhancements

Synthetics

Synthetic diamonds are diamonds manufactured in a laboratory, as opposed to diamonds mined from the Earth. The gemological and industrial uses of diamond have created a large demand for rough stones. This demand has been satisfied in large part by synthetic diamonds, which have been manufactured by various processes for more than half a century. However, in recent years it has become possible to produce gem-quality synthetic diamonds of significant size. It is possible to make colorless synthetic gemstones that, on a molecular level, are identical to natural stones and so visually similar that only a gemologist with special equipment can tell the difference.

Synthetic diamonds of various colors grown by the high-pressure high-temperature technique

The majority of commercially available synthetic diamonds are yellow and are produced by so-called *high-pressure high-temperature* (HPHT) processes. The yellow color is caused by nitrogen impurities. Other colors may also be reproduced such as blue, green or pink, which are a result of the addition of boron or from irradiation after synthesis.

Colorless gem cut from diamond grown by chemical vapor deposition

Another popular method of growing synthetic diamond is chemical vapor deposition (CVD). The growth occurs under low pressure (below atmospheric pressure). It involves feeding a mixture of gases (typically 1 to 99 methane to hydrogen) into a chamber and splitting them to chemically active radicals in a plasma ignited by microwaves, hot filament, arc discharge, welding torch or laser. This method is mostly used for coatings, but can also produce single crystals several millimeters in size.

As of 2010, nearly all 5,000 million carats (1,000 tonnes) of synthetic diamonds produced per year are for industrial use. Around 50% of the 133 million carats of natural diamonds mined per year end up in industrial use. Mining companies' expenses average $40 to $60 per carat for natural colorless diamonds, while synthetic manufacturers' expenses average $2,500 per carat for synthetic, gem-quality colorless diamonds. However, a purchaser is more likely to encounter a synthetic when looking for a fancy-colored diamond because nearly all synthetic diamonds are fancy-colored, while only 0.01% of natural diamonds are.

Simulants

Gem-cut synthetic silicon carbide set in a ring

A diamond simulant is a non-diamond material that is used to simulate the appearance of a diamond, and may be referred to as diamante. Cubic zirconia is the most common. The gemstone moissanite (silicon carbide) can be treated as a diamond simulant, though more costly to produce than cubic zirconia. Both are produced synthetically.

Enhancements

Diamond enhancements are specific treatments performed on natural or synthetic diamonds (usually those already cut and polished into a gem), which are designed to better the gemological characteristics of the stone in one or more ways. These include laser drilling to remove inclusions, application of sealants to fill cracks, treatments to improve a white diamond's color grade, and treatments to give fancy color to a white diamond.

Coatings are increasingly used to give a diamond simulant such as cubic zirconia a more "diamond-like" appearance. One such substance is diamond-like carbon—an amorphous carbonaceous material that has some physical properties similar to those of the diamond. Advertising suggests that such a coating would transfer some of these diamond-like properties to the coated stone, hence enhancing the diamond simulant. Techniques such as Raman spectroscopy should easily identify such a treatment.

Identification

Early diamond identification tests included a scratch test relying on the superior hardness of diamond. This test is destructive, as a diamond can scratch another diamond, and is rarely used nowadays. Instead, diamond identification relies on its superior thermal conductivity. Electronic thermal probes are widely used in the gemological centers to separate diamonds from their imitations. These probes consist of a pair of battery-powered thermistors mounted in a fine copper tip. One thermistor functions as a heating device while the other measures the temperature of the copper tip: if the stone being tested is a diamond, it will conduct the tip's thermal energy rapidly enough to produce a measurable temperature drop. This test takes about 2–3 seconds.

Whereas the thermal probe can separate diamonds from most of their simulants, distinguishing between various types of diamond, for example synthetic or natural, irradiated or non-irradiated, etc., requires more advanced, optical techniques. Those techniques are also used for some diamonds simulants, such as silicon carbide, which pass the thermal conductivity test. Optical techniques can distinguish between natural diamonds and synthetic diamonds. They can also identify the vast majority of treated natural diamonds. "Perfect" crystals (at the atomic lattice level) have never been found, so both natural and synthetic diamonds always possess characteristic imperfections, arising from the circumstances of their crystal growth, that allow them to be distinguished from each other.

Laboratories use techniques such as spectroscopy, microscopy and luminescence under shortwave ultraviolet light to determine a diamond's origin. They also use specially made instruments to aid them in the identification process. Two screening instruments are the *DiamondSure* and the *DiamondView*, both produced by the DTC and marketed by the GIA.

Several methods for identifying synthetic diamonds can be performed, depending on the method of production and the color of the diamond. CVD diamonds can usually be identified by an orange fluorescence. D-J colored diamonds can be screened through the Swiss Gemmological Institute's Diamond Spotter. Stones in the D-Z color range can be examined through the DiamondSure UV/visible spectrometer, a tool developed by De Beers. Similarly, natural diamonds usually have minor imperfections and flaws, such as inclusions of foreign material, that are not seen in synthetic diamonds.

Screening devices based on diamond type detection can be used to make a distinction between diamonds that are certainly natural and diamonds that are potentially synthetic. Those potentially synthetic diamonds require more investigation in a specialized lab. Examples of commercial screening devices are D-Screen (WTOCD / HRD Antwerp) and Alpha Diamond Analyzer (Bruker / HRD Antwerp).

Stolen Diamonds

Occasionally large thefts of diamonds take place. In February 2013 armed robbers carried out a raid at Brussels Airport and escaped with gems estimated to be worth $50m (£32m; 37m euros). The gang broke through a perimeter fence and raided the cargo hold of a Swiss-bound plane. The gang have since been arrested and large amounts of cash and diamonds recovered.

The identification of stolen diamonds presents a set of difficult problems. Rough diamonds will have a distinctive shape depending on whether their source is a mine or from an alluvial environment such as a beach or river - alluvial diamonds have smoother surfaces than those that have been mined. Determining the provenance of cut and polished stones is much more complex.

The Kimberley Process was developed to monitor the trade in rough diamonds and prevent their being used to fund violence. Before exporting, rough diamonds are certificated by the government of the country of origin. Some countries, such as Venezuela, are not party to the agreement. The Kimberley Process does not apply to local sales of rough diamonds within a country.

Diamonds may be etched by laser with marks invisible to the naked eye. Lazare Kaplan, a US-based company, developed this method. However, whatever is marked on a diamond can readily be removed.

Hyalite

A sample of hyalite

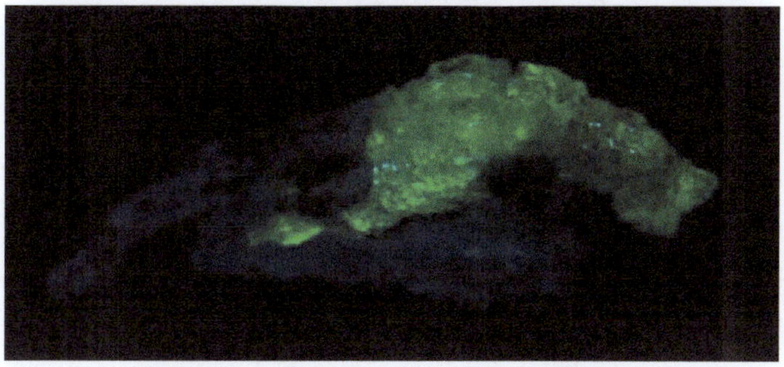

Fluorescent hyalite under ultraviolet light

Hyalite is a form of opal with a glassy and clear appearance which may exhibit an internal play of colors if natural inclusions are present. It is also called *Muller's glass, water opal* and *jalite*. The name Müller's glass derived from the name of its discoverer, Franz-Joseph Müller von Reichenstein.

Properties

Hyalite's Mohs hardness is 5.5 to 6 and has a specific gravity of 1.9 - 2.1. It has no planes of cleavage but fractures conchoidally, is clear or translucent and has a globular structure. Its luster is vitreous and its streak is white. Hyalite is an amorphous form of silica (SiO_2) formed as a volcanic sublimate in volcanic or pegmatic rock and is therefore considered a mineraloid. It contains 3 - 8% water, either as a silanol group or in molecular form.

Uses

Opalescent hyalite is used in jewellery, and well-formed samples are of interest to collectors due to their unusual appearance, mode of formation and relative rarity. It is sometimes mistaken for resin opal or silica glass since they both may appear clear and globular, but it can be identified under ultraviolet light due to its bright green fluorescence.

Slocum Stone

Slocum stone (sometimes sold as "Slocum opal") is an early opal simulant which was briefly popular prior to the introduction of synthetics and less expensive simulants. It was named after its inventor, John S. Slocum of Rochester, Michigan.

opal simulant ("Slocum Stone")	
 a "Slocum Stone" cabochon	
Category	**Mineral**
Identification	
Color	varies
Fracture	conchoidal
Mohs scale hardness	6
Streak	white
Specific gravity	2.4–2.51
Refractive index	1.489–1.520

John Slocum experimented during the 1960s with various methods. In 1971, he began marketing his first commercial simulant under the "Opal Essence" trade name. Not satisfied, he continued experimenting and in 1974 brought out the more realistic "Slocum Stone" type. However, Gilson synthetic opals and other imitation opals such as "Opalite" came onto the market soon thereafter and greatly diluted the market for "Slocum Stone" By 1976, this product was being manufactured by MDI Corporation.

Slocum Stone is a silicate glass which shows traces of sodium, magnesium, aluminum and titanium. It was manufactured in several base colors, and the opalescence is produced by very thin layers of metallic film (estimated at 30 nanometres in thickness), in the form of translucent flakes, which produce a thin-film interference effect. These flakes themselves lend color, along with colorant within the glass base. Bubbles and swirls typical of glass are other typical inclusions which may be noticed under magnification. In later examples, built-up laminations are visible when viewed from the side.

Opal

Opal is a hydrated amorphous form of silica ($SiO_2 \cdot nH_2O$); its water content may range from 3 to 21% by weight, but is usually between 6 and 10%. Because of its amorphous character, it is classed as a mineraloid, unlike crystalline forms of silica, which are classed as minerals. It is deposited at a relatively low temperature and may occur in the fissures of almost any kind of rock, being most commonly found with limonite, sandstone, rhyolite, marl, and basalt. Opal is the national gemstone of Australia.

The internal structure of precious opal makes it diffract light. Depending on the conditions in which it formed, it can take on many colors. Precious opal ranges from clear through white, gray, red, orange, yellow, green, blue, magenta, rose, pink, slate, olive, brown, and black. Of these hues, the black opals are the most rare, whereas white and greens are the most common. Opals vary in optical density from opaque to semitransparent.

Precious Opal

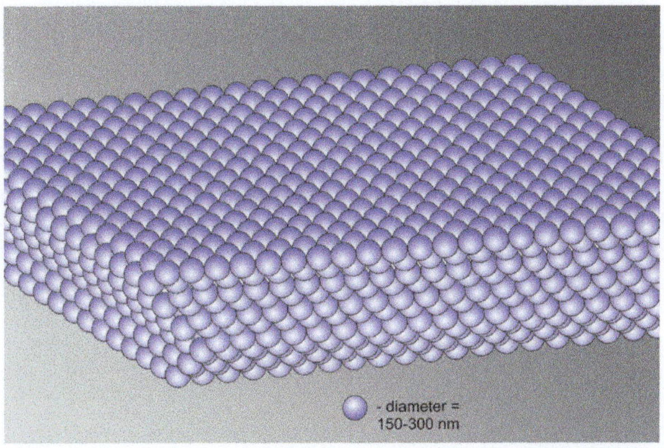

Precious opal consists of spheres of silicon dioxide molecules arranged in regular, closely packed planes. (Idealized diagram)

Multicolor rough crystal opal from Coober Pedy, South Australia, expressing nearly every color of the visible spectrum

Precious opal shows a variable interplay of internal colors, and though it is a mineraloid, it has an internal structure. At microscopic scales, precious opal is composed of silica spheres some 150 to

300 nm in diameter in a hexagonal or cubic close-packed lattice. It was shown by J. V. Sanders in the mid-1960s, that these ordered silica spheres produce the internal colors by causing the interference and diffraction of light passing through the microstructure of the opal. The regularity of the sizes and the packing of these spheres determines the quality of precious opal. Where the distance between the regularly packed planes of spheres is around half the wavelength of a component of visible light, the light of that wavelength may be subject to diffraction from the grating created by the stacked planes. The colors that are observed are determined by the spacing between the planes and the orientation of planes with respect to the incident light. The process can be described by Bragg's law of diffraction.

Precious opal replacing ichthyosaur backbone; display specimen, South Australian Museum

Visible light of diffracted wavelengths cannot pass through large thicknesses of the opal. This is the basis of the optical band gap in a photonic crystal. The notion that opals are photonic crystals for visible light was expressed in 1995 by Vasily Astratov's group. In addition, microfractures may be filled with secondary silica and form thin lamellae inside the opal during solidification. The term opalescence is commonly and erroneously used to describe this unique and beautiful phenomenon, which is correctly termed play of color. Contrarily, opalescence is correctly applied to the milky, turbid appearance of common or potch opal. Potch does not show a play of color.

For gemstone use, most opal is cut and polished to form a cabochon. "Solid" opal refers to polished stones consisting wholly of precious opal. Opals too thin to produce a "solid" may be combined with other materials to form attractive gems. An opal doublet consists of a relatively thin layer of precious opal, backed by a layer of dark-colored material, most commonly ironstone, dark or black common opal (potch), onyx, or obsidian. The darker backing emphasizes the play of color, and results in a more attractive display than a lighter potch. An opal triplet is similar to a doublet, but has a third layer, a domed cap of clear quartz or plastic on the top. The cap takes a high polish and acts as a protective layer for the opal. The top layer also acts as a magnifier, to emphasize the play of color of the opal beneath, which is often of lower quality. Triplet opals therefore have a more artificial appearance, and are not classed as precious opal. Jewelry applications of precious opal can be somewhat limited by opal's sensitivity to heat due primarily to its relatively high water content and predisposition to scratching. Combined with modern techniques of polishing, doublet opal produces a similar effect to black or boulder opal at a fraction of the price. Doublet opal also has the added benefit of having genuine opal as the top visible and touchable layer, unlike triplet opals.

Common Opal

Besides the gemstone varieties that show a play of color, the other kinds of common opal include the milk opal, milky bluish to greenish (which can sometimes be of gemstone quality); resin opal, which is honey-yellow with a resinous luster; wood opal, which is caused by the replacement of the organic material in wood with opal; menilite, which is brown or grey; hyalite, a colorless glass-clear opal sometimes called Muller's glass; geyserite, also called siliceous sinter, deposited around hot springs or geysers; and diatomite or diatomaceous earth, the accumulations of diatom shells or tests.

A piece of milky raw opal from Andamooka, South Australia

An opal "triplet" from Andamooka showing blue and green fire

Fire opal from Mexico

A rock showing striations of opal throughout

A cabochon cut from a piece of opalized wood

Precious opal replacing calcite of bivalve shells, from Coober Pedy; display specimen, South Australian Museum

Other Varieties of Opal

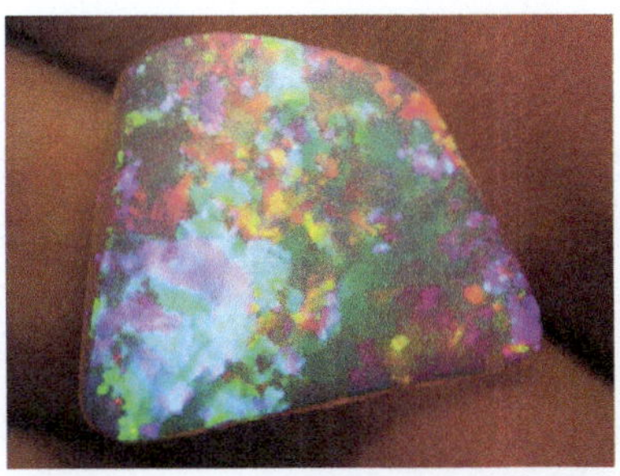

Brightness of the fire in opal ranges on a scale of 1 to 5 (5 being the brightest)

Australian opal doublet, a slice of precious opal with a backing of ironstone

Faceted yellow fire opal, 4.76 carats (0.952 g; 0.0336 oz), Mexico

Fire opal is a transparent to translucent opal, with warm body colors of yellow to orange to red. Although it does not usually show any play of color, occasionally a stone will exhibit bright green flashes. The most famous source of fire opals is the state of Querétaro in Mexico; these opals are commonly called Mexican fire opals. Fire opals that do not show play of color are sometimes referred to as jelly opals. Mexican opals are sometimes cut in their rhyolitic host material if it is hard enough to allow cutting and polishing. This type of Mexican opal is referred to as a Cantera opal. Also, a type of opal from Mexico, referred to as Mexican water opal, is a colorless opal which exhibits either a bluish or golden internal sheen.

Girasol opal is a term sometimes mistakenly and improperly used to refer to fire opals, as well as a type of transparent to semitransparent type milky quartz from Madagascar which displays an asterism, or star effect, when cut properly. However, the true girasol opal is a type of hyalite opal that exhibits a bluish glow or sheen that follows the light source around. It is not a play of color as seen in precious opal, but rather an effect from microscopic inclusions. It is also sometimes referred to as water opal, too, when it is from Mexico. The two most notable locations of this type of opal are Oregon and Mexico.

Peruvian opal (also called blue opal) is a semiopaque to opaque blue-green stone found in Peru, which is often cut to include the matrix in the more opaque stones. It does not display pleochroism. Blue opal also comes from Oregon in the Owyhee region, as well as from Nevada around the Virgin Valley.

Sources of Opal

Gem opal from Brazil

Polished opal from Yowah (Yowah Nut), Queensland

Multicolored solid black opal cabochon from Lightning Ridge, New South Wales

Boulder opal, Carisbrooke Station near Winton, Queensland

Australian opal has often been cited as accounting for 95–97% of the world's supply of precious opal, with the state of South Australia accounting for 80% of the world's supply. Recent data suggests that the world supply of precious opal may have changed. In 2012, Ethiopian opal production was estimated to be 14,000 kg (31,000 lb) by the United States Geological Survey. USGS data from the same period (2012), reveals that Australian opal production to be $41 million. Because of the units of measurement, it is not possible to directly compare Australian and Ethiopian opal production, but these data and others suggest that the traditional percentages given for Australian opal production may be overstated. Yet, the validity of data in the USGS report appears to conflict with that of Laurs et al. and Mesfin, who estimated the 2012 Ethiopian opal output (from Wegal Tena) to be only 750 kg (1,650 lb).

Australian Opal

The town of Coober Pedy in South Australia is a major source of opal. The world's largest and most valuable gem opal "Olympic Australis" was found in August 1956 at the "Eight Mile" opal field in Coober Pedy. It weighs 17,000 carats (3.4 kg; 7.5 lb) and is 11 inches (280 mm) long, with a height of 4 $^3/_4$ in (120 mm) and a width of 4 $^1/_2$ in (110 mm). The Mintabie Opal Field located about 250 km (160 mi) north west of Coober Pedy has also produced large quantities of crystal opal and the rarer black opal. Over the years, it has been sold overseas incorrectly as Coober Pedy opal. The black opal is said to be some of the best examples found in Australia.

Andamooka in South Australia is also a major producer of matrix opal, crystal opal, and black opal. Another Australian town, Lightning Ridge in New South Wales, is the main source of black opal, opal containing a predominantly dark background (dark gray to blue-black displaying the play of color). Boulder opal consists of concretions and fracture fillings in a dark siliceous ironstone matrix. It is found sporadically in western Queensland, from Kynuna in the north, to Yowah and Koroit in the south. Its largest quantities are found around Jundah and Quilpie (known as the "home of the boulder opal") in South West Queensland. Australia also has opalized fossil remains, including dinosaur bones in New South Wales, and marine creatures in South Australia. The rarest type of Australian opal is "pipe" opal, closely related to boulder opal, which forms in sandstone with some iron oxide content, usually as fossilized tree roots.

Ethiopian Opal

Gem-grade precious Ethiopian Welo opal pendant

Although it has been reported that Northern African opal was used to make tools as early as 4000 BC, the first published report of gem opal from Ethiopia appeared in 1994, with the discovery of precious opal in the Menz Gishe District, North Shewa Province. The opal, found mostly in the form of nodules, was of volcanic origin and was found predominantly within weathered layers of rhyolite. This Shewa Province opal was mostly dark brown in color and had a tendency to crack. These qualities made it unpopular in the gem trade. In 2008, a new opal deposit was found near the town of Wegel Tena, in Ethiopia's Wollo Province. The Wollo Province opal was different from the previous Ethiopian opal finds in that it more closely resembled the sedimentary opals of Australia and Brazil, with a light background and often vivid play-of-color. Wollo Province opal, more commonly referred to as "Welo" or "Wello" opal, has become the dominant Ethiopian opal in the gem trade.

Virgin Valley, Nevada

Multicolored rough opal specimen from Virgin Valley, Nevada, US

The Virgin Valley opal fields of Humboldt County in northern Nevada produce a wide variety of precious black, crystal, white, fire, and lemon opal. The black fire opal is the official gemstone of Nevada. Most of the precious opal is partial wood replacement. The precious opal is hosted and found within a subsurface horizon or zone of bentonite in place which is considered a "lode" deposit. Opals which have weathered out of the in-place deposits are alluvial and considered placer deposits. Miocene-age opalised teeth, bones, fish, and a snake head have been found. Some of the opal has high water content and may desiccate and crack when dried. The largest producing mines of Virgin Valley have been the famous Rainbow Ridge, Royal Peacock, Bonanza, Opal Queen, and WRT Stonetree/Black Beauty mines. The largest unpolished black opal in the Smithsonian Institution, known as the "Roebling opal", came out of the tunneled portion of the Rainbow Ridge Mine in 1917, and weighs 2,585 carats (517.0 g; 18.24 oz). The largest polished black opal in the Smithsonian Institution comes from the Royal Peacock opal mine in the Virgin Valley, weighing 160 carats (32 g; 1.1 oz), known as the "Black Peacock".

Other Locations

Another source of white base opal or creamy opal in the United States is Spencer, Idaho. A high percentage of the opal found there occurs in thin layers.

Other significant deposits of precious opal around the world can be found in the Czech Republic,

Canada, Slovakia, Hungary, Turkey, Indonesia, Brazil (in Pedro II, Piauí), Honduras (more precisely in Erandique), Guatemala and Nicaragua.

In late 2008, NASA announced it had discovered opal deposits on Mars.

Synthetic Opal

Opals of all varieties have been synthesized experimentally and commercially. The discovery of the ordered sphere structure of precious opal led to its synthesis by Pierre Gilson in 1974. The resulting material is distinguishable from natural opal by its regularity; under magnification, the patches of color are seen to be arranged in a "lizard skin" or "chicken wire" pattern. Furthermore, synthetic opals do not fluoresce under ultraviolet light. Synthetics are also generally lower in density and are often highly porous.

Two notable producers of synthetic opal are Kyocera and Inamori of Japan. Most so-called synthetics, however, are more correctly termed "imitation opal", as they contain substances not found in natural opal (such as plastic stabilizers). The imitation opals seen in vintage jewelry are often foiled glass, glass-based "Slocum stone", or later plastic materials.

Other research in macroporous structures have yielded highly ordered materials that have similar optical properties to opals and have been used in cosmetics.

Local Atomic Structure of Opals

The lattice of spheres of opal that cause the interference with light are several hundred times larger than the fundamental structure of crystalline silica. As a mineraloid, no unit cell describes the structure of opal. Nevertheless, opals can be roughly divided into those that show no signs of crystalline order (amorphous opal) and those that show signs of the beginning of crystalline order, commonly termed cryptocrystalline or microcrystalline opal. Dehydration experiments and infrared spectroscopy have shown that most of the H_2O in the formula of $SiO_2 \cdot nH_2O$ of opals is present in the familiar form of clusters of molecular water. Isolated water molecules, and silanols, structures such as SiOH, generally form a lesser proportion of the total and can reside near the surface or in defects inside the opal.

The crystal structure of crystalline α-cristobalite. Locally, the structures of some opals, opal-C, are similar to this.

The structure of low-pressure polymorphs of anhydrous silica consist of frameworks of fully corner bonded tetrahedra of SiO_4. The higher temperature polymorphs of silica cristobalite and tridymite are frequently the first to crystallize from amorphous anhydrous silica, and the local structures of microcrystalline opals also appear to be closer to that of cristobalite and tridymite than to quartz. The structures of tridymite and cristobalite are closely related and can be described as hexagonal and cubic close-packed layers. It is therefore possible to have intermediate structures in which the layers are not regularly stacked.

Microcrystalline Opal

Lussatite (opal-CT)

Opal-CT has been interpreted as consisting of clusters of stacking of cristobalite and tridymite over very short length scales. The spheres of opal in opal-CT are themselves made up of tiny microcrystalline blades of cristobalite and tridymite. Opal-CT has occasionally been further subdivided in the literature. Water content may be as high as 10 wt%. Lussatite is a synonym. Opal-C, also called lussatine, is interpreted as consisting of localized order of α-cristobalite with a lot of stacking disorder. Typical water content is about 1.5 wt%.

Noncrystalline Opal

Two broad categories of noncrystalline opals, sometimes just referred to as "opal-A", have been proposed. The first of these is opal-AG consisting of aggregated spheres of silica, with water filling the space in between. Precious opal and potch opal are generally varieties of this, the difference being in the regularity of the sizes of the spheres and their packing. The second "opal-A" is opal-AN or water-containing amorphous silica-glass. Hyalite is another name for this.

Noncrystalline silica in siliceous sediments is reported to gradually transform to opal-CT and then opal-C as a result of diagenesis, due to the increasing overburden pressure in sedimentary rocks, as some of the stacking disorder is removed.

Etymology

The word 'opal' is adapted from the Latin term *opalus*, but the origin of this word is a matter of debate. However, most modern references suggest it is adapted from the Sanskrit word *úpala*.

References to the gem are made by Pliny the Elder. It is suggested to have been adapted from Ops, the wife of Saturn and goddess of fertility. The portion of Saturnalia devoted to Ops was "Opalia", similar to *opalus*.

Another common claim that the term is adapted from the Ancient Greek word, *opallios*. This word has two meanings, one is related to "seeing" and forms the basis of the English words like "opaque"; the other is "other" as in "alias" and "alter". It is claimed that *opalus* combined these uses, meaning "to see a change in color". However, historians have noted the first appearances of *opallios* do not occur until after the Romans had taken over the Greek states in 180 BC, and they had previously used the term *paederos*.

However, the argument for the Sanskrit origin is strong. The term first appears in Roman references around 250 BC, at a time when the opal was valued above all other gems. The opals were supplied by traders from the Bosporus, who claimed the gems were being supplied from India. Before this the stone was referred to by a variety of names, but these fell from use after 250 BC.

Historical Superstitions

In the Middle Ages, opal was considered a stone that could provide great luck because it was believed to possess all the virtues of each gemstone whose color was represented in the color spectrum of the opal. It was also said to confer the power of invisibility if wrapped in a fresh bay leaf and held in the hand. Following the publication of Sir Walter Scott's *Anne of Geierstein* in 1829, opal acquired a less auspicious reputation. In Scott's novel, the Baroness of Arnheim wears an opal talisman with supernatural powers. When a drop of holy water falls on the talisman, the opal turns into a colorless stone and the Baroness dies soon thereafter. Due to the popularity of Scott's novel, people began to associate opals with bad luck and death. Within a year of the publishing of Scott's novel in April 1829, the sale of opals in Europe dropped by 50%, and remained low for the next 20 years or so.

Even as recently as the beginning of the 20th century, it was believed that when a Russian saw an opal among other goods offered for sale, he or she should not buy anything more, as the opal was believed to embody the evil eye.

Opal is considered the birthstone for people born in October.

Famous Opals

- The Olympic Australis, the world's largest and most valuable gem opal, found in Coober Pedy
- The Andamooka Opal, presented to Queen Elizabeth II, also known as the Queen's Opal
- The Addyman Plesiosaur from Andamooka, "the finest known opalised skeleton on Earth"
- The Burning of Troy, the now-lost opal presented to Joséphine de Beauharnais by Napoleon I of France and the first named opal
- The Flame Queen Opal

- The Halley's Comet Opal, the world's largest uncut black opal
- Although the clock faces above the information stand in Grand Central Terminal Manhattan, New York, are often said to be opal, they are in fact opalescent glass
- The Roebling Opal, Smithsonian Institution
- The Galaxy Opal, listed as the "World's Largest Polished Opal" in the 1992 Guinness Book of Records
- The Rainbow Virgin, "the finest opal ever unearthed"

References

- McCall, G. J. H. (2001) Tektites in the Geological Record: Showers of Glass from the Sky. The Geological Society Publishing House, Bath, United Kingdom. 256 pp. ISBN 1-86239-085-1.
- Montanari, A., and C. Koeberl (2000) Impact Stratigraphy. The Italian Record. Lecture Notes in Earth Sciences Series no. 93. Springer-Verlag, New York, New York. 364 pp. ISBN 3540663681.
- O'Keefe, J. A., (1978) Tektites and Their Origin. Developments in petrology vol. 4. Elsevier Scientific Publishing Company, New York, New York. 254 pp. ISBN 9780444413505.
- Peter Roger Stuart Moorey (1999). Ancient mesopotamian materials and industries: the archaeological evidence. Eisenbrauns. pp. 108–. ISBN 978-1-57506-042-2.
- Pliny the Elder (translated by John Bostock, Henry Thomas Riley) (1857). The natural history of Pliny. 6. H G Bohn. ISBN 1851099301.
- M Martini; M Milazzo; M Piacentini; Società Italiana di Fisica (2004). Physics Methods in Archaeometry. 154. IOS Press. ISBN 1586034243.
- Iain Morley and Colin Renfrew. The Archaeology of Measurement: Comprehending Heaven, Earth and Time in Ancient Societies, Cambridge University Press, 2010 ISBN 0521119901.
- L Romano. 6 ICAANE, Otto Harrassowitz Verlag, 2010 Volume 3 of Proceedings of the 6th International Congress of the Archaeology of the Ancient Near East: 5–10 May 2009 ISBN 3447062177.
- P R S Moorey. Ancient Mesopotamian Materials and Industries: The Archaeological Evidence Eisenbrauns, 1999 ISBN 1575060426.
- D Schmandt-Besserat. Volume 3 of Invited lectures on the Middle East at the University of Texas at Austin, Undena Publications, 1979, ISBN 0890030316.

Production of Glass

Glass production has mainly two main methods; the float glass process and glassblowing. The most famous technique of forming glass is glassblowing; it involves inflating molten glass and converting them into a bubble with the help of blowpipe. The aspects that have been explained in the chapter are cylinder blown sheet, machine drawn cylinder sheet, silicon dioxide, polished plate glass and glazier. The aspects elucidated in this section are of vital importance and provide a better understanding of the production of glass.

Glass Production

Glass bottles (for cucumber slices) on shipping pallets

Glass production involves two main methods – the float glass process that produces sheet glass, and glassblowing that produces bottles and other containers.

Glass Container Production

Glass Container Factories

Broadly, modern glass container factories are three-part operations: the *batch house,* the *hot end,* and the *cold end*. The *batch house* handles the raw materials; the *hot end* handles the manufacture proper — the forehearth, annealing ovens, and forming machines; and the *cold end* handles the product-inspection and -packaging equipment.

Hot End

The following table lists common viscosity fixpoints, applicable to large-scale glass production and experimental glass melting in the laboratory:

$\log_{10}$(η, Pa·s)	$\log_{10}$(η, P)	Description
1	2	Melting Point (glass melt homogenization and fining)
3	4	Working Point (pressing, blowing, gob forming)
4	5	Flow Point
6.6	7.6	Littleton Softening Point (Glass deforms visibly under its own weight. Standard procedures ASTM C338, ISO 7884-3)
8–10	9-11	Dilatometric Softening Point, T_d, depending on load
10.5	11.5	Deformation Point (Glass deforms under its own weight on the μm-scale within a few hours.)
11–12.3	12–13.3	Glass Transition Temperature, T_g
12	13	Annealing Point (Stress is relieved within several minutes.)
13.5	14.5	Strain Point (Stress is relieved within several hours.)

Batch Processing System (Batch House)

Batch processing is one of the initial steps of the glass-making process. The batch house simply houses the raw materials in large silos (fed by truck or railcar) and holds anywhere from 1–5 days of material. Some batch systems include material processing such as raw material screening/sieve, drying, or pre-heating (i.e. cullet). Whether automated or manual, the batch house measures, assembles, mixes, and delivers the glass raw material recipe (batch) via an array of chutes, conveyors, and scales to the furnace. The batch enters the furnace at the 'dog house' or 'batch charger'. Different glass types, colors, desired quality, raw material purity / availability, and furnace design will affect the batch recipe.

Furnace

Batch feed (doghouse) of a glass furnace

The *hot end* of a glassworks is where the molten glass is formed into glass products, beginning when the batch is fed into the furnace at a slow, controlled rate by the batch processing system (batch house). The furnaces are natural gas- or fuel oil-fired, and operate at temperatures up to 1,575 °C (2,867 °F). The temperature is limited only by the quality of the furnace's superstructure material and by the glass composition. Types of furnaces used in container glass making include 'end-port' (end-fired), 'side-port', and 'oxy-fuel'. Typically, furnace "size" is classified by metric tons per day (MTPD) production capability.

Forming Process

Glass container forming

There are currently two primary methods of making glass containers: the *blow & blow* method for narrow-neck containers only, and the *press & blow* method used for jars and tapered narrow-neck containers.

In both methods, a stream of molten glass, at its plastic temperature (1,050–1,200 °C [1,920–2,190 °F]), is cut with a shearing blade to form a solid cylinder of glass, called a *gob*. The gob is of predetermined weight just sufficient to make a bottle. Both processes start with the gob falling, by gravity, and guided, through troughs and chutes, into the blank moulds, two halves of which are clamped shut and then sealed by the "baffle" from above.

In the blow and blow process, the glass is first blown through a valve in the baffle, forcing it down into the three-piece "ring mould" which is held in the "neckring arm" below the blanks, to form the "finish", [The term "finish" describes the details (such as cap sealing surface, screw threads, retaining rib for a tamper-proof cap, etc.) at the open end of the container.] The compressed air is blown through the glass, which results in hollow and partly formed container. Compressed air is then blown again at the second stage to give final shape.

Containers are made in two major stages. The first stage moulds all the details ("finish") around the opening, but the body of the container is initially made much smaller than its final size. These partly manufactured containers are called parisons, and quite quickly, they are blow-molded into final shape.

Referring to the mechanism, the "rings" are sealed from below by a short plunger. After the "settleblow" finishes, the plunger retracts slightly, to allow the skin that's formed to soften. "Counterblow" air then comes up through the plunger, to create the parison. The baffle rises and the blanks open. The parison is inverted in an arc to the "mould side" by the "neckring arm", which holds the parison by the "finish".

As the neckring arm reaches the end of its arc, two mould halves close around the parison. The neckring arm opens slightly to release its grip on the "finish", then reverts to the blank side. *Final blow*, applied through the "blowhead", blows the glass out, expanding into the mould, to make the final container shape.

In the press and blow process, the parison is formed by a long metal plunger which rises up and presses the glass out, in order to fill the ring and blank moulds. The process then continues as before, with the parison being transferred to the final-shape mould, and the glass being blown out into the mould.

The container is then picked up from the mould by the "take-out" mechanism, and held over the "deadplate", where air cooling helps cool down the still-soft glass. Finally, the bottles are swept onto a conveyor by the "push out paddles" that have air pockets to keep the bottles standing after landing on the "deadplate"; they're now ready for annealing.

Forming Machines

IS machine during bottle production

The forming machines hold and move the parts that form the container. The machine consist of basic 19 mechanisms in operation to form a bottle and generally powered by compressed air (high pressure - 3.2 bar and low pressure - 2.8 bar), the mechanisms are electronically timed to coordinate all movements of the mechanisms. The most widely used forming machine arrangement is the *individual section* machine (or IS machine). This machine has a bank of 5–20 identical sections, each of which contains one complete set of mechanisms to make containers. The sections are in a row, and the gobs feed into each section via a moving chute, called the *gob distributor*. Sections make either one, two, three or four containers simultaneously. (Referred to as *single*, *double*, *triple* and *quad* gob). In the case of multiple gobs, the *shears* cut the *gobs* simultaneously, and they fall into the blank moulds in parallel.

Internal Treatment

After the forming process, some containers—particularly those intended for alcoholic spirits—undergo a treatment to improve the chemical resistance of the inside, called *internal treatment* or dealkalization. This is usually accomplished through the injection of a sulfur- or fluorine-containing gas mixture into bottles at high temperatures. The gas is typically delivered to the container either in the air used in the forming process (that is, during the final blow of the container), or through a nozzle directing a stream of the gas into the mouth of the bottle after forming. The treatment renders the container more resistant to alkali extraction, which can cause increases in product pH, and in some cases container degradation.

Annealing

As glass cools, it shrinks and solidifies. Uneven cooling causes weak glass due to stress. Even cooling is achieved by annealing. An annealing oven (known in the industry as a Lehr) heats the container to about 580 °C (1,076 °F), then cools it, depending on the glass thickness, over a 20 – 60 minute period.

Cold End

The role of the *cold end is to spray on a polyethylene coating for abrasion resistance and increased lubricity, inspect the containers for defects, package the containers for shipment, and label the containers.*

Inspection Equipment

Glass containers are 100% inspected; automatic machines, or sometimes persons, inspect every container for a variety of faults. Typical faults include small cracks in the glass called *checks* and foreign inclusions called *stones* which are pieces of the refractory brick lining of the melting furnace that break off and fall into the pool of molten glass, or more commonly oversized silica granules (sand) that have failed to melt and which subsequently are included in the final product. These are especially important to select out due to the fact that they can impart a destructive element to the final glass product. For example, since these materials can withstand large amounts of thermal energy, they can cause the glass product to sustain thermal shock resulting in explosive destruction when heated. Other defects include bubbles in the glass called *blisters* and excessively thin walls. Another defect common in glass manufacturing is referred to as a *tear*. In the *press and blow* forming, if a plunger and mould are out of alignment, or heated to an incorrect temperature, the glass will stick to either item and become torn. In addition to rejecting faulty containers, inspection equipment gathers statistical information and relays it to the forming machine operators in the hot end. Computer systems collect fault information and trace it to the mould that produced the container. This is done by reading the mould number on the container, which is encoded (as a numeral, or a binary code of dots) on the container by the mould that made it. Operators carry out a range of checks manually on samples of containers, usually visual and dimensional checks.

Secondary Processing

Sometimes container factories will offer services such as labelling. Several labelling technologies are available. Unique to glass is the *Applied Ceramic Labelling* process (ACL). This is screen-printing of the decoration onto the container with a vitreous enamel paint, which is then baked on. An example of this is the original Coca-Cola bottle. Absolut Vodka Bottles have various added services such as: Etching (Absolut Citron/) Coating (Absolut Raspberry/Ruby Red)and *Applied Ceramic Labelling* (Absolut Blue/Pears/Red/Black)

Packaging

Glass containers are packaged in various ways. Popular in Europe are bulk pallets with between 1000 and 4000 containers each. This is carried out by automatic machines (palletisers) which arrange and stack containers separated by layer sheets. Other possibilities include boxes and even hand-sewn sacks. Once packed, the new "stock units" are labelled and warehoused.

Coatings

Glass containers typically receive two surface coatings, one at the *hot end*, just before annealing and one at the *cold end* just after annealing. At the *hot end* a very thin layer of tin(IV) oxide is applied either using a safe organic compound or inorganic stannic chloride. Tin based systems are not the only ones used, although the most popular. Titanium tetrachloride or organo titanates can also be used. In all cases the coating renders the surface of the glass more adhesive to the *cold end* coating. At the *cold end* a layer of typically, polyethylene wax, is applied via a water based emulsion. This makes the glass slippery, protecting it from scratching and stopping containers from sticking together when they are moved on a conveyor. The resultant invisible combined coating gives a virtually unscratchable surface to the glass. Due to reduction of in-service surface damage, the coatings often are described as strengtheners, however a more correct definition might be strength-retaining coatings.

Ancillary Processes – compressors & Cooling

Forming machines are largely powered by compressed air and a typical glass works will have several large compressors (totaling 30k-60k cfm) to provide the needed compressed air. Furnaces, compressors and forming machine generate quantities of waste heat which is generally cooled by water. Hot glass which is not used in the forming machine is diverted and this diverted glass (called *cullet*) is generally cooled by water, and sometimes even processed and crushed in a water bath arrangement. Often cooling requirements are shared over banks of cooling towers arranged to allow for backup during maintenance.

Marketing

Glass container manufacture in the developed world is a mature market business. World demand for flat glass was approximately 52 million tonnes in 2009. The United States, Europe and China account for 75% of demand, with China's consumption having increased from 20% in the early 1990s to 50%. Glass container manufacture is also a geographical business; the product is heavy and large in volume, and the major raw materials (sand, soda ash and limestone) are generally readily available, therefore production facilities need to be located close to their markets. A typical glass furnace holds hundreds of tonnes of molten glass, and so it is simply not practical to shut it down every night, or in fact in any period short of a month. Factories therefore run 24 hours a day 7 days a week. This means that there is little opportunity to either increase or decrease production rates by more than a few percent. New furnaces and forming machines cost tens of millions of dollars and require at least 18 months of planning. Given this fact, and the fact that there are usually more products than machine lines means that products are sold from stock. The marketing/production challenge is therefore to be able to predict demand both in the short 4–12 week term and over the 24–48 month long term. Factories are generally sized to service the requirements of a city; in developed countries there is usually a factory per 1–2 million people. A typical factory will produce 1–3 million containers a day. Despite its positioning as a mature market product, glass does enjoy a high level of consumer acceptance and is perceived as a "premium" quality packaging format.

Lifecycle Impact

Glass containers are wholly recyclable and the glass industries in many countries retain a policy, sometimes required by government regulations, to maintain a high price on cullet to ensure high return rates. Return rates of 95% are not uncommon in the Nordic countries (Sweden, Norway, Denmark and Finland). Return rates of less than 50% are usual in other countries. Of course glass containers can also be reused, and in developing countries this is common, however the environmental impact of washing the container as against remelting them is uncertain. Factors to consider here are the chemicals and fresh water used in the washing, and the fact that a single-use container can be made much lighter, using less than half the glass (and therefore energy content) of a multiuse container. Also, a significant factor in the developed world's consideration of reuse are producer concerns over the risk and consequential product liability of using a component (the reused container) of unknown and unqualified safety. How glass containers compare to other packaging types (plastic, cardboard, aluminium) is hard to say; conclusive lifecycle studies are yet to be produced.

Float Glass Process

Float glass is a sheet of glass made by floating molten glass on a bed of molten metal, typically tin, although lead and various low melting point alloys were used in the past. This method gives the sheet uniform thickness and very flat surfaces. Modern windows are made from float glass. Most float glass is soda-lime glass, but relatively minor quantities of specialty borosilicate and flat panel display glass are also produced using the float glass process. The float glass process is also known as the Pilkington process, named after the British glass manufacturer Pilkington, who pioneered the technique (invented by Sir Alastair Pilkington) in the 1950s.

Environmental Impacts

Local Impacts

As with all highly concentrated industries, glassworks suffer from moderately high local environmental impacts. Compounding this is that because they are mature market businesses, they often have been located on the same site for a long time and this has resulted in residential encroachment. The main impacts on residential housing and cities are noise, fresh water use, water pollution, NOx and SOx air pollution, and dust.

Noise is created by the forming machines. Operated by compressed air, they can produce noise levels of up to 106dBA. How this noise is carried into the local neighborhood depends heavily on the layout of the factory. Another factor in noise production is truck movements. A typical factory will process 600T of material a day. This means that some 600T of raw material has to come onto the site and the same off the site again as finished product.

Water is used to cool the furnace, compressor and unused molten glass. Water use in factories varies widely; it can be as little as one tonne water used per melted tonne of glass. Of the one tonne, roughly half is evaporated to provide cooling, the rest forms a wastewater stream.

Most factories use water containing an emulsified oil to cool and lubricate the *gob* cutting *shear blades*. This oil-laden water mixes with the water outflow stream, thus polluting it. Factories usually have some kind of water processing equipment that removes this emulsified oil to various degrees of effectiveness.

The oxides of nitrogen are a natural product of the burning of gas in air and are produced in large quantities by gas-fired furnaces. Some factories in cities with particular air pollution problems will mitigate this by using liquid oxygen, however the logic of this given the cost in carbon of (1) not using regenerators and (2) having to liquefy and transport oxygen is highly questionable. The oxides of sulfur are produced as a result of the glass melting process. Manipulating the batch formula can effect some limited mitigation of this; alternatively exhaust plume scrubbing can be used.

The raw materials for glass-making are all dusty material and are delivered either as a powder or as a fine-grained material. Systems for controlling dusty materials tend to be difficult to maintain, and given the large amounts of material moved each day, only a small amount has to escape for there to be a dust problem. *Cullet* is also moved about in a glass factory and tends to produce fine glass particles when shovelled or broken.

Glassblowing

Glassblowing is a glassforming technique that involves inflating molten glass into a bubble (or parison), with the aid of a blowpipe (or blow tube). A person who blows glass is called a glassblower, glassmith, or gaffer. A lampworker manipulates glass with the use of a torch on a smaller scale, such as in producing precision laboratory glassware out of borosilicate glass.

Glassblower Romeo Lefebvre in his workshop in Montreal, 1942

A stage in the manufacture of a Bristol blue glass ship's decanter.
The blowpipe is being held in the glassblower's left hand. The glass is glowing yellow.

Technology

Principles

As a novel glass forming technique created in the middle of the last century BC, glassblowing exploited a working property of glass which was previously unknown to the glassworkers: inflation. Inflation refers to the expansion of a molten blob of glass by introducing a small amount of air to it. This property is based on the liquid structure of glass where the atoms are held together by strong chemical bonds in a disordered and random network, therefore molten glass is viscous enough to be blown and gradually hardens as it loses heat.

In order to increase the stiffness of the molten glass, which in turn facilitates the process of blowing, there was a subtle change in the composition of glass. With reference to their studies of the ancient glass assemblages from Sepphoris of Israel, Fischer and McCray postulated that the concentration of natron, which acts as flux in glass, is slightly lower in blown vessels than those manufactured by casting. Lower concentration of natron would have allowed the glass to be stiffer for blowing.

During blowing, thinner layers of glass cool faster than thicker ones and become more viscous than the thicker layers. This effect allows production of blown glass with uniform thickness, instead of causing blow-through of the thinned layers.

A full range of glassblowing techniques was developed within decades of its invention. The two major methods of glassblowing are free-blowing and mold-blowing.

Free-blowing

This method held a pre-eminent position in glassforming ever since its introduction in the middle of the 1st century BC until the late 19th century, and is still widely used nowadays as a glassforming technique, especially for artistic purposes. The process of free-blowing involves the blowing of short puffs of air into a molten portion of glass called a '"gather" which has been spooled at one end of the blowpipe. This has the effect of forming an elastic skin on the interior of the glass blob that matches the exterior skin caused by the removal of heat from the furnace. The glassworker can then quickly inflate the molten glass to a coherent blob and work it into a desired shape.

Researchers at the Toledo Museum of Art attempted to reconstruct the ancient free-blowing technique by using clay blowpipes. The result proved that short clay blowpipes of about 30–60 cm (12–24 in) facilitate free-blowing because they are simple to handle, easy to manipulate and can be re-used several times. Skilled workers are capable of shaping almost any vessel forms by rotating the pipe, swinging it and controlling the temperature of the piece while they blow. They can produce a great variety of glass objects, ranging from drinking cups to window glass.

An outstanding example of the free-blowing technique is the Portland Vase, which is a cameo manufactured during the Roman period. An experiment was carried out by Gudenrath and Whitehouse with the aim of re-creating the Portland Vase. A full amount of blue glass required for the body of the vase was gathered on the end of the blowpipe and was subsequently dipped into a pot of hot white glass. Inflation occurred when the glassworker blew the molten glass into a sphere which was then stretched or elongated into a vase with a layer of white glass overlying the blue body.

Mold-blowing

Glassblower Jean-Pierre Canlis sculpting a section of his piece "Insignificance"

Mold-blowing was an alternate glassblowing method that came after the invention of free-blowing, during the first part of the second quarter of the 1st century AD. A glob of molten glass is placed on the end of the blowpipe, and is then inflated into a wooden or metal carved mold. In this way, the shape and the texture of the bubble of glass is determined by the design on the interior of the mold rather than the skill of the glassworker.

Two types of molds, namely single-piece mold and multi-piece mold, are frequently used to produce mold-blown vessels. The former allows the finished glass object to be removed in one movement by pulling it upwards from the single-piece mold and is largely employed to produce tableware and utilitarian vessels for storage and transportation. Whereas the latter is made in multi-paneled mold segments that join together, thus permitting the development of more sophisticated surface modeling, texture and design.

The Roman leaf beaker which is now on display in the J. Paul Getty Museum was blown in a three-part mold decorated with the foliage relief frieze of four vertical plants. Meanwhile, Taylor and Hill tried to reproduce mold-blown vessels by using three-part molds made of different materials. The result suggested that metal, in particular bronze, molds are more effective in producing high-relief design on glass than plaster molds and wooden molds.

The development of the mold-blowing technique has enabled the speedy production of glass objects in large quantity, thus encouraging the mass production and widespread distribution of glass objects.

Modern Glassblowing

Use of a glory hole to reheat a piece on the end of a blowpipe

Glass created from complex murrine and zanfirico cane can possess a great deal of detail.

Glass can be made with precise striped patterns through a process called cane which involves the use of rods of colored glass.

The transformation of raw materials into glass takes place around 2,400 °F (1,320 °C); the glass emits enough heat to appear almost white hot. The glass is then left to "fine out" (allowing the bubbles to rise out of the mass), and then the working temperature is reduced in the furnace to around 2,000 °F (1,090 °C). At this stage, the glass appears to be a bright orange color. Though most glassblowing is done between 1,600 and 1,900 °F (870 and 1,040 °C), "soda-lime" glass remains somewhat plastic and workable as low as 1,350 °F (730 °C). Annealing is usually done between 700 and 900 °F (371 and 482 °C).

Glassblowing involves three furnaces. The first, which contains a crucible of molten glass, is simply referred to as the furnace. The second is called the glory hole, and is used to reheat a piece in between steps of working with it. The final furnace is called the lehr or annealer, and is used to slowly cool the glass, over a period of a few hours to a few days, depending on the size of the pieces. This keeps the glass from cracking or shattering due to thermal stress. Historically, all three furnaces were contained in one structure, with a set of progressively cooler chambers for each of the three purposes. Many glassblowing studios in Mexico and South America still employ this method.

The major tools used by a glassblower are the blowpipe (or blow tube), punty (or punty rod, pontil, or mandrel), bench, marver, blocks, jacks, paddles, tweezers, paper, and a variety of shears.

The tip of the blowpipe is first preheated; then dipped in the molten glass in the furnace. The molten glass is "gathered" onto the end of the blowpipe in much the same way that viscous honey is picked up on a honey dipper. This glass is then rolled on the marver, which was traditionally a flat slab of marble, but today is more commonly a fairly thick flat sheet of steel. This process, called marvering, forms a cool skin on the exterior of the molten glass blob, and shapes it. Then air is blown into the pipe, creating a bubble. Next, the glassworker can gather more glass over that bubble to create a larger piece. Once a piece has been blown to its approximate final size, the bottom is finalized. Then, the molten glass is attached to a stainless steel or iron rod called a punty for shaping and transferring the hollow piece from the blowpipe to provide an opening and/or to finalize the top.

The bench is a glassblower's workstation, and has a place for the glassblower to sit, a place for the handheld tools, and two rails that the pipe or punty rides on while the blower works with the piece. Blocks are ladle-like tools made from water-soaked fruitwood, and are used similarly to the marver to shape and cool a piece in the early steps of creation. Jacks are tools shaped somewhat like large tweezers with two blades, which are used for forming shape later in the creation of a piece. Paddles are flat pieces of wood or graphite used for creating flat spots such as a bottom. Tweezers are used to pick out details or to pull on the glass. There are two important types of shears, straight shears and diamond shears. Straight shears are essentially bulky scissors, used for making linear cuts. Diamond shears have blades that form a diamond shape when partially open. These are used for cutting off masses of glass.

There are many ways to apply patterns and color to blown glass, including rolling molten glass in powdered color or larger pieces of colored glass called frit. Complex patterns with great detail can be created through the use of cane (rods of colored glass) and murrine (rods cut in cross-sections to reveal patterns). These pieces of color can be arranged in a pattern on a flat surface, and then "picked up" by rolling a bubble of molten glass over them. One of the most exacting and complicated caneworking techniques is "reticello", which involves creating two bubbles from cane, each twisted in a different direction and then combining them and blowing out the final form.

A lampworker, usually operating on a much smaller scale, historically used alcohol lamps and breath or bellows-driven air to create a hot flame at a workbench to manipulate preformed glass rods and tubes. These stock materials took form as laboratory glassware, beads, and durable scientific "specimens"—miniature glass sculpture. The craft, which was raised to an art form in the late 1960s by Hans Godo Frabel (later followed by lampwork artists such as Milon Townsend and Robert Mickelson), is still practised today. The modern lampworker uses a flame of oxygen and propane or natural gas. The modern torch permits working both the soft glass from the furnace worker and the borosilicate glass (low-expansion) of the scientific glassblower. This latter worker may also have multiple headed torches and special lathes to help form the glass or fused quartz used for special projects.

History

Origins

The earliest evidence of glassblowing was found by Roman Ghirshman in Chogha Zanbil, where many glass bottles were found in the excavations of the 2nd millennium BC site. Later evidence comes from a collection of waste from a glass shop, including fragments of glass tubes, glass rods and tiny blown bottles, which was dumped in a mikvah, a ritual bath in the Jewish Quarter of the Old City of Jerusalem, dated from 37 to 4 BC. Some of the glass tubes recovered are fire-closed at one end and are partially inflated by blowing through the open end while still hot to form a small bottle; thus they are considered as a rudimentary form of blowpipe.

Hence, tube blowing not only represents the initial attempts of experimentation by glassworkers at blowing glass, it is also a revolutionary step that induced a change in conception and a deep understanding of glass. Such inventions swiftly eclipsed all other traditional methods, such as casting and core-forming, in working glass.

In the Roman Empire

Roman blown glass hydria from Baelo Claudia (4th century AD)

A glassworks in England in 1858. During the Industrial Revolution, techniques for mass-produced glassware were improved.

Glassblowing production methods in England in 1858

The invention of glassblowing coincided with the establishment of the Roman Empire in the 1st century BC, which enhanced the spread and dominance of this new technology. Glassblowing was greatly supported by the Roman government (although Roman citizens could not be "in trade", in particular under the reign of Augustus), and glass was being blown in many areas of the Roman world. On the eastern borders of the Empire, the first large glass workshops were set up by the Phoenicians in the birthplace of glassblowing in contemporary Lebanon and Israel as well as in the neighbouring province of Cyprus.

Ennion for example, was among one of the most prominent glassworkers from Lebanon of the time. He was renowned for producing the multi-paneled mold-blown glass vessels that were complex in their shapes, arrangement and decorative motifs. The complexity of designs of these mold-blown glass vessels illustrated the sophistication of the glassworkers in the eastern regions of the Roman Empire. Mold-blown glass vessels manufactured by the workshops of Ennion and other contemporary glassworkers such as Jason, Nikon, Aristeas, and Meges, constitutes some of the earliest evidence of glassblowing found in the eastern territories.

Eventually, the glassblowing technique reached Egypt and was described in a fragmentary poem printed on papyrus which was dated to 3rd century AD. The Roman hegemony over the Mediterranean areas resulted in the substitution of glassblowing for earlier Hellenistic casting, core-forming and mosaic fusion techniques. The earliest evidence of blowing in Hellenistic work consists of small blown bottles for perfume and oil retrieved from the glass workshops on the Greek island of Samothrace and at Corinth in mainland Greece which were dated to the 1st century AD.

Later, the Phoenician glassworkers exploited their glassblowing techniques and set up their workshops in the western territories of the Roman Empire, first in Italy by the middle of the 1st century AD. Rome, the heartland of the Empire, soon became a major glassblowing center, and more glassblowing workshops were subsequently established in other provinces of Italy, for example Campania, Morgantina and Aquileia. A great variety of blown glass objects, ranging from unguentaria (toiletry containers for perfume) to cameo, from tableware to window glass, were produced.

From there, escaping craftsmen (who had been forbidden to travel) otherwise advanced to the rest of Europe by building their glassblowing workshops in the north of the Alps (which is now Switzerland), and then at sites in northern Europe in present-day France and Belgium.

One of the most prolific glassblowing centers of the Roman period was established in Cologne on the river Rhine in Germany by late 1st century BC. Stone base molds and terracotta base molds were discovered from these Rhineland workshops, suggesting the adoption and the application of mold-blowing technique by the glassworkers. Besides, blown flagons and blown jars decorated with ribbing, as well as blown perfume bottles with letters CCAA or CCA which stand for Colonia Claudia Agrippiniensis, were produced from the Rhineland workshops. Remains of blown blue-green glass vessels, for example bottles with a handle, collared bowls and indented beakers, were found in abundance from the local glass workshops at Poetovio and Celeia in Slovenia.

Surviving physical evidence, such as blowpipes and molds which are indicative of the presence of blowing, is fragmentary and limited. Pieces of clay blowpipes were retrieved from the late 1st century AD glass workshop at Avenches in Switzerland. Clay blowpipes, also known as mouthblowers, were made by the ancient glassworkers due to the accessibility and availability of the resources before the introduction of the metal blowpipes. Hollow iron rods, together with blown vessel fragments and glass waste dating to approximately 4th century AD, were recovered from the glass workshop in Mérida of Spain, as well as in Salona in Croatia.

Middle Ages

The glass blowing tradition was carried on in Europe from the medieval period through the Middle Ages to the Renaissance in the demise of the Roman Empire in the 5th century AD. During the early medieval period, the Franks manipulated the technique of glassblowing by creating the simple corrugated molds and developing the claws decoration techniques. Blown glass objects, such as the drinking vessels that imitated the shape of the animal horn were produced in the Rhine and Meuse valleys, as well as in Belgium. The Byzantine glassworkers made mold-blown glass decorated with Christian and Jewish symbols in Jerusalem between late 6th century and the middle of the 7th century AD. Mold-blown vessels with facets, relief and linear-cut decoration were discovered at Samarra in the Islamic lands.

Renaissance Europe witnessed the revitalization of glass industry in Italy. Glassblowing, in particular the mold-blowing technique, was employed by the Venetian glassworkers from Murano to produce the fine glassware which is also known as cristallo. The technique of glassblowing, coupled with the cylinder and crown methods, was used to manufacture sheet or flat glass for window panes in the late 17th century. The applicability of glassblowing was so widespread that glass was being blown in many parts of the world, for example, in China, Japan and the Islamic Lands.

The Nøstetangen Museum at Hokksund, Norway shows how glass was made according to ancient tradition. The Nøstetangenglassworks had operated there from 1741 to 1777, producing table-glass and chandeliers in the German and English style.

Industrial Revolution

Recent Developments

The "studio glass movement" began in 1962 when Harvey Littleton, a ceramics professor, and Dominick Labino, a chemist and engineer, held two workshops at the Toledo Museum of Art, during which they started experimenting with melting glass in a small furnace and creating blown glass art. Littleton promoted the use of small furnaces in individual artists studios(). This approach to glassblowing blossomed into a worldwide movement, producing such flamboyant and prolific artists as Dale Chihuly, Dante Marioni, Fritz Driesbach and Marvin Lipofsky as well as scores of other modern glass artists. Today there are many different institutions around the world that offer glassmaking resources for training and sharing equipment.

Working with large or complex pieces requires a team of several glassworkers, in a complex choreography of precisely timed movements. This practical requirement has encouraged collaboration among glass artists, in both semi-permanent and temporary working groups.

In Literature

The writer Daphne du Maurier was descended from a family of glass-blowers in 18th century France, and she wrote about her forbears in the 1963 historical novel *The Glass-Blowers*.

The subject of mystery novelist Donna Leon's *Through a Glass, Darkly* is the investigation of a crime in a Venetian glassworks on the island of Murano.

Cylinder Blown Sheet Glass

Cylinder blown sheet is a type of hand-blown window glass. It is created with a similar process to broad sheet, but larger cylinders are produced by swinging the cylinder in a trench. The glass is then allowed to cool before the cylinder is cut. The glass is then re-heated and flattened. The result is much larger panes and improved surface quality over broad sheet.

In this manufacturing process glass is blown into a cylindrical iron mould. The ends are cut off and a cut is made down the side of the cylinder. The cut cylinder is then placed in an oven where the cylinder unrolls into a flat glass sheet. William J. Blenko used this method in the early 1900s to make stained glass. These imperfect panes have led to the misconception that glass is actually a high-viscosity fluid at room temperature, which is not the case.

Other methods of producing hand-blown window glass included broad sheet, blown plate, crown glass and polished plate. These methods of manufacture lasted at least until the end of the 19th century. The early 20th century marks the move away from hand-blown to machine manufactured glass such as rolled plate, machine drawn cylinder sheet, the Fourcault process of flat drawn sheet, single and twin ground polished plate and most common, float glass.

Cylinder blown sheet glass was manufactured in the UK in the mid 19th century. It had been manufactured in France and Germany (and imported to the UK) since the 18th century.

Machine Drawn Cylinder Sheet Glass

Machine drawn cylinder sheet was the first mechanical method for "drawing" window glass. Cylinders of glass 40 feet (12 m) high are drawn vertically from a circular tank. The glass is then annealed and cut into 7 to 10 foot (2 to 3 m) cylinders. These are cut lengthways, reheated, and flattened.

This process was invented in the USA in 1903. This type of glass was manufactured in the early 20th century (it was manufactured in the UK by Pilkingtons from 1910 to 1933).

Other historical methods for making window glass included broad sheet, blown plate, crown glass, polished plate and cylinder blown sheet. These methods of manufacture lasted at least until the end of the 19th century. The early 20th century marks the move away from hand-blown to machine manufactured glass such as rolled plate, flat drawn sheet, single and twin ground polished plate and float glass.

Silicon Dioxide

FSilicon dioxide, also known as silica (from the Latin *silex*), is a chemical compound that is an oxide of silicon with the chemical formula SiO_2. It has been known since ancient times. Silica is most commonly found in nature as quartz, as well as in various living organisms. In many parts of the world, silica is the major constituent of sand. Silica is one of the most complex and most abundant families of materials, existing both as several minerals and being produced synthetically. Notable examples include fused quartz, crystal, fumed silica, silica gel, and aerogels. Applications range from structural materials to microelectronics to components used in the food industry.

Production

Silicon dioxide is mostly obtained by mining and purification of quartz. Quartz comprises more than 10% by mass of the earth's crust. This product is suitable for many purposes while for others chemical processing is required to make a purer or otherwise more suitable (e.g. more reactive or fine-grained) product.

Fumed Silica

Pyrogenic silica (sometimes called fumed silica or silica fume) is a very fine particulate or colloidal form of silicon dioxide. It is prepared by burning $SiCl_4$ in an oxygen rich hydrocarbon flame to produce a "smoke" of SiO_2.

$$SiCl_4 + 2\ H_2 + O_2 \rightarrow SiO_2 + 4\ HCl.$$

Silica Fume

This product is obtained as byproduct from hot processes like ferro-silicon production. It is less pure than fumed silica and should not be confused with that product. The production process, particle characteristics and fields of application of fumed silica are all different from those of silica fume.

Precipitated Silica

Amorphous silica, silica gel, is produced by the acidification of solutions of sodium silicate. The gelatinous precipitate is first washed and then dehydrated to produce colorless microporous silica. Idealized equation involving a trisilicate and sulfuric acid is shown:

$$Na_2Si_3O_7 + H_2SO_4 \rightarrow 3\ SiO_2 + Na_2SO_4 + H_2O$$

Approximately one billion kilograms/year (1999) of silica was produced in this manner, mainly for use for polymer composites – tires and shoe soles.

On Microchips

Thin films of silica grow spontaneously on silicon wafers via thermal oxidation. This route gives a very shallow layer (approximately 1 nm or 10 Å) of so-called native oxide. Higher temperatures and alternative environments are used to grow well-controlled layers of silicon dioxide on silicon, for example at temperatures between 600 and 1200 °C, using so-called dry or wet oxidation with O_2 or H_2O, respectively. The depth of the layer of silicon replaced by the dioxide is 44% of the depth of the silicon dioxide layer produced.

$$Si + O_2 \rightarrow SiO_2$$

or H_2O, respectively.

$$Si + 2\ H_2O \rightarrow SiO_2 + 2\ H_2$$

The depth of the layer of silicon replaced by the dioxide is 44% of the depth of the silicon dioxide layer produced.

The native oxide layer can be beneficial in microelectronics, where it acts as electric insulators with high chemical stability. In electrical applications, it can protect the silicon, store charge, block current, and even act as a controlled pathway to limit current flow.

Laboratory or Special Methods

From Silicate Esters

Many routes to silicon dioxide start with silicate esters, the best known being tetraethyl orthosilicate (TEOS). Simply heating TEOS at 680–730 °C gives the dioxide:

$$Si(OC_2H_5)_4 \rightarrow SiO_2 + 2\ O(C_2H_5)_2$$

Similarly TEOS combusts around 400 °C:

$$Si(OC_2H_5)_4 + 12\ O_2 \rightarrow SiO_2 + 10\ H_2O + 8\ CO_2$$

TEOS undergoes hydrolysis via the so-called sol-gel process. The course of the reaction and nature of the product are affected by catalysts, but the idealized equation is:

$$Si(OC_2H_5)_4 + 2\ H_2O \rightarrow SiO_2 + 4\ HOCH_2CH_3$$

Other Methods

Being highly stable, silicon dioxide arises from many methods. Conceptually simple, but of little practical value, combustion of silane gives silicon dioxide. This reaction is analogous to the combustion of methane:

$$SiH_4 + 2\ O_2 \rightarrow SiO_2 + 2\ H_2O.$$

Uses

An estimated 95% of silicon dioxide produced is consumed in the construction industry, e.g. for the production of Portland cement. Other major applications are listed below.

Precursor to Glass and Silicon

Silica is used primarily in the production of glass for windows, drinking glasses, beverage bottles, and many other uses. The majority of optical fibers for telecommunication are also made from silica. It is a primary raw material for many ceramics such as earthenware, stoneware, and porcelain.

Silicon dioxide is used to produce elemental silicon. The process involves carbothermic reduction in an electric arc furnace:

$$SiO_2 + 2\ C \rightarrow Si + 2\ CO$$

Major Component Used in Sand Casting

Silica, in the form of sand is used as the main ingredient in sand casting for the manufacture of a large number of metallic components in engineering and other applications. The high melting point of silica enables it to be used in such applications.

Food and Pharmaceutical Applications

Silica is a common additive in the production of foods, where it is used primarily as a flow agent in powdered foods, or to adsorb water in hygroscopic applications. It is the primary component of diatomaceous earth. Colloidal silica is also used as a wine, beer, and juice fining agent.

In pharmaceutical products, silica aids powder flow when tablets are formed.

Other

A silica-based aerogel was used in the Stardust spacecraft to collect extraterrestrial particles. Silica is also used in the extraction of DNA and RNA due to its ability to bind to the nucleic acids under the presence of chaotropes. Hydrophobic silica is used as a defoamer component. In hydrated form, it is used in toothpaste as a hard abrasive to remove tooth plaque.

In its capacity as a refractory, it is useful in fiber form as a high-temperature thermal protection fabric. In cosmetics, it is useful for its light-diffusing properties and natural absorbency. It is also used as a thermal enhancement compound in ground source heat pump industry.

Structure

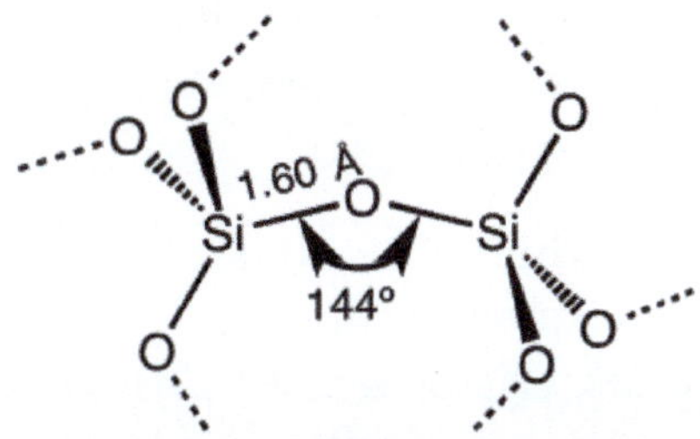

Structural motif found in α-quartz, but also found in almost all forms of silicon dioxide.

In the majority of silicates, the Si atom shows tetrahedral coordination, with 4 oxygen atoms surrounding a central Si atom. The most common example is seen in the quartz crystalline form of silica SiO_2. In each of the most thermodynamically stable crystalline forms of silica, on average, all 4 of the vertices (or oxygen atoms) of the SiO_4 tetrahedra are shared with others, yielding the net chemical formula: SiO_2.

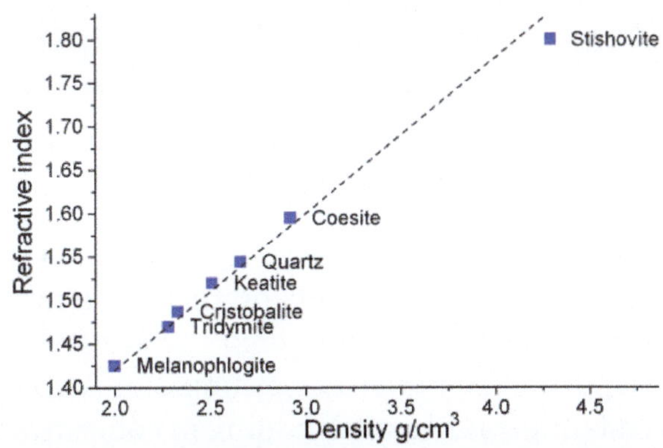

Relation between refractive index and density for some SiO_2 forms.

For example, in the unit cell of α-quartz, the central tetrahedron shares all 4 of its corner O atoms, the 2 face-centered tetrahedra share 2 of their corner O atoms, and the 4 edge-centered tetrahedra share just one of their O atoms with other SiO_4 tetrahedra. This leaves a net average of 12 out of 24 total vertices for that portion of the 7 SiO_4 tetrahedra that are considered to be a part of the unit cell for silica.

SiO_2 has a number of distinct crystalline forms (polymorphs) in addition to amorphous forms. With the exception of stishovite and fibrous silica, all of the crystalline forms involve tetrahedral

SiO_4 units linked together by shared vertices in different arrangements. Silicon–oxygen bond lengths vary between the different crystal forms, for example in α-quartz the bond length is 161 pm, whereas in α-tridymite it is in the range 154–171 pm. The Si-O-Si angle also varies between a low value of 140° in α-tridymite, up to 180° in β-tridymite. In α-quartz the Si-O-Si angle is 144°.

Fibrous silica has a structure similar to that of SiS_2 with chains of edge-sharing SiO_4 tetrahedra. Stishovite, the higher-pressure form, in contrast has a rutile-like structure where silicon is 6-coordinate. The density of stishovite is 4.287 g/cm³, which compares to α-quartz, the densest of the low-pressure forms, which has a density of 2.648 g/cm³. The difference in density can be ascribed to the increase in coordination as the six shortest Si-O bond lengths in stishovite (four Si-O bond lengths of 176 pm and two others of 181 pm) are greater than the Si-O bond length (161 pm) in α-quartz. The change in the coordination increases the ionicity of the Si-O bond. But more important is the observation that any deviations from these standard parameters constitute microstructural differences or variations, which represent an approach to an amorphous, vitreous or glassy solid.

The only stable form under normal conditions is α-quartz and this is the form in which crystalline silicon dioxide is usually encountered. In nature impurities in crystalline α-quartz can give rise to colors. The high temperature minerals, cristobalite and tridymite, have both a lower density and index of refraction than quartz. Since the composition is identical, the reason for the discrepancies must be in the increased spacing in the high temperature minerals. As is common with many substances, the higher the temperature the farther apart the atoms due to the increased vibration energy.

The transformation from α-quartz to beta-quartz takes place abruptly at 573 °C. Since the transformation is accompanied by a significant change in volume it can easily induce fracturing of ceramics or rocks passing through this temperature limit.

The high-pressure minerals, seifertite, stishovite, and coesite, on the other hand, have a higher density and index of refraction when compared to quartz. This is probably due to the intense compression of the atoms that must occur during their formation, resulting in a more condensed structure.

Faujasite silica is another form of crystalline silica. It is obtained by dealumination of a low-sodium, ultra-stable Y zeolite with a combined acid and thermal treatment. The resulting product contains over 99% silica, has high crystallinity and high surface area (over 800 m²/g). Faujasite-silica has very high thermal and acid stability. For example, it maintains a high degree of long-range molecular order (or crystallinity) even after boiling in concentrated hydrochloric acid.

Molten silica exhibits several peculiar physical characteristics that are similar to the ones observed in liquid water: negative temperature expansion, density maximum (at temperatures ~5000 °C), and a heat capacity minimum. Its density decreases from 2.08 g/cm³ at 1950 °C to 2.03 g/cm³ at 2200 °C. When molecular silicon monoxide, SiO, is condensed in an argon matrix cooled with helium along with oxygen atoms generated by microwave discharge, molecular SiO_2 is produced with a linear structure. Dimeric silicon dioxide, $(SiO_2)_2$ has been prepared by reacting O_2 with matrix isolated dimeric silicon monoxide, (Si_2O_2). In dimeric silicon dioxide there are two oxygen atoms bridging between the silicon atoms with an Si-O-Si angle of 94°

and bond length of 164.6 pm and the terminal Si-O bond length is 150.2 pm. The Si-O bond length is 148.3 pm, which compares with the length of 161 pm in α-quartz. The bond energy is estimated at 621.7 kJ/mol.

Fused Quartz

When molten silicon dioxide SiO_2 is rapidly cooled, it does not crystallize but solidifies as a glass. The geometry of the silicon and oxygen centers in glass is similar to that in quartz and most other crystalline forms of the same composition, i.e., silicon is surrounded by a regular tetrahedra of oxygen centers. The difference between the glass and the crystalline forms arise from the connectivity of these tetrahedral units. Although there is no long range periodicity in the glassy network there remains significant ordering at length scales well beyond the SiO bond length. One example of this ordering is found in the preference of the network to form rings of 6-tetrahedra.

The glass transition temperature of pure SiO_2 is about 1475 K.

Chemical Reactions

Manufactured silica fume at maximum surface area of 380 m^2/g

Silica is converted to silicon by reduction with carbon.

Fluorine reacts with silicon dioxide to form SiF_4 and O_2 whereas the other halogen gases (Cl_2, Br_2, I_2) are essentially unreactive.

Silicon dioxide is attacked by hydrofluoric acid (HF) to produce hexafluorosilicic acid:

$$SiO_2 + 6\,HF \rightarrow H_2SiF_6 + 2\,H_2O.$$

HF is used to remove or pattern silicon dioxide in the semiconductor industry.

Silicon dioxide acts as a Lux-Flood acid, being able to react with bases under certain conditions. As it does not contain any hydrogen, it cannot act as a Brønsted–Lowry acid. While not soluble in water, some strong bases will react with glass and have to be stored in plastic bottles as a result.

Silicon dioxide dissolves in hot concentrated alkali or fused hydroxide, as described in this idealized equation:

$$SiO_2 + 2\ NaOH \rightarrow Na_2SiO_3 + H_2O.$$

Silicon dioxide will neutralise basic metal oxides (e.g. sodium oxide, potassium oxide, lead(II) oxide, zinc oxide, or mixtures of oxides, forming silicates and glasses as the Si-O-Si bonds in silica are broken successively). As an example the reaction of sodium oxide and SiO_2 can produce sodium orthosilicate, sodium silicate, and glasses, dependent on the proportions of reactants:

$$2\ Na_2O + SiO_2 \rightarrow Na_4SiO_4;$$

$$Na_2O + SiO_2 \rightarrow Na_2SiO_3;$$

$$(0.25–0.8)Na_2O + SiO_2 \rightarrow \text{glass}.$$

Examples of such glasses have commercial significance, e.g. soda-lime glass, borosilicate glass, lead glass. In these glasses, silica is termed the network former or lattice former. The reaction is also used in blast furnaces to remove sand impurities in the ore by neutralisation with calcium oxide, forming calcium silicate slag.

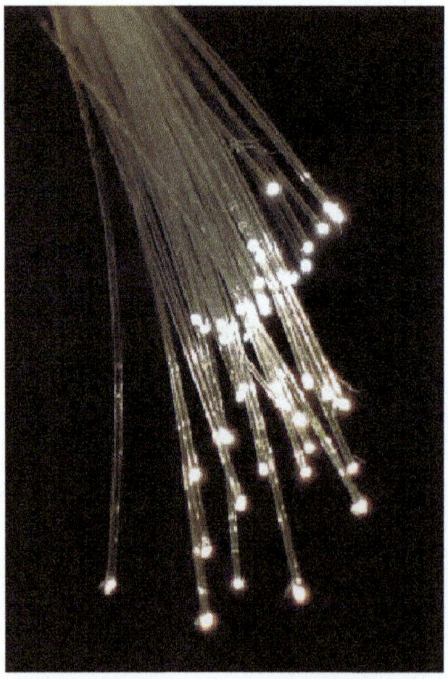

Bundle of optical fibers composed of high purity silica.

Silicon dioxide reacts in heated reflux under dinitrogen with ethylene glycol and an alkali metal base to produce highly reactive, pentacoordinate silicates which provide access to a wide variety of new silicon compounds. The silicates are essentially insoluble in all polar solvent except methanol.

Silicon dioxide reacts with elemental silicon at high temperatures to produce SiO:

$$SiO_2 + Si \rightarrow 2\ SiO$$

Solubility in Water

The solubility of silicon dioxide in water strongly depends on its crystalline form and is 3–4 times higher for silica than quartz; as a function of temperature, it peaks at about 340 °C. This property is used to grow single crystals of quartz in a hydrothermal process where natural quartz is dissolved in superheated water in a pressure vessel that is cooler at the top. Crystals of 0.5–1 kg can be grown over a period of 1–2 months. These crystals are a source of very pure quartz for use in electronic applications.

Occurrence

Biology

Even though it is poorly soluble, silica occurs widely in many plants. Plant materials with high silica phytolith content appear to be of importance to grazing animals, from chewing insects to ungulates. Studies have shown that it accelerates tooth wear, and high levels of silica in plants frequently eaten by herbivores may have developed as a defense mechanism against predation.

It is also the primary component of rice husk ash, which is used, for example, in filtration and cement manufacturing.

Silicification in and by cells has been common in the biological world for well over a billion years. In the modern world it occurs in bacteria, single-celled organisms, plants, and animals (invertebrates and vertebrates). Prominent examples include:

- Tests or frustules (i.e. shells) of diatoms, Radiolaria and testate amoebae.
- Silica phytoliths in the cells of many plants, including Equisetaceae, practically all grasses, and a wide range of dicotyledons.
- The spicules forming the skeleton of many sponges.

Crystalline minerals formed in the physiological environment often show exceptional physical properties (e.g., strength, hardness, fracture toughness) and tend to form hierarchical structures that exhibit microstructural order over a range of scales. The minerals are crystallized from an environment that is undersaturated with respect to silicon, and under conditions of neutral pH and low temperature (0–40 °C).

Formation of the mineral may occur either within the cell wall of an organism (such as with phytoliths), or outside the cell wall, as typically happens with tests. Specific biochemical reactions exist for mineral deposition. Such reactions include those that involve lipids, proteins, and carbohydrates.

It is unclear in what ways silica is important in the nutrition of animals. This field of research is challenging because silica is ubiquitous and in most circumstances dissolves in trace quantities only. All the same it certainly does occur in the living body, leaving us with the problem that it is hard to create proper silica-free controls for purposes of research. This makes it difficult to be sure when the silica present has had operative beneficial effects, and when its presence is coincidental, or even harmful. The current consensus is that it certainly seems important in the growth, strength, and management of many connective tissues. This is true not only for hard connective tissues such as bone and tooth but possibly in the biochemistry of the subcellular enzyme-containing structures as well.

Health Effects

Silica ingested orally is essentially nontoxic, with an LD_{50} of 5000 mg/kg (5 g/kg). On the other hand, inhaling finely divided crystalline silica dust can lead to silicosis, bronchitis, or cancer, as the dust becomes lodged in the lungs and continuously irritates them, reducing lung capacities. Studies of workers with exposure to crystalline silica have shown 10-fold higher than expected rates of lupus and other systemic autoimmune diseases compared to expected rates in the general population. Prior to new rules issued in 2013, OSHA allowed 100 μg per cubic meter of air. The new regulations reduce the amount to 50 μg/m³. The exposure limit for the construction industry is also set at 50 μg/m³ down from 250 μg/m³.

Quartz sand (silica) as main raw material for commercial glass production

In the body, crystalline silica particles do not dissolve over clinically relevant periods. Silica crystals inside the lungs can activate the NLRP3 inflammasome inside macrophages and dendritic cells and thereby result in processing of pro-Interleukin 1 beta into its mature form. Chronic exposure to silica may thereby account for some of its health hazards, as interleukin-1 is a highly pro-inflammatory cytokine in the immune system. This effect can create an occupational hazard for people working with sandblasting equipment, products that contain powdered crystalline silica and so on. Children, asthmatics of any age, allergy sufferers, and the elderly (all of whom have reduced lung capacity) can be affected in much less time. Amorphous silica, such as fumed silica is not associated with development of silicosis, but may cause irreversible lung damage in some cases. Laws restricting silica exposure *with respect to the silicosis hazard* specify that they are concerned only with silica that is both crystalline and dust-forming.

A study that followed subjects for 15 years found that higher levels of silica in water appeared to decrease the risk of dementia. The study found an association between an increase of 10 milligram-per-day of the intake of silica in drinking water with a decreased risk of dementia of 11%.

Crystalline silica is used in hydraulic fracturing of formation which contain tight oil and shale gas, a use which presents a health hazard to workers. In 2013 OSHA announced tightened restrictions on the amount of crystalline silica which could be present and required "green completion" of fracked wells to reduce exposure. Crystalline silica is an occupational hazard for those working with stone countertops, because the process of cutting and installing the countertops creates large amounts of airborne silica.

Crystalline Forms

SiO_2, more so than almost any material, exists in many crystalline forms (called polymorphs).

Crystalline forms of SiO_2				
Form	**Crystal symmetry Pearson symbol, group No.**	**ρ g/cm³**	**Notes**	**Structure**
α-quartz	rhombohedral (trigonal) hP9, $P3_121$ No.152	2.648	Helical chains making individual single crystals optically active; α-quartz converts to β-quartz at 846 K	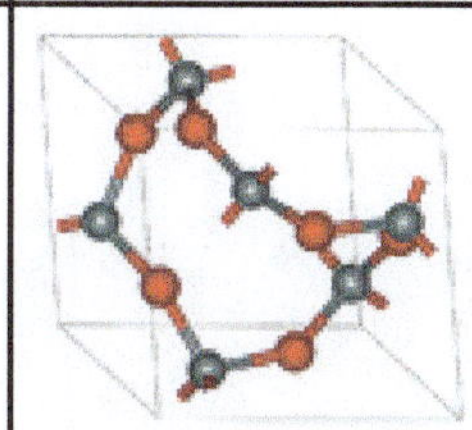
β-quartz	hexagonal hP18, $P6_222$, No. 180	2.533	Closely related to α-quartz (with an Si-O-Si angle of 155°) and optically active; β-quartz converts to β-tridymite at 1140 K	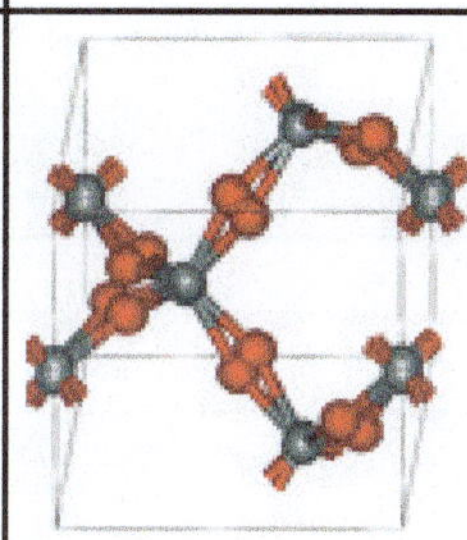
α-tridymite	orthorhombic oS24, $C222_1$, No.20	2.265	Metastable form under normal pressure	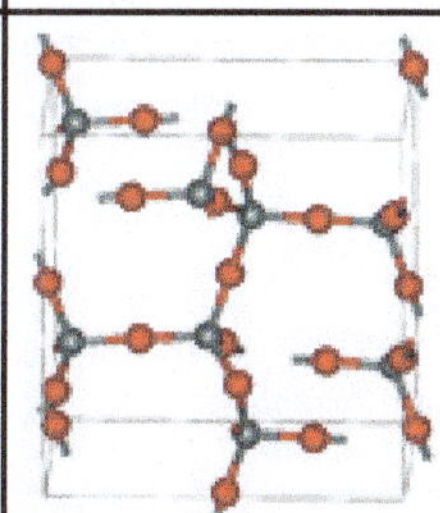
β-tridymite	hexagonal hP12, $P6_3/mmc$, No. 194		Closely related to α-tridymite; β-tridymite converts to β-cristobalite at 2010 K	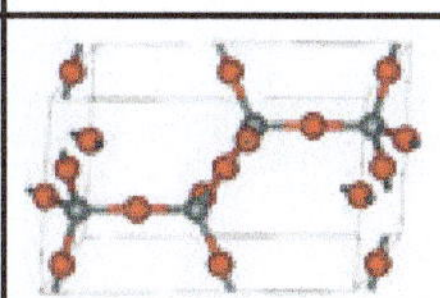
α-cristobalite	tetragonal tP12, $P4_12_12$, No. 92	2.334	Metastable form under normal pressure	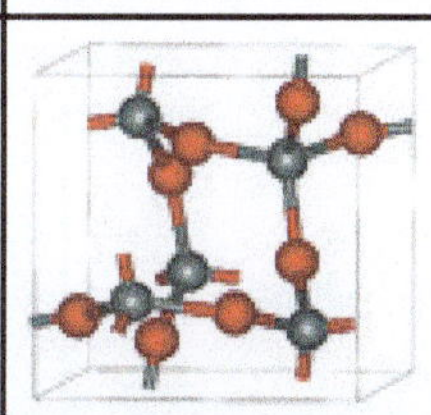

β-cristobalite	cubic cF104, Fd3m, No.227		Closely related to α-cristobalite; melts at 1978 K	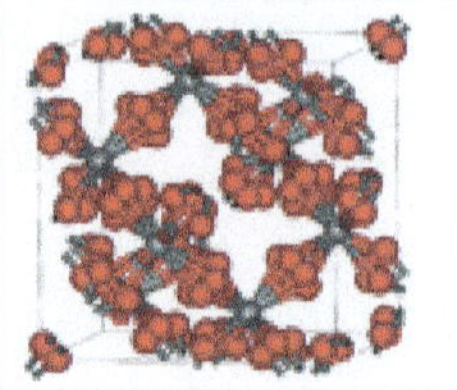
keatite	tetragonal tP36, $P4_12_12$, No. 92	3.011	Si_5O_{10}, Si_4O_{14}, Si_8O_{16} rings; synthesised from glassy silica and alkali at 600–900 K and 40–400 MPa	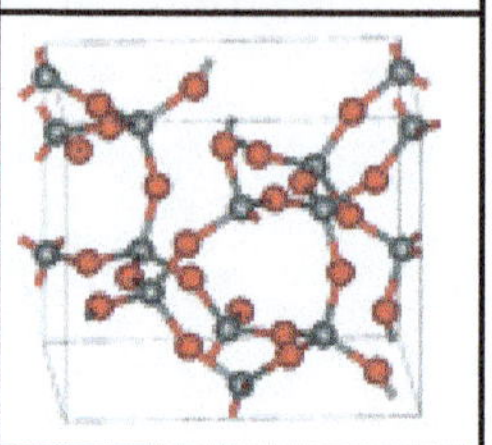
moganite	monoclinic mS46, C2/c, No.15		Si_4O_8 and Si_6O_{12} rings	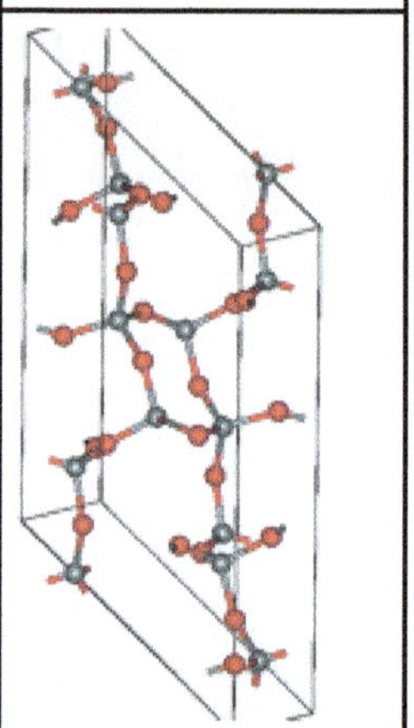
coesite	monoclinic mS48, C2/c, No.15	2.911	Si_4O_8 and Si_8O_{16} rings; 900 K and 3–3.5 GPa	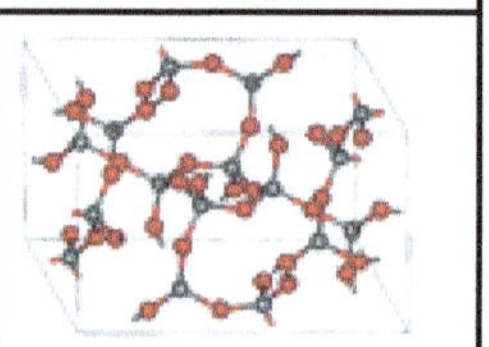
stishovite	Tetragonal tP6, $P4_2/mnm$, No.136	4.287	One of the densest (together with seifertite) polymorphs of silica; rutile-like with 6-fold coordinated Si; 7.5–8.5 GPa	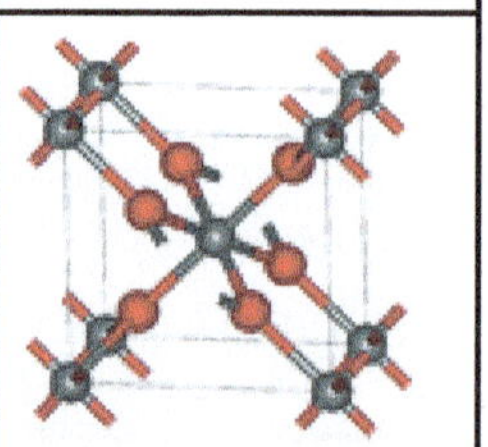
seifertite	orthorhombic oP, Pbcn	4.294	One of the densest (together with stishovite) polymorphs of silica; is produced at pressures above 40 GPa.	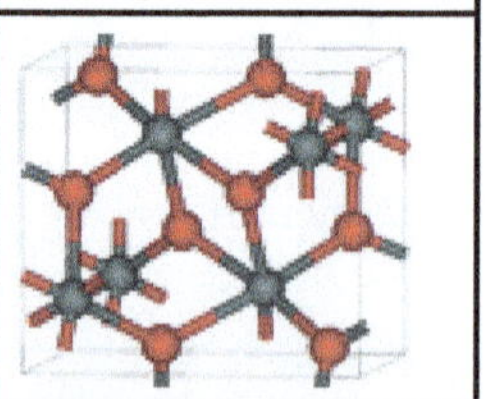
melanophlogite	cubic (cP*, $P4_232$, No.208) or tetragonal ($P4_2/nbc$)	2.04	Si_5O_{10}, Si_6O_{12} rings; mineral always found with hydrocarbons in interstitial spaces - a clathrasil	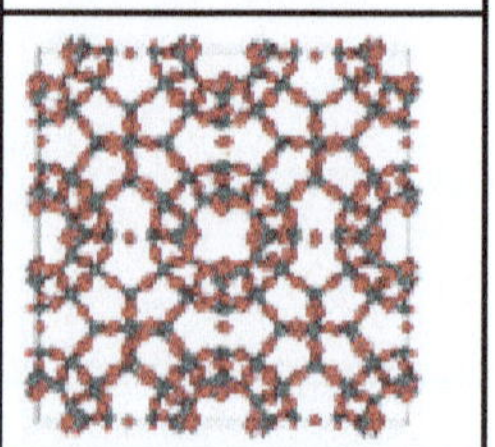

fibrous W-silica	orthorhombic oI12, Ibam, No.72	1.97	Like SiS_2 consisting of edge sharing chains, melts at ~1700 K	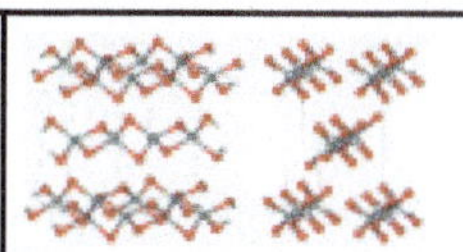
2D silica	hexagonal		Sheet-like bilayer structure	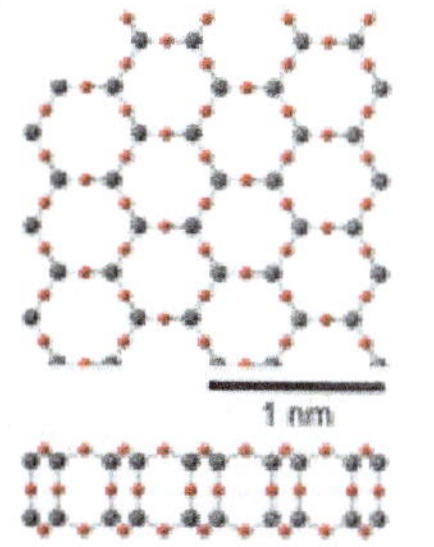

Polished Plate Glass

Polished plate is a type of hand-made glass. It is produced by casting glass onto a table and then subsequently grinding and polishing the glass. This was originally done by hand, and then later by machine. It was an expensive process requiring a large capital investment.

Other methods of producing hand-blown window glass included: broad sheet, blown plate, crown glass and cylinder blown sheet. These methods of manufacture lasted at least until the end of the 19th century. The early 20th century marks the move away from hand-blown to machine manufactured glass such as rolled plate, machine drawn cylinder sheet, flat drawn sheet, single and twin ground polished plate, and float glass.

The Frenchman, Louis Lucas de Nehou, in 1688, in conjunction with A. Thevart, succeeded in perfecting the process of casting plate-glass. Mirror plates previous to the invention had been made from blown "sheet" glass, and were consequently very limited in size. De Nehou's process of rolling molten glass poured on an iron table rendered the manufacture of very large plates possible.

In 1773 English polished plate (by the French process) was produced at Ravenhead.

By 1800 a steam engine was used to carry out the grinding and polishing of the cast glass.

Fourcault Process

The Fourcault Process is a method of manufacturing flat glass. First developed in Belgium by Émile Fourcault during the early 1900s, the process was used globally. Fourcault is an example of a "vertical draw" process, in that the glass is drawn against gravity in an upward direction. Gravity forces influence parts of the process.

Process

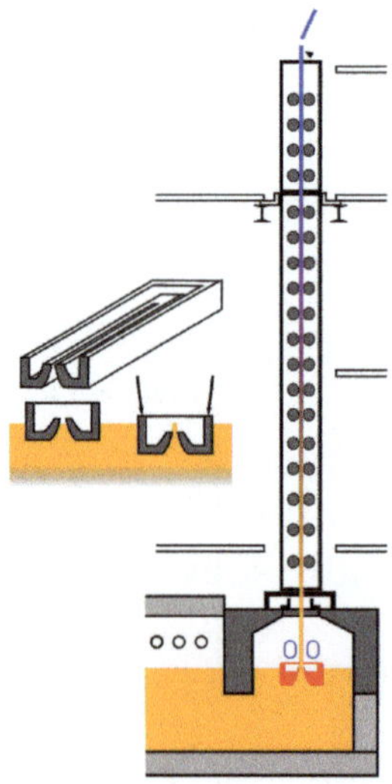

Fourcault drawing line with detail of the Debiteuse (red)

The Fourcault process requires a "pit" or drawing area and an assembly of machines to draw up the ribbon of glass while performing actions upon it that ensure desired quality and process yields. Today most glass manufacture has a "hot end" where the products are made. Fourcault is no exception.

The action in Fourcault happens "at the draw", or area where the glass is taken from a liquid state into the start of the process needed to make it into flat glass.

At the bottom of the draw is the "pit" or place where the molten glass is sufficiently cooled to be close to forming temperature. The cooling process uses a device known as a "canal". As the name describes, a canal is a box shaped structure which conveys the glass from the refining area to the pit.

The canal links the pit with the "refining" area, a section of the glass furnace that removes gas bubbles and other sources of imperfection. Since refining requires much higher temperatures to release gas bubbles than those required to form the glass it is not possible to draw directly from the refining area, hence the need for canals.

Forming

The Fourcault Process uses a ceramic die to shape fused (or molten) glass into a ribbon of rectangular cross section. The die, known as a Debiteuse, floats in the molten glass inside of the pit to a prescribed depth which slightly pushes a part of the molten glass slightly above the top surface of the die. A slot is cut through the center of the Debiteuse, which is shaped to produce the best quality of glass.

The Debiteuse is the starting point of the vertical draw, where the glass begins to change from a hot syrupy mass into useful flat glass. We will call the glass from the point of the Debiteuse until it is cut a "ribbon".

The base of the ribbon is shielded from heat radiation from the fused glass so that it continues to hold the shape imparted to it by the Debiteuse. This cooling preserves the rectangular cross section of the drawn glass by cooling the ribbon glass below the temperature where it would collapse into a column or break back into the melted glass. It is especially important to shield the outside edges of the ribbon from heat so that they are firmer and will hold the rest of the ribbon in a proper shape. In some cases manufacturers will allow the edges to form thicker "bulbs", which are removed after final cutting.

Immediately after being drawn the ribbon is cooled using mechanical coolers so that it maintains its rectangular shape in two dimensions, but assumes a ribbon like structure that extends down into the Debi and upwards into a drawing assembly. This mechanical cooling allows the ribbon to hold its integrity. In the author's experience the mechanical coolers used water, contained in specially shaped radiators, to remove heat radiated by the ribbon.

Sometimes a mild vacuum is applied to the ribbon in this early part of the process since mechanical cooling can induce air currents which impact upon surface quality.

Quenching

Some manufacturers also will apply sulfur dioxide gas during the draw in order to change the chemistry of the glass on the surface. By changing the chemistry it is possible to affect the surface characteristics of the glass, improving its quality and durability.

Glass rollers hold the ribbon throughout various parts of the process, supporting its weight and continuing the drawing process.

The process continues as the ribbon is drawn upwards into a chimney like structure, where it is quenched or rapidly cooled. When the ribbon reaches the end of the process it is scored, or cut, and then removed for further processing in discrete sheets of flat glass.

The "bulb edges" are recycled as cullet (flawed glass which is remelted) or were resold for shelving or displays. Sometimes flawed parts of the sheets were removed, leaving behind decent quality flat glass.

Operations

Time, speed and spacing of the various phases of the process are critical factors in the Fourcault process. Fourcault process machine operators require experience in order to judge placement of the die, location of various parts of the process, and rates of draw. These must be balanced against glass quality and the age of the draw.

As the draw continues the glass in the pit grows cooler and cooler, eventually leading to failures or diminished quality. The draw must be stopped, the pit must be "heated back" and then the process can continue anew.

Glass chemistry has a huge impact on the process since it controls the melting, forming and annealing temperatures, liquidus temperature (point where various chemicals that make up the glass start to crystallize out of the glass) and rates of change of characteristics of the glass itself.

Occasionally the ribbon will break or crack, leading to failure of the drawing process. Such breaks, known as "checks" can be alleviated by using proper operating parameters. Sometimes an expedient measure, using a portable source of heat, can be used to make the checks migrate to the edge of the ribbon where they disappear. The author has even seen crude torches made of wood which can migrate the checks.

The resultant product is a form of flat glass which is suitable for lesser quality uses. Due to process instabilities Fourcault process glass can have waves, seeds (small gas bubbles) or stones (undissolved materials).

In terms of economics and product quality the Fourcault process has been supplanted in many countries by the Pilkington developed "Float" process. The Float process lets the molten glass settle on top of a pool of liquid tin, so that gravity creates a flat sheet. Due to various chemistry and physical aspects of window glass the Pilkington Float process produces a vastly superior product.

Broad Sheet Glass

Among the Glass-Workers, an 1871 engraving by Harry Fenn

Broad sheet is a type of hand-blown glass. It was first made in Sussex in 1226.

Production

It is made by blowing molten glass into an elongated balloon shape with a blowpipe. Then, while the glass is still hot, the ends are cut off and the resulting cylinder is split with shears and flattened on an iron plate. The quality of broad sheet glass is not good, with many imperfections and mostly translucent. Due to the relatively small sizes blown, broad sheet was typically made into leadlights. The centerpiece was used for decoration in places where looking through the glass wasn't vital. If the piece was large, it was possible to see bubble tracks and strain lines.

Other methods for making hand-blown glass included blown plate glass, crown glass, polished plate glass and cylinder blown sheet glass. These methods of manufacture lasted at least until the end of the 19th century. The early 20th century marks the move away from hand-blown to machine manufactured glass such as rolled plate glass, machine drawn cylinder sheet glass, flat drawn sheet glass, single and twin ground polished plate glass and float glass.

Broad sheet glass was first made in the UK in Sussex in 1226. This glass was of poor quality and fairly opaque. Manufacture slowly decreased and ceased by the early 16th century. French glass-makers and others were making broad sheet glass earlier than this.

Blown Plate Glass

Blown plate is a hand-blown glass. There is a record of blown plate being produced in London in 1620.

Production

Blown plate was made by hand-grinding broad sheet glass. As the process was labour-intensive, and expensive, blown plate was mainly used for carriages and mirrors rather than in windows for buildings. Other methods for making hand-blown glass included: broad sheet, crown glass, polished plate and cylinder blown sheet. These methods of manufacture lasted at least until the end of the 19th century. The early 20th century marked the move away from hand-blown to machine manufactured glass such as rolled plate, machine drawn cylinder sheet, flat drawn sheet, single and twin ground polished plate and float glass.

Glazier

A glazier at work, 1946.

This Deutsche Bundespost postage stamp, issued in 1986, commemorates glaziers.

A glazier is a skilled tradesman responsible for cutting, installing, and removing glass (and materials used as substitutes for glass, such as some plastics). Glaziers may work with glass in various surfaces and settings, such as windows, doors, shower doors, skylights, storefronts, display cases, mirrors, facades, interior walls, ceilings, and tabletops.

Duties and Tools

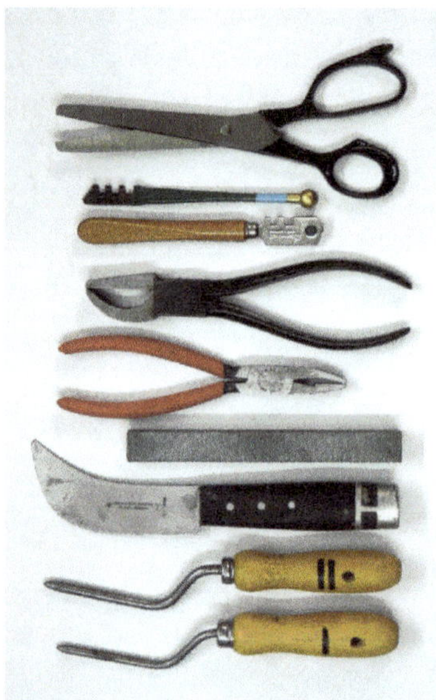
A set of glazier tools

The *Occupational Outlook Handbook* of the U.S. Department of Labor lists the following as typical tasks for a glazier:

- Follow blueprints or specifications
- Remove any old or broken glass before installing replacement glass
- Cut glass to the specified size and shape
- Make or install sashes or moldings for glass installation

- Fasten glass into sashes or frames with clips, moldings, or other types of fasteners
- Add weather seal or putty around pane edges to seal joints.

The National Occupational Analysis recognized by the Canadian Council of Directors of Apprenticeship separates the trade into 5 blocks of skills, each with a list of skills, and a list of tasks and subtasks a journeyman is expected to be able to accomplish:

Block A – Occupational Skills

1. Uses and maintains tools and equipment
2. Organizes work
3. Performs routine activities

Block B – Commercial Window and Door Systems

4. Fabricates commercial window and door systems
5. Installs commercial window and door systems

Block C – Residential Window and Door Systems

6. Installs residential window systems
7. Installs residential door systems

Block D – Specialty Glass and Products

8. Fabricates and installs specialty glass and products
9. Installs glass systems on vehicles

Block E – Servicing

10. Services commercial window and door systems
11. Services residential window and door systems
12. Services specialty glass and products.

Tools used by glaziers "include cutting boards, glass-cutting blades, straightedges, glazing knives, saws, drills, grinders, putty, and glazing compounds."

Some glaziers work specifically with glass in motor vehicles; other work specifically with the safety glass used in aircraft.

Education and Training

Glaziers are typically educated at the high school diploma or equivalent level and learn the skills of the trade through an apprenticeship program, which in the U.S. is typically four years.

In the U.S., apprenticeship programs are offered through the National Glass Association as well as trade associations and local contractors' associations. Construction-industry glaziers are frequently members of the International Union of Painters and Allied Trades.

In Ontario, Canada, apprenticeships are offered at the provincial level and certified through the Ontario College of Trades.

Other provinces manage their own apprenticeship programs.

The Trade of Glazier is a designated Red Seal Trade in Canada.

Occupational Hazards

Occupational hazards encountered by glaziers include the risks of being cut by glass or tools and falling from scaffolds or ladders. The use of heavy equipment may also cause injury: the National Institute for Occupational Safety and Health (NIOSH) reported in 1990 that a journeyman glazier died in an industrial accident in Indiana after attempting to use a manlift to carry a thousand-pound case of glass which the manlift did not have capacity to carry.

In the United States

According to the *Occupational Outlook Handbook*, there are some 45,300 glaziers in the United States, with median pay of $38,410 per year in 2014. Two-thirds of Glaziers work in the foundation, structure, and building exterior contractors industry, with smaller numbers working in building material and supplies dealing, building finishing contracting, automotive repair and maintenance, and glass and glass product manufacturing.

Among the 50 states, only Connecticut and Florida require glaziers to hold a license.

Glass Coloring and Color Marking

Glass coloring and color marking may be obtained by 1) addition of *coloring ions*, by 2) precipitation of nanometer sized colloides (so-called *striking glasses* such as "gold ruby" or red "selenium ruby"), 3) by *colored inclusions* (as in milk glass and smoked glass), 4) by *light scattering* (as in phase separated glass), 5) by *dichroic coatings*, or 6) by *colored coatings.*

Coloring Ions

Iron(II) oxide glass

Ordinary soda-lime glass appears colorless to the naked eye when it is thin, although iron oxide impurities produce a green tint which can be viewed in thick pieces or with the aid of scientific instruments. Further metals and metal oxides can be added to glass during its manufacture to change its color which can enhance its aesthetic appeal. Examples of these additives are listed below:

Amber Glass

Uranium glass glowing under UV light

Cobalt glass for decoration

- Iron(II) oxide may be added to glass resulting in bluish-green glass which is frequently used in beer bottles. Together with chromium it gives a richer green color, used for wine bottles.
- Sulfur, together with carbon and iron salts, is used to form iron polysulfides and produce amber glass ranging from yellowish to almost black. In borosilicate glasses rich in boron, sulfur imparts a blue color. With calcium it yields a deep yellow color.
- Manganese can be added in small amounts to remove the green tint given by iron, or in higher concentrations to give glass an amethyst color. Manganese is one of the oldest glass additives, and purple manganese glass was used since early Egyptian history.
- Manganese dioxide, which is black, is used to remove the green color from the glass; in a very slow process this is converted to sodium permanganate, a dark purple compound. In New England some houses built more than 300 years ago have window glass which is lightly tinted violet because of this chemical change; and such glass panes are prized as antiques. This process is widely confused with the formation of "desert amethyst glass", in which glass exposed to desert sunshine with a high ultraviolet component develops a delicate violet tint. Details of the process and the composition of the glass vary and so do the results, because it is not a simple matter to obtain or produce properly controlled specimens.
- Small concentrations of cobalt (0.025 to 0.1%) yield blue glass. The best results are achieved when using glass containing potash. Very small amounts can be used for decolorizing.
- 2 to 3% of copper oxide produces a turquoise color.
- Nickel, depending on the concentration, produces blue, or violet, or even black glass. Lead crystal with added nickel acquires purplish color. Nickel together with a small amount of cobalt was used for decolorizing of lead glass.
- Chromium is a very powerful colorizing agent, yielding dark green or in higher concentrations even black color. Together with tin oxide and arsenic it yields emerald

green glass. Chromium aventurine, in which aventurescence was achieved by growth of large parallel chromium(III) oxide plates during cooling, was also made from glass with added chromium oxide in amount above its solubility limit in glass.

- Cadmium together with sulphur forms cadmium sulfide and results in deep yellow color, often used in glazes. However, cadmium is toxic. Together with selenium and sulphur it yields shades of bright red and orange.
- Adding titanium produces yellowish-brown glass. Titanium, rarely used on its own, is more often employed to intensify and brighten other colorizing additives.
- Uranium (0.1 to 2%) can be added to give glass a fluorescent yellow or green color. Uranium glass is typically not radioactive enough to be dangerous, but if ground into a powder, such as by polishing with sandpaper, and inhaled, it can be carcinogenic. When used with lead glass with very high proportion of lead, produces a deep red color.
- Didymium gives green color (used in UV filters) or lilac red.

Striking Glasses

Cranberry glass bowl

Photochromic eyeglass lens. The coloring is caused by silver-nanoparticles.

- Selenium, like manganese, can be used in small concentrations to decolorize glass, or in higher concentrations to impart a reddish color, caused by selenium nanoparticles dispersed in glass. It is a very important agent to make pink and red glass. When used together with cadmium sulfide, it yields a brilliant red color known as "Selenium Ruby".

- Pure metallic copper produces a very dark red, opaque glass, which is sometimes used as a substitute for gold in the production of ruby-colored glass.
- Metallic gold, in very small concentrations (around 0.001%, or 10 ppm), produces a rich ruby-colored glass ("Ruby Gold" or "Rubino Oro"), while lower concentrations produces a less intense red, often marketed as "cranberry". The color is caused by the size and dispersion of gold particles. Ruby gold glass is usually made of lead glass with added tin.
- Silver compounds such as silver nitrate and silver halides can produce a range of colors from orange-red to yellow. The way the glass is heated and cooled can significantly affect the colors produced by these compounds. Also photochromic lenses and photosensitive glass are based on silver.

Colored Inclusions

Tin oxide with antimony and arsenic oxides produce an opaque white glass (milk glass), first used in Venice to produce an imitation porcelain. Similarly, some smoked glasses may be based on dark-colored inclusions, but with ionic coloring it is also possible to produce dark colors.

Color Caused by Scattering

The Tyndall effect in opalescent glass: it appears blue from the side, but orange light shines through.

Porous glass pore-size gradient (large pores on the right); coloring based on the Tyndall effect.

Glass containing two or more phases with different refractive indices shows coloring based on the Tyndall effect and explained by the Mie theory, if the dimensions of the phases are similar or larger than the wavelength of visible light. The scattered light is blue and violet as seen in the image, while the transmitted light is yellow and red.

Dichroic Glass

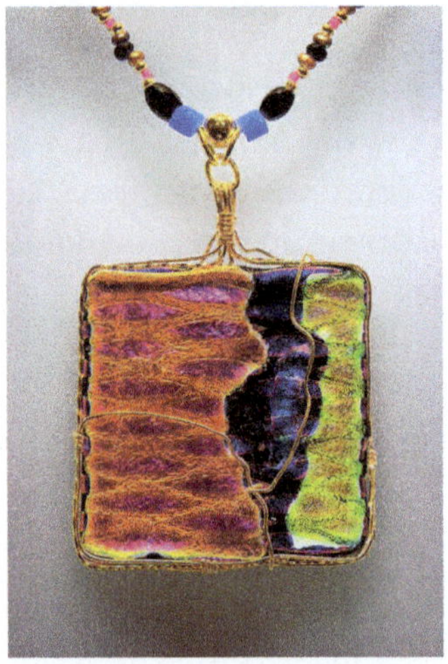

A pendant made from dichroic glass

Dichroic glass has one or several coatings in the nanometer-range (for example metals, metal oxides, or nitrides) which give the glass dichroic optical properties. Also the blue appearance of some automobile windshields is caused by dichroism.

References

- Werner Vogel: "Glass Chemistry"; Springer-Verlag Berlin and Heidelberg GmbH & Co. K; 2nd revised edition (November 1994), ISBN 3-540-57572-3
- B. H. W. S. de Jong, "Glass"; in "Ullmann's Encyclopedia of Industrial Chemistry"; 5th edition, vol. A12, VCH Publishers, Weinheim, Germany, 1989, ISBN 3-527-20112-2, pp. 365–432.
- Freestone, I. (1991). "Looking into Glass". In S. Bowman (ed.) Science and the Past. pp.37–56. University of Toronto Press: Toronto & Buffalo. ISBN 0-7141-2071-5
- Israeli, Y. (1991). "The Invention of Blowing". In M. Newby & K. Painter (eds.) Roman Glass: Two Centuries of Art and Invention. pp. 46–55. The Society of Antiquaries of London: London ISBN 0-85431-255-2
- Tatton-Brown, V. (1991). "Early Medieval Europe AD 400 – 1066". In H. Tait (ed.) Five Thousand Years of Glass. pp. 98–111. British Museum Press: London. ISBN 0-8122-1888-4
- Vose, R.H. (1989). "From Dark Ages to the Fall of Constantinople". In D. Klein & W. Lloyd (eds.) The History of Glass. pp. 39–66. Macdonald & Co. (Publishers) Ltd.: ISBN 0-85613-516-X
- Tait, H. (1994). "Europe from the Middle Ages to Industrial Revolution". In H. Tait (ed.) Five Thousand Years of Glass. pp. 145–187. British Museum Press: London ISBN 0-8122-1888-4
- Haynes, William M., ed. (2011). CRC Handbook of Chemistry and Physics (92nd ed.). Boca Raton, FL: CRC Press. ISBN 1439855110.
- Greenwood, Norman N.; Earnshaw, Alan (1984). Chemistry of the Elements. Oxford: Pergamon Press. pp. 393–99. ISBN 0-08-022057-6.
- Doering, Robert & Nishi, Yoshio (2007). Handbook of Semiconductor Manufacturing Technology. CRC Press. ISBN 1-57444-675-4.

6

Applications of Glass

The applications of glass that have been discussed in this section are glass tile, glass-bottom boat, fiberglass, windshield, storm glass etc. Glass-bottom boats are boats that have some portion that has a section of glass. This section helps the passengers in observing the underwater environment from within the boat. The chapter serves as a source to understand the applications of glass.

Glass Tile

Glass mosaic (detail). Glass tiles can be observed.

Glass tiles are pieces of glass formed into consistent shapes.

Early History

Glass was used in mosaics as early as 2500 BC, but it took until the 3rd century BC before innovative artisans in Greece, Persia and India created glass tiles.

Whereas clay tiles are dated as early as 8000 BC, there were significant barriers to the development of glass tiles, included the high temperatures required to melt glass, and the complexities of mastering various annealing curves for glass.

In recent years, glass tiles have become popular for both field and accent tiles. This trend can be attributed to recent technological breakthroughs, as well as the tiles' inherent properties, in particular their potential to impart intense color and reflect light, and their imperviousness to water.

Glass tile introduces complexities to the installer. Since glass is more rigid than ceramic or porcelain tile, glass tiles break more readily under the duress of substrate shifts.

Smalti Tiles

Smalti tile, sometimes referred to as *Byzantine glass mosaic tile*, are typically opaque glass tiles that were originally developed for use in mosaics created during the time of the Byzantine empire.

Smalti is made by mixing molten glass with metal oxides for color in a furnace; the result is a cloudy mixture that is poured into flat slabs that are cooled and broken into individual pieces. The molten mixture can also be topped with gold leaf, followed by a thin glass film to protect against tarnishing. During the Byzantine era, Constantinople became the center of the mosaic craft, and the use of gold leaf glass mosaic reached perhaps its greatest artistic expression in the former seat of the Orthodox patriarch of Constantinople, the Hagia Sophia.

Traditional smalti tiles are still found today in many European churches and ornamental objects; the method is also used by some present-day artisans, both in installations and fine art. In the 1920s, mass production methods were applied to Smalti tile manufacturing, which enabled these tiles to find their way into many middle-class homes. Instead of the old method of rolling the colored glass mixture out, cooling, and cutting, the new method called for molten liquid to be poured and cooled in trays, usually resulting in 3/4 inch chicklet-type pieces.

Modern era

Since the 1990s a variety of modern glass tile technologies, including methods to take used glass and recreate it as 'green' tiles, has resulted in a resurgence of interest in glass tile as a floor and wall cladding. It is now most commonly used in pools, kitchens, spas, and bathrooms. And while smalti tiles are still popular, small and large format glass products are now commonly formed using cast and fused glass methods. The plasticity of these last two methods has resulted in a wide variety of looks and applications, including floor tiles.

In the late 1990s, special glass tiles have been coated on the back side with a receptive white coating. This has allowed impregnation of heat-transfer dyes by a printing process reproducing high resolution pictures and designs. Custom printed glass tile and glass tile murals exhibit the toughness of glass on the wearing surface with photo-like pictures. These are especially practical in kitchens and showers, where cleanser and moisture resistance are important.

Glass-bottom Boat

A glass-bottom boat is a boat with sections of glass, or other suitable transparent material, below the waterline allowing passengers to observe the underwater environment from within the boat. The view through the glass bottom is better than simply looking into the water from above, because one does not have to look through optically erratic surface disturbances. The effect is similar to that achieved by a diving mask, while the passengers are able to stay dry, out of the water.

The interior of a glass bottom boat

A glass bottom boat showing undersea windows on the Great Barrier Reef.

Use

Glass bottom boats are used almost exclusively for giving tours, as they are usually designed to allow the maximum number of tourists to view out the glass bottom and are not really suitable for other uses. Glass bottom boats are in use in nearly every seaside tourist destination. However many of them are gradually being replaced by semi submarines. That type of vessels offer a better view of the marine life.

History

Glass bottom boats were first used near Santa Catalina Island off the coast of Southern California. They also became popular in Florida at several areas of natural springs that became tourist attractions, for example, Silver Springs, Wakulla Springs, Rainbow Springs, and Weeki Wachee Springs.

Typical tours in these boats include views of underwater flora and fauna, reefs, shipwrecks, and other underwater sights.

Fiberglass

Fiberglass (or fibreglass) is a type of fiber-reinforced plastic where the reinforcement fiber is specifically glass fiber. The glass fiber may be randomly arranged, flattened into a sheet (called a chopped strand mat), or woven into a fabric. The plastic matrix may be a thermosetting plastic – most often epoxy, polyester resin – or vinylester, or a thermoplastic.

The glass fibers are made of various types of glass depending upon the fiberglass use. These glasses all contain silica or silicate, with varying amounts of oxides of calcium, magnesium, and sometimes boron. To be used in fiberglass, glass fibers have to be made with very low levels of defects.

Fiberglass is a strong lightweight material and is used for many products. Although it is not as strong and stiff as composites based on carbon fiber, it is less brittle, and its raw materials are much cheaper. Its bulk strength and weight are also better than many metals, and it can be more readily molded into complex shapes. Applications of fiberglass include aircraft, boats, automobiles, bath tubs and enclosures, swimming pools, hot tubs, septic tanks, water tanks, roofing, pipes, cladding, casts, surfboards, and external door skins.

Other common names for fiberglass are glass-reinforced plastic (GRP), glass-fiber reinforced plastic (GFRP) or GFK (from German: *Glasfaserverstärkter Kunststoff*). Because glass fiber itself is sometimes referred to as "fiberglass", the composite is also called "fiberglass reinforced plastic." This article will adopt the convention that "fiberglass" refers to the complete glass fiber reinforced composite material, rather than only to the glass fiber within it.

History

Glass fibers have been produced for centuries, but mass production of glass strands was accidentally discovered in 1932 when Games Slayter, a researcher at Owens-Illinois, directed a jet of compressed air at a stream of molten glass and produced fibers. A patent for this method of producing glass wool was first applied for in 1933. Owens joined with the Corning company in 1935 and the method was adapted by Owens Corning to produce its patented "fibreglas" (one "s") in 1936. Originally, fibreglas was a glass wool with fibers entrapping a great deal of gas, making it useful as an insulator, especially at high temperatures.

A suitable resin for combining the "fibreglass" with a plastic to produce a composite material was developed in 1936 by du Pont. The first ancestor of modern polyester resins is Cyanamid's resin of 1942. Peroxide curing systems were used by then. With the combination of fiberglass and resin the gas content of the material was replaced by plastic. This reduced the insulation properties to values typical of the plastic, but now for the first time the composite showed great strength and promise as a structural and building material. Confusingly, many glass fiber composites continued to be called "fiberglass" (as a generic name) and the name was also used for the low-density glass wool product containing gas instead of plastic.

Ray Greene of Owens Corning is credited with producing the first composite boat in 1937, but did not proceed further at the time due to the brittle nature of the plastic used. In 1939 Russia was reported to have constructed a passenger boat of plastic materials, and the United States a fuselage and wings of an aircraft. The first car to have a fiber-glass body was a 1946 prototype of the Stout Scarab, but the model did not enter production.

Fiber

Glass reinforcements used for fiberglass are supplied in different physical forms, microspheres, chopped or woven.

Unlike glass fibers used for insulation, for the final structure to be strong, the fiber's surfaces must be almost entirely free of defects, as this permits the fibers to reach gigapascal tensile strengths. If a bulk piece of glass were defect-free, it would be equally as strong as glass fibers; however, it is generally impractical to produce and maintain bulk material in a defect-free state outside of laboratory conditions.

Production

The process of manufacturing fiberglass is called pultrusion. The manufacturing process for glass fibers suitable for reinforcement uses large furnaces to gradually melt the silica sand, limestone, kaolin clay, fluorspar, colemanite, dolomite and other minerals to liquid form. It is then extruded through bushings, which are bundles of very small orifices (typically 5–25 micrometres in diameter for E-Glass, 9 micrometres for S-Glass). These filaments are then *sized* (coated) with a chemical solution. The individual filaments are now bundled in large numbers to provide a roving. The diameter of the filaments, and the number of filaments in the roving, determine its *weight*, typically expressed in one of two measurement systems:

- *yield*, or yards per pound (the number of yards of fiber in one pound of material; thus a smaller number means a heavier roving). Examples of standard yields are 225yield, 450yield, 675yield.

- *tex*, or grams per km (how many grams 1 km of roving weighs, inverted from yield; thus a smaller number means a lighter roving). Examples of standard tex are 750tex, 1100tex, 2200tex.

These rovings are then either used directly in a composite application such as pultrusion, filament winding (pipe), gun roving (where an automated gun chops the glass into short lengths and drops it into a jet of resin, projected onto the surface of a mold), or in an intermediary step, to manufacture fabrics such as *chopped strand mat* (CSM) (made of randomly oriented small cut lengths of fiber all bonded together), woven fabrics, knit fabrics or uni-directional fabrics.

Chopped Strand Mat

Chopped strand mat or CSM is a form of reinforcement used in fiberglass. It consists of glass fibers laid randomly across each other and held together by a binder.

It is typically processed using the hand lay-up technique, where sheets of material are placed in a mold and brushed with resin. Because the binder dissolves in resin, the material easily conforms to different shapes when wetted out. After the resin cures, the hardened product can be taken from the mold and finished.

Using chopped strand mat gives a fiberglass with isotropic in-plane material properties.

Sizing

A coating or primer is applied to the roving to:

- Help protect the glass filaments for processing and manipulation.
- Ensure proper bonding to the resin matrix, thus allowing for transfer of shear loads from the glass fibers to the thermoset plastic. Without this bonding, the fibers can 'slip' in the matrix, causing localized failure..

Properties

An individual structural glass fiber is both stiff and strong in tension and compression—that is, *along* its axis. Although it might be assumed that the fiber is weak in compression, it is actually only the long aspect ratio of the fiber which makes it seem so; i.e., because a typical fiber is long and narrow, it buckles easily. On the other hand, the glass fiber is weak in shear—that is, *across* its axis. Therefore, if a collection of fibers can be arranged permanently in a preferred direction within a material, and if they can be prevented from buckling in compression, the material will be preferentially strong in that direction.

Furthermore, by laying multiple layers of fiber on top of one another, with each layer oriented in various preferred directions, the material's overall stiffness and strength can be efficiently controlled. In fiberglass, it is the plastic matrix which permanently constrains the structural glass fibers to directions chosen by the designer. With chopped strand mat, this directionality is essentially an entire two dimensional plane; with woven fabrics or unidirectional layers, directionality of stiffness and strength can be more precisely controlled within the plane.

A fiberglass component is typically of a thin "shell" construction, sometimes filled on the inside with structural foam, as in the case of surfboards. The component may be of nearly arbitrary shape, limited only by the complexity and tolerances of the mold used for manufacturing the shell.

The mechanical functionality of materials is heavily relied on the combined performances of both the resin (AKA matrix) and fibres. For example, in severe temperature condition (over 180 °C) resin component of the composite may lose its functionality partially because of bond deterioration of resin and fibre. However, GFRPs can show still significant residual strength after experiencing high temperature (200 °C).

Types of Glass Fiber Used

Composition: the most common types of glass fiber used in fiberglass is E-glass, which is alumino-borosilicate glass with less than 1% w/w alkali oxides, mainly used for glass-reinforced plastics. Other types of glass used are A-glass (Alkali-lime glass with little or no boron oxide), E-CR-glass (Electrical/Chemical Resistance; alumino-lime silicate with less than 1% w/w alkali oxides, with high acid resistance), C-glass (alkali-lime glass with high boron oxide content, used for glass staple fibers and insulation), D-glass (borosilicate glass, named for its low Dielectric constant), R-glass (alumino silicate glass without MgO and CaO with high mechanical requirements as Reinforcement), and S-glass (alumino silicate glass without CaO but with high MgO content with high tensile strength).

Naming and use: pure silica (silicon dioxide), when cooled as fused quartz into a glass with no true melting point, can be used as a glass fiber for fiberglass, but has the drawback that it must be worked at very high temperatures. In order to lower the necessary work temperature, other materials are introduced as "fluxing agents" (i.e., components to lower the melting point). Ordinary A-glass ("A" for "alkali-lime") or soda lime glass, crushed and ready to be remelted, as so-called cullet glass, was the first type of glass used for fiberglass. E-glass ("E" because of initial Electrical application), is alkali free, and was the first glass formulation used for continuous filament formation. It now makes up most of the fiberglass production in the world, and also is the single largest consumer of boron minerals globally. It is susceptible to chloride ion attack and is a poor choice for marine applications. S-glass ("S" for "stiff") is used when tensile strength (high modulus) is important, and is thus an important building and aircraft epoxy composite (it is called R-glass, "R" for "reinforcement" in Europe). C-glass ("C" for "chemical resistance") and T-glass ("T" is for "thermal insulator"—a North American variant of C-glass) are resistant to chemical attack; both are often found in insulation-grades of blown fiberglass.

Table of Some Common Fiberglass Types

Material	Specific gravity	Tensile strength MPa (ksi)	Compressive strength MPa (ksi)
Polyester resin (Not reinforced)	1.28	55 (7.98)	140 (20.3)
Polyester and Chopped Strand Mat Laminate 30% E-glass	1.4	100 (14.5)	150 (21.8)
Polyester and Woven Rovings Laminate 45% E-glass	1.6	250 (36.3)	150 (21.8)
Polyester and Satin Weave Cloth Laminate 55% E-glass	1.7	300 (43.5)	250 (36.3)

Polyester and Continuous Rovings Laminate 70% E-glass	1.9	800 (116)	350 (50.8)
E-Glass Epoxy composite	1.99	1,770 (257)	
S-Glass Epoxy composite	1.95	2,358 (342)	

Applications

A cryostat made of fiberglass

Fiberglass is an immensely versatile material due to its light weight, inherent strength, weather-resistant finish and variety of surface textures.

The development of fiber-reinforced plastic for commercial use was extensively researched in the 1930s. It was of particular interest to the aviation industry. A means of mass production of glass strands was accidentally discovered in 1932 when a researcher at Owens-Illinois directed a jet of compressed air at a stream of molten glass and produced fibers. After Owens merged with the Corning company in 1935, Owens Corning adapted the method to produce its patented "Fiberglas" (one "s"). A suitable resin for combining the "Fiberglas" with a plastic was developed in 1936 by du Pont. The first ancestor of modern polyester resins is Cyanamid's of 1942. Peroxide curing systems were used by then.

During World War II, fiberglass was developed as a replacement for the molded plywood used in aircraft radomes (fiberglass being transparent to microwaves). Its first main civilian application was for the building of boats and sports car bodies, where it gained acceptance in the 1950s. Its use has broadened to the automotive and sport equipment sectors. In production of some products, such as aircraft, carbon fiber is now used instead of fiberglass, which is stronger by volume and weight.

Advanced manufacturing techniques such as pre-pregs and fiber rovings extend fiberglass's applications and the tensile strength possible with fiber-reinforced plastics.

Fiberglass is also used in the telecommunications industry for shrouding antennas, due to its RF permeability and low signal attenuation properties. It may also be used to conceal other equipment where no signal permeability is required, such as equipment cabinets and steel support structures, due to the ease with which it can be molded and painted to blend with existing structures and surfaces. Other uses include sheet-form electrical insulators and structural components commonly found in power-industry products.

Because of fiberglass's light weight and durability, it is often used in protective equipment such as helmets. Many sports use fiberglass protective gear, such as goaltenders' and catchers' masks.

Storage Tanks

Several large fiberglass tanks at an airport

Storage tanks can be made of fiberglass with capacities up to about 300 tonnes. Smaller tanks can be made with chopped strand mat cast over a thermoplastic inner tank which acts as a preform during construction. Much more reliable tanks are made using woven mat or filament wound fiber, with the fiber orientation at right angles to the hoop stress imposed in the side wall by the contents. Such tanks tend to be used for chemical storage because the plastic liner (often polypropylene) is resistant to a wide range of corrosive chemicals. Fiberglass is also used for septic tanks.

House Building

A fiberglass dome house in Davis, California

Glass-reinforced plastics are also used to produce house building components such as roofing laminate, door surrounds, over-door canopies, window canopies and dormers, chimneys, coping systems, and heads with keystones and sills. The material's reduced weight and easier handling, compared to wood or metal, allows faster installation. Mass-produced fiberglass brick-effect panels can be used in the construction of composite housing, and can include insulation to reduce heat loss.

Piping

GRP and GRE pipe can be used in a variety of above- and below-ground systems, including those for:

- Desalination
- Water treatment
- Water distribution networks
- Chemical process plants
- Firewater
- Hot and Cold water
- Drinking water
- Wastewater/sewage, Municipal waste
- Natural gas, LPG

Examples of Fiberglass Use

Kayaks made of fiberglass

- DIY bows / youth recurve; longbows
- Pole vaulting poles
- Equipment handles(Hammers, axes, etc.)
- Traffic lights

- Ship hulls
- Rowing shells and oars
- Waterpipes
- Helicopter rotor blades
- Surfboards, tent poles
- Gliders, kit cars, microcars, karts, bodyshells, kayaks, flat roofs, lorries
- Pods, domes and architectural features where a light weight is necessary
- High-end bicycles
- Auto body parts (for instance, body kits, hoods, spoilers, etc.), and entire auto bodies (e.g. Lotus Elan, Anadol, Reliant, Quantum Quantum Coupé, Chevrolet Corvette and Studebaker Avanti, and DeLorean DMC-12 underbody)
- Antenna covers and structures, such as radomes, UHF broadcasting antennas, and pipes used in hex beam antennas for amateur radio communications
- FRP tanks and vessels: FRP is used extensively to manufacture chemical equipment and tanks and vessels. BS4994 is a British standard related to this application.
- Most commercial velomobiles
- Most printed circuit boards consist of alternating layers of copper and fiberglass FR-4
- Large commercial wind turbine blades
- RF coils used in MRI scanners
- Drum Sets
- Sub-sea installation protection covers
- Reinforcement of asphalt pavement, as a fabric or mesh interlayer between lifts
- Helmets and other protective gear used in various sports
- Orthopedic casts
- Fiberglass grating is used for walkways on ships and oil rigs, and in factories
- Fiber-reinforced composite columns
- Water slides

Construction Methods

Filament Winding

Filament winding is a fabrication technique mainly used for manufacturing open (cylinders) or closed end structures (pressure vessels or tanks). The process involves winding filaments under tension over a male mandrel. The mandrel rotates while a wind eye on a carriage moves horizontally, laying down fibers in the desired pattern. The most common filaments are carbon or glass fiber and are coated with synthetic resin as they are wound. Once the mandrel is completely covered to the desired thickness, the resin is cured, often the mandrel is placed in an oven to achieve this, though sometimes radiant heaters are used with the mandrel still turning in the machine. Once the resin has cured, the mandrel is removed, leaving the hollow final product. For some products such as gas bottles the 'mandrel' is a permanent part of the finished product forming a liner to prevent gas leakage or as a barrier to protect the composite from the fluid to be stored.

Filament winding is well suited to automation, and there are many applications, such as pipe and small pressure vessel that are wound and cured without any human intervention. The controlled variables for winding are fiber type, resin content, wind angle, tow or bandwidth and thickness of the fiber bundle. The angle at which the fiber has an effect on the properties of the final product. A high angle "hoop" will provide circumferential or "burst" strength, while lower angle patterns (polar or helical) will provide greater longitudinal tensile strength.

Products currently being produced using this technique range from pipes, golf clubs, Reverse Osmosis Membrane Housings, oars, bicycle forks, bicycle rims, power and transmission poles, pressure vessels to missile casings, aircraft fuselages and lamp posts and yacht masts.

Fiberglass Hand Lay-up Operation

A release agent, usually in either wax or liquid form, is applied to the chosen mold to allow finished product to be cleanly removed from the mold. Resin—typically a 2-part polyester, vinyl or epoxy—is mixed with its hardener and applied to the surface. Sheets of fiberglass matting are laid into the mold, then more resin mixture is added using a brush or roller. The material must conform to the mold, and air must not be trapped between the fiberglass and the mold. Additional resin is applied and possibly additional sheets of fiberglass. Hand pressure, vacuum or rollers are used to be sure the resin saturates and fully wets all layers, and that any air pockets are removed. The work must be done quickly, before the resin starts to cure, unless high temperature resins are used which will not cure until the part is warmed in an oven. In some cases, the work is covered with plastic sheets and vacuum is drawn on the work to remove air bubbles and press the fiberglass to the shape of the mold.

Fiberglass Spray Lay-up Operation

The fiberglass spray lay-up process is similar to the hand lay-up process, but differs in the application of the fiber and resin to the mold. Spray-up is an open-molding composites fabrication process where resin and reinforcements are sprayed onto a mold. The resin and glass may be applied separately or simultaneously "chopped" in a combined stream from a chopper gun. Workers roll out the spray-up to compact the laminate. Wood, foam or other core material may then be added, and a secondary spray-up layer imbeds the core between the laminates. The part is then cured, cooled and removed from the reusable mold.

Pultrusion Operation

Pultrusion is a manufacturing method used to make strong, lightweight composite materials. In pultrusion, material is pulled through forming machinery using either a hand-over-hand method or a continuous-roller method (as opposed to extrusion, where the material is pushed through dies). In fiberglass pultrusion, fibers (the glass material) are pulled from spools through a device that coats them with a resin. They are then typically heat-treated and cut to length. Fiberglass produced this way can be made in a variety of shapes and cross-sections, such as W or S cross-sections.

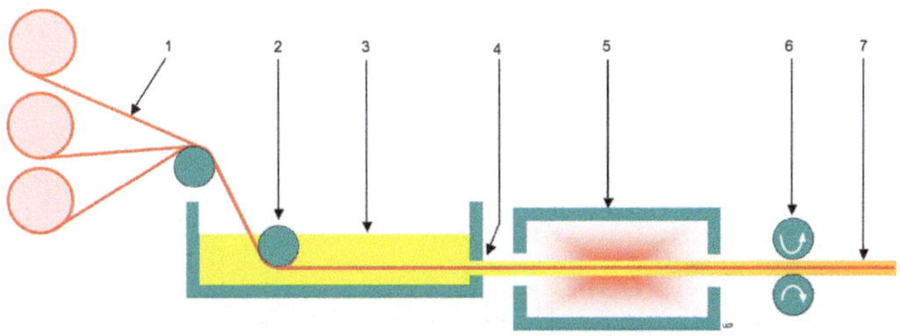

Diagram of the pultrusion process.

Warping

One notable feature of fiberglass is that the resins used are subject to contraction during the curing process. For polyester this contraction is often 5–6%; for epoxy, about 2%. Because the fibers do not contract, this differential can create changes in the shape of the part during curing. Distortions can appear hours, days or weeks after the resin has set.

While this distortion can be minimised by symmetric use of the fibers in the design, a certain amount of internal stress is created; and if it becomes too great, cracks form.

Health Problems

In June 2011, the National Toxicology Program (NTP) removed from its Report on Carcinogens all biosoluble glass wool used in home and building insulation and for non-insulation products. However, NTP considers fibrous glass dust to be "reasonably anticipated [as] a human carcinogen (Certain Glass Wool Fibers (Inhalable))". Similarly, California's Office of Environmental Health Hazard Assessment ("OEHHA") published a November, 2011 modification to its Proposition 65 listing to include only "Glass wool fibers (inhalable and biopersistent)." The actions of U.S. NTP and California's OEHHA mean that a cancer warning label for biosoluble fiber glass home and building insulation is no longer required under federal or California law. All fiberglass wools commonly used for thermal and acoustical insulation were reclassified by the International Agency for Research on Cancer ("IARC") in October 2001 as Not Classifiable as to carcinogenicity to humans (Group 3).

People can be exposed to fiberglass in the workplace by breathing it in, skin contact, or eye contact. The Occupational Safety and Health Administration (OSHA) has set the legal limit (permissible exposure limit) for fiberglass exposure in the workplace as 15 mg/m^3 total and 5 mg/m^3 in respiratory exposure over an 8-hour workday. The National Institute for Occupational Safety and Health (NIOSH) has set a recommended exposure limit (REL) of 3 fibers/cm^3 (less than 3.5 micrometers in diameter and greater than 10 micrometers in length) as a time-weighted average over an 8-hour workday, and a 5 mg/m^3 total limit.

The European Union and Germany classify synthetic vitreous fibers as possibly or probably carcinogenic, but fibers can be exempt from this classification if they pass specific tests. Evidence for these classifications is primarily from studies on experimental animals and mechanisms of carcinogenesis. The glass wool epidemiology studies have been reviewed by a panel of international experts convened by the IARC. These experts concluded: "Epidemiologic studies published during the 15 years since the previous IARC monographs review of these fibers in 1988 provide no evidence of increased risks of lung cancer or mesothelioma (cancer of the lining of the body cavities) from occupational exposures during the manufacture of these materials, and inadequate evidence overall of any cancer risk." Similar reviews of the epidemiology studies have been conducted by the Agency for Toxic Substances and Disease Registry ("ATSDR"), the National Toxicology Program, the National Academy of Sciences and Harvard's Medical and Public Health Schools which reached the same conclusion as IARC that there is no evidence of increased risk from occupational exposure to glass wool fibers.

Fiberglass will irritate the eyes, skin, and the respiratory system. Potential symptoms include irritation of eyes, skin, nose, throat, dyspnea (breathing difficulty); sore throat, hoarseness and cough. Scientific evidence demonstrates that fiber glass is safe to manufacture, install and use when recommended work practices are followed to reduce temporary mechanical irritation.

While the resins are cured, styrene vapors are released. These are irritating to mucous membranes and respiratory tract. Therefore, the Hazardous Substances Ordinance in Germany dictates a maximum occupational exposure limit of 86 mg/m^3. In certain concentrations may even occur a potentially explosive mixture. Further manufacture of GRP components (grinding, cutting, sawing) creates fine dusts and chips containing glass filaments, as well as tacky dust, in quantities substantial enough to affect people's health and the functionality of machines and equipment. The installation of effective extraction and filtration equipment is required to ensure safety and efficiency.

Windshield

Panoramic (wrap-around) windshield on a 1959 Edsel Corsair

The windshield (North America) or windscreen (EU & Commonwealth countries) of an aircraft, car, bus, motorbike or tram is the front window. Modern windshields are generally made of laminated safety glass, a type of treated glass, which consists of two (typically) curved sheets of glass with a plastic layer laminated between them for safety, and are bonded into the window frame. Motorbike windshields are often made of high-impact acrylic plastic.

Usage

Split and raked windshield on a 1952 DeSoto. Note the panes of glass are flat.

Windshields protect the vehicle's occupants from wind and flying debris such as dust, insects, and rocks, and provide an aerodynamically formed window towards the front. UV coating may be applied to screen out harmful ultraviolet radiation. However, this is usually unnecessary since most auto windshields are made from laminated safety glass. The majority of UV-B is absorbed by the glass itself, and any remaining UV-B together with most of the UV-A is absorbed by the PVB bonding layer.

On motorbikes their main function is to shield the rider from wind, though not as completely as in a car, whereas on sports and racing motorcycles the main function is reducing drag when the rider assumes the optimal aerodynamic configuration with his or her body in unison with the machine, and does not shield the rider from wind when sitting upright.

Safety

Automobile windshield displaying "spiderweb" cracking typical of laminated safety glass.

Early windshields were made of ordinary window glass, but that could lead to serious injuries in the event of a crash. A series of crashes led up to the development of stronger windshields. The most notable example of this is the *Pane vs. Ford* case of 1917 that decided against Pane in that he was only injured through reckless driving. They were replaced with windshields made of toughened glass and were fitted in the frame using a rubber or neoprene seal. The hardened glass shattered into many mostly harmless fragments when the windshield broke. use windshields, hovever, could shatter from a simple stone chip. In 1919, Henry Ford solved the problem of flying debris by using the new French technology of glass laminating. Windshields made using this process were two layers of glass with a cellulose inner layer. This inner layer held the glass together when it fractured. Between 1919 and 1929, Ford ordered the use of laminated glass on all of his vehicles.

Modern, glued-in windshields contribute to the vehicle's rigidity, but the main force for innovation has historically been the need to prevent injury from sharp glass fragments. Almost all nations now require windshields to stay in one piece even if broken, except if pierced by a strong force. Properly installed automobile windshields are also essential to safety; along with the roof of the car, they provide protection to the vehicle's occupants in the case of a roll-over accident.

Today's windshields are a safety device just like seat belts and airbags. The installation of the auto glass is done with an automotive grade urethane designed specifically for automobiles. The adhesive creates a molecular bond between the glass and the vehicle. If the adhesive bond fails at any point on the glass it can reduce the effectiveness of the air bag and substantially compromise the structural integrity of the roof.

Other Aspects

In many places, laws restrict the use of heavily tinted glass in vehicle windshields; generally, laws specify the maximum level of tint permitted. Some vehicles have noticeably more tint in the uppermost part of the windshield to block sunglare.

In aircraft windshields, an electric current is applied through a conducting layer of tin(IV) oxide to generate heat to prevent icing. A similar system for automobile windshields, introduced on Ford vehicles as "Quickclear" in Europe ("InstaClear" in North America) in the 1980s and through the early 1990s, used this conductive metallic coating applied to the inboard side of the outer layer of glass. Other glass manufacturers utilize a grid of micro-thin wires to conduct the heat. These systems are more typically utilized by European auto manufacturers such as Jaguar and Porsche.

The use of thermal glass prevents some navigation systems from functioning correctly, as the embedded metal blocks the satellite signal.The RF signal tends to flow along the metal wires or layer so very little radiation can pass. This can be resolved by using an external antenna. Mobile telephones can also have problems; thermal glass typically allows only 0.0001 (1‰, or one per mille) of the signal to pass, whereas a concrete wall with rebars allows up to 0.0100 (10%, or 100‰) of the signal to pass.

Terminology

Singled aero screen on Bentley Blower No.1

The term *windshield* is used generally throughout North America. The term *windscreen* is the usual term in the British Isles and Australasia for all vehicles. In the US *windscreen* refers to the mesh or foam placed over a microphone to minimize wind noise, while a *windshield* refers to the front window of a car. In the UK, the terms are reversed, although generally, the foam screen is referred to as a microphone shield, and not a windshield.

Sports or racing cars would sometimes have aero screens, which were small semi-circular or rectangular windshields. These were often mounted in pairs behind a foldable flat windshield. Aero screens are usually less than 20 cm (8 in) in height. They are known as aero screens because they only deflect the wind. The twin aeroscreen setup (often called Brooklands) was popular among older sports and modern cars in vintage style.

A *wiperless windshield* is a windshield that uses a mechanism other than wipers to remove snow and rain from the windshield. The concept car Acura TL features a wiperless windshield using a series of jet nozzles in the cowl to blow pressurized air onto the windshield.

Repair of Stone-chip and Crack Damage

According to the US National Windshield Repair Association many types of stone damage can be successfully repaired. Whether the windshield can be repaired always depends upon four factors: the size, type, depth, and location of the damage.

Size and Depth

Repair of cracks up to 61 cm (24 in) is within permissible limits; automobile glass with more severe damage needs to be replaced. However, this is dependent on state and local laws. Unless it goes to the edge of the window. It would then compromise the structural integrity of the windshield.

Type

Circular bullseyes, linear cracks, crack chips, dings, pits, and star-shaped breaks can be repaired without removing the glass, eliminating the risk of leaking or bonding problems sometimes associated with replacement.

Location

If the damage is in the driver's line of sight or very near the windshield's edge, usually it cannot be repaired.

Some damages are very difficult to repair, or cannot be repaired:

- on inside of the windshield
- deep damage on both layers of glass
- damage over rain sensor or internal radio antenna
- damages that reach into the driver's critical viewing area
- complex multiple cracks

- very long cracks (i.e. over 18–24 inches (46–61 cm) long)
- contaminated cracks
- edge cracks and chips

In cracked windshield repair, air is removed from the damaged area on the windshield with a specified vacuum injection pump. Then using the injection pump, clear adhesive resin is injected to replace the air in the windshield crack. The resin is then cured with an ultraviolet light. When done properly, the damaged area's strength is restored, as is 90–95% of the clarity.

Windshield Replacement

Windshields that cannot be repaired have to be replaced. Replacement of a windshield typically takes less than 1 hour. To ensure the vehicle is safe to drive, time values called the Safe Drive Away Time have been established. Windshields which have been replaced must cure or bond sufficiently until they are able to withstand the forces of a crash. Knowing the minimum time needed to cure the glass bonding adhesives is therefore important. This safe drive away time (SDAT) or minimum drive away time (MDAT) refers to the time required until a windshield installation or glass replacement is considered safe to drive again. Criteria are specified in U.S. Federal Motor Vehicle Safety Standards 212/208 to ensure the reliability of adhesive systems. Typically the SDAT is verified with crash tests as well as with high speed laboratory test methods.

Consumers may be unaware that the MDAT or SDAT time is focused on safety and not necessarily on the quality, durability, or warranty of the installation. Care must be taken not to drive the vehicle prior to the SDAT/MDAT. If a vehicle is released to be driven before the SDAT and the adhesive used to set the new windshield has not had appropriate cure time, the occupants will not be properly protected in the event of a collision.

Airbags deploy at speeds up to 200 mph (320 km/h; 89 m/s) and in some cases exert tremendous force on the windshield. Occupants can impact the airbag just 50 ms after initial deployment. Depending on vehicle design, airbag deployment and/or occupant impact into the airbag may increase forces on the windshield, dramatically in some cases. Forces of occupants on the airbags - and hence the potential forces on the windshield - are lower for belted occupants. As consequence, adhesive suppliers usually inform their customers about the level of security achieved:

1) Example: Security exceeding FMVSS 212/208 belted

2) Example: Security exceeding FMVSS 212/208 unbelted

With the advent of quick cure adhesives, mobile windshield replacements have become more prevalent. Often the temperature and humidity cannot be controlled for mobile installations. For most common glass adhesives the ideal environment is 70 °F (21 °C) and 50% humidity. Variations from the ideal curing environment can increase the time needed for a sufficiently safe bond to form. A controlled environment always ensures the performance and safety of the windshield installation.

If not properly cured, the adhesive bond could fail during a crash, allowing occupants to be injured, ejected, or causing other harm. Because of the variables and difficulties involved in mobile windshield replacement, many vehicle manufacturers do not recommend this method of installation.

Disposal

Waste disposal of laminated glass is no longer permitted in landfill in most European countries as the End of Life Vehicles Directive (ELV) is implemented. A study by Surrey University and Pilkington Glass proposes that waste laminated glass be placed into a separating device such as a rolling mill where the glass is fragmented and the larger cullet is mechanically detached from the inner film. The application of heat then melts the laminating plastic, usually Polyvinyl Butyral "PVB" enabling both the glass and the interior film to be recycled. The PVB recycling process is a simple procedure of melting and reshaping it.

Storm Glass

Crystals forming in a storm glass prior to a storm

A storm glass is a type of weather forecasting device, composed of a sealed glass container filled with liquid. The appearance of the liquid purportedly predicts the weather, but most modern experiments have failed to confirm this.

Description

The liquid within the glass is a mixture of several ingredients, most commonly distilled water, ethanol, potassium nitrate, ammonium chloride, and camphor. This specific mixture was developed by Admiral Robert FitzRoy and used on HMS *Beagle*.

During the historic voyage, FitzRoy carefully documented how the storm glass would predict the weather:

- If the liquid in the glass is clear, the weather will be bright and clear.
- If the liquid is cloudy, the weather will be cloudy as well, perhaps with precipitation.
- If there are small dots in the liquid, humid or foggy weather can be expected.
- A cloudy glass with small stars indicates thunderstorms.
- If the liquid contains small stars on sunny winter days, then snow is coming.
- If there are large flakes throughout the liquid, it will be overcast in temperate seasons or snowy in the winter.
- If there are crystals at the bottom, this indicates frost.
- If there are threads near the top, it will be windy.

In 1859, violent storms struck the British Isles. In response, the British Crown distributed storm glasses, then known as "FitzRoy's storm barometers," to many small fishing communities around the British Isles that were to be consulted by ships at port before setting sail.

Crystals in FitzRoy Stormglass

Ingredients

- 2.5 g potassium nitrate
- 2.5 g ammonium chloride
- 33 ml distilled water
- 40 ml ethanol
- 10 g camphor

The potassium nitrate and ammonium chloride are dissolved in warm water and the ethanol and camphor are then added. The resulting solution is placed in a glass container and sealed.

Photochromic Lens

A photochromic eyeglass lens (polymer film), after exposure to sunlight with part of the lens covered by paper. Note the two levels of coloration around the dark-light interface, arising from the fact that the photochromic molecules are located in two thin films (on the front and back surfaces of the lens).

Photochromic lenses are optical lenses that darken on exposure to specific types of light of sufficient intensity, most commonly ultraviolet (UV) radiation. In the absence of activating light the lenses return to their clear state. Photochromic lenses may be made of glass, polycarbonate, or another plastic. They are principally used in eyeglasses that are dark in bright sunlight, but clear in low ambient light conditions. They darken significantly within about a minute of exposure to bright light, and take somewhat longer to clear. A range of clear and dark transmittances are available; one manufacturer makes one glass with transmittance reducing from 87% to 20%, and another reducing from 45% to 9%.

In one sort of technology, molecules of silver chloride or another silver halide are embedded in photochromatic lenses. They are transparent to visible light without significant ultraviolet component, which is normal for artificial lighting. In another sort of technology, organic photochromic molecules, when exposed to ultraviolet (UV) rays as in direct sunlight, undergo a chemical process that causes them to change shape and absorb a significant percentage of the visible light, i.e., they darken. These processes are reversible; once the lens is removed from strong sources of UV rays the photochromic compounds return to their transparent state.

Invention

Photochromic lenses were developed by Roger Araujo at the Corning Glass Works Inc. in the 1960s, and the process was used in the first mass-produced variable tint lenses.

Technical Details

The glass version of these lenses achieve their photochromic properties through the embedding of microcrystalline silver halides (usually silver chloride) in a glass substrate. Plastic photochromic lenses use organic photochromic molecules (for example oxazines and naphthopyrans) to achieve the reversible darkening effect. These lenses darken when exposed to ultraviolet light of the intensity present in sunlight, but not in artificial light.

With the photochromic material dispersed in the glass substrate, the degree of darkening depends on the thickness of glass, which poses problems with variable-thickness lenses in prescription glasses. With plastic lenses, the material is typically embedded into the surface layer of the plastic in a uniform thickness of up to 150 μm.

Typically, photochromic lenses darken substantially in response to UV light in less than one minute, and continue to darken a little more over the next fifteen minutes. The lenses begin to clear in the absence of UV light, and will be noticeably lighter within two minutes, mostly clear within five minutes, and fully back to their non-exposed state in about fifteen minutes. A report by the

Institute of Ophthalmology at the University College London suggested that at their clearest photochromic lenses can absorb up to 20% of ambient light.

Because photochromic compounds fade back to their clear state by a thermal process, the higher the temperature, the less dark photochromic lenses will be. This thermal effect is called "temperature dependency" and prevents these devices from achieving true sunglass darkness in very hot weather. Conversely, photochromic lenses will get very dark in cold weather conditions. Once inside, away from the triggering UV light, the cold lenses take longer to regain their transparency than warm lenses.

A number of sunglass manufacturers and suppliers including Tifosi, Intercast, Oakley, Serengeti Eyewear, and Persol provide tinted lenses that use photochromism to go from a dark to a darker state. They are typically used for outdoor sunglasses rather than as general-purpose lenses.

Picture Framing Glass

Picture framing glass ("glazing," "conservation glass," "museum quality glass") usually refers to flat glass or acrylic ("plexi") used for framing artwork and for presenting art objects in a display box (also, "conservation framing").

Purpose

The primary purpose of glazing in art framing is to physically protect the work of art from damaging factors such as humidity, heat, and soiling. Laminated glass and some acrylic may be used to protect against physical damage from glass breakage and to offer protection from a malicious attack. Regular glass as well as some glass surface treatments can also filter some of the damaging ultra-violet radiation (UV) and heat (NIR). Artworks that require protective glazing are those rendered on paper or fabrics (including photographs), which contain pigments and dyes that absorb UV and are susceptible to discoloration. In the case if the framed object or artwork is UV resistant, UV protection can still serve the purpose of preserving the integrity and colors of non-conservation grade framing materials susceptible to UV damage, such as mat board (passe partout).

Although protection is a primary purpose of glazing, *displaying* an artwork is the primary purpose of framing it. Therefore, the least visible glazing best displays the artwork behind it. Visible light transmission is the primary measure of glass' *invisibility*, since the viewer actually sees the light, reflected from the artwork. Light transmission of glass is especially important in art framing, since light passes through the glass twice – once to illuminate the artwork, and then again, reflected from the artwork, as colors - before reaching the viewer.

Light transmission through glass is diminished either by light reflection or light absorption of the glazing material. The total light transferred through the glazing material (light transmission) is reduced by reflection and/or absorption. In art framing, light reflection causes glare, while light absorption also may cause the transmitted colors to be dulled or distorted. While type of the glass substrate will affect the light absorption of the glazing, the surface treatment can affect light scattering, light reflection and in some cases, light absorption. There are various glazing options to achieve this goal as explained in the following sections on Types of Picture Framing

Glass.

Types of Picture Framing Glass

Regular (or "Clear")

Due to widespread availability and low cost, Soda Lime Glass is most commonly used for picture framing glass. Glass thicknesses typically range from thin 2.0 mm, to 2.5 mm. Clear glass has light transmission of ~ 90%, absorption of ~ 2%, and reflection of ~ 8%. Whereas absorption can be reduced by using low-iron glass, reflection can only be reduced by an anti-reflective surface treatment.

Low-iron (or "Extra-Clear," "Water White," etc)

Low-iron, or water white glass, is made using special iron free silica, and is generally only available in 2.0 mm thicknesses for picture framing applications. Because low iron glass light absorption can be as low as 0.5% vs. ~ 2% for clear glass, the light transmission will be significantly better vs. clear glass. Low iron glass has light transmission of ~91.5% and reflection of 8%.

Laminated Glass

Laminated glass offers shatter-resistance and protection from malicious breakage to art glazing. The most commonly used configuration is Glass + PVB Foil + Glass. Some variations of foils and glass thicknesses can offer shatter- and breakage-resistance or even bullet-resistance. The absorption of laminated glass depends on the glass substrates and foils used in the laminating process. Reflection of laminated glass is similar to monolithic glass, unless surface treatments are applied to reduce reflection.

Acrylic

Some types of acrylic glass can have the high light transmission and optical quality of glass. Acrylic is also light weight, compared to glass, and is shatter-resistant, making acrylic an attractive choice for framing large, oversized works of art. In general, acrylic sheet scratches easily and retains a static charge, which can be problematic when framing pastels or charcoals. Some manufacturers, add dyes to acrylic glass to filter the UV light transmittance, and its surface can also be treated with both anti-static and anti-reflective coatings.

Glass Surface Treatments and Coatings

Due to the change in the refractive index, as a light beam travels from air (refractive index of ~1) into glass or acrylic (refractive index of ~1.5) and then back into air, these transitions cause part of the light to be reflected. While "anti-glare" (a.k.a. "non-glare" or matte finish) glass treatments focus on scattering the light, "anti-reflective" coatings actually reduce the amount of light, which is reflected from each glazing surface, which has the benefit of increasing the amount of light transmitted through the glazing.

Matte (Etched, "Non-Glare," or "Anti-Glare")

The main purpose of matte glass is to transform the specular reflection into reflection haze. So-called "scattering" of the reflected light renders reflected images blurry, so that distinct reflected shapes and sources of light do not distract from the art viewing experience. Scattering the light does not reduce the reflection or absorption, which remain at the level of the glass substrate. There are several ways of making the glass surface matte – from pressing the pattern when the glass is still soft to fine etching of the glass surface by acid. The quality of matte glass is usually determined by its gloss factor or haze factor.

Anti-reflective Coatings

Single-layer

Single-layer anti-reflective coatings aim to achieve the refractive index of 1.25 (halfway between air and glass), and can be made either by single-layer micro-porous structures achieved by etching, hybrid materials and other processes suitable for producing large-area coatings for art framing purposes. Single layer coatings have been used as a lower-cost alternative to multi-layer anti-reflective coatings. Single-layer anti-reflective coatings can reduce light reflection to as low as 1.5%.

Multi-layer

The lowest reflection can be achieved with multi-layer anti-reflective coatings, which can be applied by either magnetron sputtering, evaporation or sol-gel process (or other processes, which can control the uniformity of deposition on nanometer-scale), and can reduce the light reflection to lower than 0.25% per side (0.5% total).

Features of Anti-reflective Coatings

- Light Reflection - the main goal of anti-reflective coatings is to reduce the light reflection that causes the so-called glare. Therefore, the lower the light reflection, the less glare reaches the viewer. The best anti-reflective products available for the picture framing market have light reflection of 0.5%. The seemingly small differences in light reflection are actually very important due to the logarithmic response of human eyes to signal intensity (Weber's law). In other words, under normal lighting conditions, the human eye perception of the intensity of a reflected light source in a 1% reflecting glass surface will be perceived as more than twice of the same light source in 0.5% reflecting glass.

- Light Absorption - light absorption of glazing is the light that is neither transmitted nor reflected by the glazing. Since light is not necessarily absorbed uniformly, some wavelengths may be transmitted more than others, causing the transmitted color to be distorted. A good way to detect light absorption of glazing is so-called *white paper test*. This test, used to detect the *transmission color* of glazing, involves placing a piece of glazing on white paper, and comparing the paper's color with and without the glass. A slight greenish tint will indicate the presence of iron oxide in the raw materials used to produce clear float glass. Additional transmitted colors may result from the absorption of any applied coatings.

- Light Transmission - the lower the light reflection and light absorption, the higher the light transmission, and therefore, the visibility of the objects displayed behind the glazing.
- Reflected Color - Uncoated glass reflects light uniformly and does not cause reflected light to be distorted (a white light source reflected in a non-coated glass pane will still appear white). However, anti-reflective coatings typically cause some wavelengths of light to be reflected more than others, causing a *shift* in the reflected color. This way, a white light source reflected in an anti-reflected glass surface may appear green or blue or red, depending on the wavelengths that are favored by a particular anti-reflective coating design.
- Intensity of Reflected Color - intensity of reflected color can be measured by its relative distance from the color neutral zone (i.e. white). Due to the variability of industrial processes, some producers design their anti-reflective coatings to have more intense colors so that the statistical deviation of outcomes falls within a specific color (green or blue, etc.). The tighter a manufacturer's control of its processes, the closer the design can be to the color neutral zone, without crossing over from a designated color.
- Reflected Color under an Angle - as a reflected light source is reflected from the glazing under a shallow angle, some anti-reflective coatings may cause the reflected color to shift. Therefore, in picture framing, a stable color under a wide viewing angle is desirable.
- Cleaning - since anti-reflective coatings render the glass surface virtually invisible, the dirt or soiling of the surface is much more visible on an anti-reflective surface. This enhanced visibility of surface smudges results in an end-user difficulty to clean the AR-coated glass. Therefore, some anti-reflective coatings have special surface treatments to improve cleanability, while others give special cleaning instructions to avoid damage to its coating.
- Handling - Some coatings are more durable than others. A scratch through an anti-reflective coating is also much more visible than a scratch through the surface of an uncoated glass due to the difference in the reflectivity of the scratched surface (glass ~8%) and the reflectivity of the anti-reflective surface around the scratch (~0.5%). Therefore, anti-reflective coatings with more scratch-resistance are preferred in art glazing. Magnetron-sputtered and sol-gel anti-reflective coatings are typically metal oxides with superior hardness compared to other application methods.

UV Filtering Coatings

In order to reduce the amount of damaging light radiation transmitted through glazing, some glass coatings are designed to either *reflect* or *absorb* the ultraviolet (UV) spectrum. The following technologies are used to reduce the amount of UV from reaching the artwork:

- Organic UV Absorbers are added to an inert, inorganic silica-based coating to yield a UV *absorbing* layer on one side of the glass. Organic UV absorbers are capable of blocking almost 100% of UV radiation between 300 nm to 380 nm, but in an industrial environment it is difficult to make a sharp UV cut-off without affecting the visible spectrum, therefore UV absorbers tend to also increase the absorption of the visible light. Chemically-deposited UV absorbers also result in a less scratch resistant surface than magnetron-sputtered or Sol-gel UV blocking layers, as evidenced by the manufacturer's recommendation for avoiding environmental and other contact with the UV-coated side.

- Interference UV Blockers are usually built into anti-reflective interference thin film stacks, and focus on maximizing the UV *reflection* below the visible light boundary. Industrially-available sol-gel processes offer up to 84% UV block, while magnetron-sputtered AR/UV-Blocking layers can block up to ~92% with no adverse effects on the transmission or absorption of visible light.

- UV Filtering of the Substrate is possible by adding UV filtering agents during the production of the substrate. While typical clear float glass blocks ~45% of UV radiation, the addition of CeOx to glass has been shown to further reduce UV transmission as well as widespread use of organic UV blocking dyes in the production of acrylic substrates. Most soda-lime glass completely absorbs short wavelength UV-B radiation below 300 nm. Low-iron glass typically blocks ~12% of UV radiation between 300 and 380 nm.

UV Protection in Art Glazing

UV Definition in Art Framing

The most widely used definition of "UV Light" in the framing industry has been defined as non-weighted average transmittance between 300 nm and 380 nm, while the ISO-DIS-21348 standard for determining irradiances defines various UV light ranges:

Name	Abbreviation	Wavelength range in nanometers	Energy per photon
Ultraviolet A, long wave, or black light	UVA	400 nm–315 nm	3.10–3.94 eV
Near	NUV	400 nm–300 nm	3.10–4.13 eV
Ultraviolet B or medium wave	UVB	315 nm–280 nm	3.94–4.43 eV

The definition of the upper limit of UV protection as 380 nm by the framing industry is not consistent with accepted standards above.

According to the Library of Congress Preservation department, the artwork damage does not stop at 380 nm though, and all radiation (UV, visible, IR) has the potential to damage art. Thus, calculating a simple average of all wavelengths between 300 nm and 380 nm does not account for the fact that different wavelengths have different artwork damage potential. At least two other methods exist, which provide a more holistic measurement of radiation damage, from both the UV and visible portions of the spectrum:

- Krochmann Damage Function (KDF) is used to rate a glazing's ability to limit fading potential. It expresses the percentage of both UV and of that portion of the visible spectrum from 300 nm to 600 nm that passes through the window and weights each wavelength in relation to the potential damage it can cause to typical materials. Lower numbers are better.

- ISO-CIE Damage-Weighted Transmission (ISO) uses a weighting function recommended by the International Commission on Illumination (CIE). Its spectral range is also weighted and extends from 300 nm to 700 nm.

For picture framing purposes, it is not appropriate to use these methods for *absolute* ratings, since "Better" ratings are obtained with lower visible light transmission, which is not aesthetically desirable in a framing glazing. However, by including more art damaging factors than UV radiation between 300 nm and 380 nm, these methods provide a more holistic *relative* ranking tool. For example, comparing a 99% and 92% UV Blocking glazing, would translate to 44% and 41%, respectively, under the KDF.

How much UV Filtering should a Glazing have

The discussion on how much UV filtering is necessary in art framing is complex and controversial, driven by conflicting corporate interests. There have so far been no independent organizations, not tied to corporate sponsors, which have presented scientifically verifiable and conclusive evidence to the amount of UV filtering necessary for a glazing to both display and at the same time protect an artwork. On one hand, the issue is complicated by the varying amount of damaging light actually present in an indoor environment (from low level indirect sources to direct daylight). On the other, by the fact that not only UV, but also visible light damages an artwork. According to the National Fenestration Rating Council, only 40% of artwork fading is caused by UV radiation. The remaining damage comes from the visible light, heat, humidity and material chemistry. This means that increasing visible light transmission by an anti-reflective coating actually increases the amount of damaging radiation on an artwork.

One of the most thorough and independent studies was conducted by the US Library of Congress in an effort to display and preserve the US Declaration of Independence. At first, it was decided to use special yellow "Plexiglass UF3", which removes both the ultra-violet as well as the blue end of the visible spectrum, with significant, but acceptable interference for viewing. Sealing the display by a chemically inert gas such as nitrogen, argon or helium also aided its preservation. In 2001, the display of the US Declaration of Independence was revised to include a multi-laminated glazing for shatter-resistance, with sol-gel interference-based multi-layer Anti-Reflective coatings on outer surfaces to improve the visibility of the document.

From the above evidence, it can be concluded that if preservation were the only goal of glazing, then only a climate-controlled, dark space would offer the best possible protection for an artwork, which can be exhibited once in every several years, while no glass at all provides a perfect displaying option. Therefore, for those artworks, which are chosen to be displayed, the ideal amount of UV blocking should be *as much as possible, without affecting the visible light transmission*.

Controlling UV Lighting Inside

While determining how much UV light should be filtered by art glazing, it may also be important to consider the amount of UV light present inside a room or a building. Note that regular window glazing filters away a significant portion of the UV light, which originates from the sun.

The relative amounts of damaging light in equal quantities of light:

Illumination	Relative Damage
Horizontal skylight, open	100%

Horizontal skylight, window glass	34%

The above indicates that the damage level of even direct sunlight coming from the horizontal skylight is reduced to 36% by regular window glass. Due to the sun's changing position, even less direct light enters through side windows and hanging an artwork away from direct sunlight reduces the exposure to potentially damaging direct sunlight even further.

Indoor lighting, especially fluorescent lighting, is considered to contain some UV Light. GELighting.com asserts that "UV exposure from sitting indoors under fluorescent lights at typical office light levels for an eight-hour workday is equivalent to just over one minute of exposure to the sun in Washington D.C. on a clear day in July. Additionally, the relative damage of incandescent light is 3 times less than that of fluorescent light. Since UV filtering picture framing glass does not protect against all damage factors, it is important to display framed artwork in a well controlled environment to reduce the effects of heat, humidity, and visible light.

References

- Department of Health and Human Services (2011), National Institute of Environmental Health Sciences, National toxicology Program, Fact Sheet, "The Report on Carcinogens," June 2011 (PDF), retrieved 2013-02-05
- North American Insulation Manufacturers Association ("NAIMA"), Insulation Facts #62 "Health and Safety Facts for Fiber Glass", Pub. No. N040, May 2012.

7

Recycling of Glass

Glass recycling is the conversion of waste glass into products that can be used. Recycled glass countertops are entirely made up of glass that is 100% recyclable. Glass wool and abrasive are also explained in the chapter. Recycling of glass is best understood in confluence with the major topics listed in the following section.

Glass Recycling

Public glass waste collection point in a neighborhood area for separating colorless, green and amber glass

Glass recycling is the processing of waste glass into usable products. Glass waste should be separated by chemical composition, and then, depending on the end use and local processing capabilities, might also have to be separated into different colors. Many recyclers collect different colors of glass separately since glass retains its color after recycling. The most common types used for consumer containers are colorless glass, green glass, and brown/amber glass. Glass is ideal for recycling since none of the material is degraded by normal use.

Glass makes up a large component of household and industrial waste due to its weight and density. The glass component in municipal waste is usually made up of bottles, broken glassware, light bulbs and other items. Adding to this waste is the fact that many manual methods of creating glass objects have a defect rate of around forty percent. Glass recycling uses less energy than manufacturing glass from sand, lime and soda.

Cullet from U.S. beer bottles (June 1972).

Every metric ton (1,000 kg) of waste glass recycled into new items saves 315 kilograms (694 lb) of carbon dioxide from being released into the atmosphere during the creation of new glass. Glass that is crushed and ready to be remelted is called cullet.

Many collection points have separate bins for clear, green and amber. Glass reprocessors that intend to make glass containers require separation by color, because glass tends to retain its color after recycling. If the recycled glass is intended for other products, or if the glass reprocessor uses newer optical sorting equipment, separation by color at the collection point may not be required. Heat-resistant glass like Pyrex or borosilicate glass should **not** be disposed of in the glass container as even a single piece of such material will alter the viscosity of the fluid in the furnace at remelt.

Recycling into Glass Containers

When glass is used for new container manufacturing, it is virtually infinitely recyclable. The processing and use of recycled glass in manufacturing conserves raw materials, and reduces energy consumption. That which is not recycled, but crushed, reduces the volume of waste sent to landfills.

Recycling into Other Products

The use of the recycled glass as aggregate in concrete has become popular in modern times, with large scale research being carried out at Columbia University in New York. This greatly enhances the aesthetic appeal of the concrete. Recent research findings have shown that concrete made with recycled glass aggregates have shown better long term strength and better thermal insulation due to its better thermal properties of the glass aggregates.

Glass aggregate, a mix of colors crushed to a small size, is substituted for pea gravel or crushed rock in many construction and utility projects, saving municipalities like the City of Tumwater, Washington Public Works, thousands of dollars (depending on the size of the project). Glass aggregate is not sharp to handle. In many cases, the state Department of Transportation has specifications for use, size and percentage of quantity for use. Common applications are as pipe bedding—placed around sewer, storm water or drinking water pipes to transfer weight from the surface and protect the pipe. Another common use would be as fill to bring the level of a concrete floor even with a foundation.

Other uses for recycled glass include:

- Fiberglass insulation products
- Ceramic sanitary ware production
- As a flux in brick manufacture
- Astroturf
- Agriculture and landscape applications, such as top dressing, root zone material or golf bunker sand
- Recycled glass countertops
- As water filtration media
- Abrasives

Mixed waste streams may be collected from materials recovery facilities or mechanical biological treatment systems. Some facilities can sort out mixed waste streams into different colours using electro-optical sorting units.

Glass Recycling by Country

Germany

In 2004, Germany recycled 2,116,000 tons of glass. Reusable glass or plastic (PET) bottles are available for many drinks, especially beer and carbonated water as well as soft drinks (*Mehrweg-flaschen*). The deposit per bottle (*Pfand*) is €0.08-€0.15, compared to €0.25 for recyclable but not reusable plastic bottles. There is no deposit for glass bottles which do not get refilled.

United Kingdom

Vehicle emptying a glass bank in Vienna

Glass collection points, known as Bottle Banks are very common near shopping centres, at civic amenity sites and in local neighborhoods in the United Kingdom. The first Bottle Bank was introduced by Stanley Race CBE, then president of the Glass Manufacturers' Federation and Ron England in Barnsley on 6 June 1977. Development work was done by the DoE at Warren Springs laboratory, Stevenage, (now AERA at Harwell) and Nazeing Glass Works, Broxboure to prove if a usable glass product could be made from over 90% recycled glass. It was found necessary to use magnets to remove unwanted metal closures in the mix.

Bottle Banks commonly stand beside collection points for other recyclable waste like paper, metals and plastics. Local, municipal waste collectors usually have one central point for all types of waste in which large glass containers are located. There are now over 50,000 bottle banks in the United Kingdom, and 752,000 tons of glass are now recycled annually.

The waste recycling industry in the UK cannot consume all of the recycled container glass that will become available over the coming years, mainly due to the colour imbalance between that which is manufactured and that which is consumed. The UK imports much more green glass in the form of wine bottles than it uses, leading to a surplus amount for recycling.

The resulting surplus of green glass from imported bottles may be exported to producing countries, or used locally in the growing diversity of secondary end uses for recycled glass. As of 2006, Cory Environmental were shipping glass cullet from the UK to Portugal.

United States of America

Rates of recycling and methods of waste collection vary substantially across the United States because laws are written on the state or local level and large municipalities often have their own unique systems. Many cities do curbside recycling, meaning they collect household recyclable waste on a weekly or bi-weekly basis that residents set out in special containers in front of their homes.

Apartment dwellers usually use shared containers that may be collected by the city or by private recycling companies which can have their own recycling rules. In some cases, glass is specifically separated into its own container because broken glass is a hazard to the people who later manually sort the co-mingled recyclables. Sorted recyclables are later sold to companies to be used in the manufacture of new products.

In 1971, the state of Oregon passed a law requiring buyers of carbonated beverages (such as beer and soda) to pay five cents per container as a deposit which would be refunded to anyone who returned the container for recycling. This law has since been copied in nine other states including New York and California. The abbreviations of states with deposit laws are printed on all qualifying bottles and cans. In states with these container deposit laws, most supermarkets automate the deposit refund process by providing machines which will count containers as they are inserted and then print credit vouchers that can be redeemed at the store for the number of containers returned. Small glass bottles (mostly beer) are broken, one-by-one, inside these deposit refund machines as the bottles are inserted. A large, wheeled hopper (very roughly 1.5m by 1.5m by 0.5m) inside the machine collects the broken glass until it can be emptied by an employee.

Recycled Glass Countertops

Recycled glass countertops are composed of 100% recycled glass (post consumer and pre-consumer) in a cement or petroleum based binder. A finished recycled glass countertop often ranges from 70 to 85 percent in recycled content.

Environmental Concerns

The glass used in recycled glass countertops is sourced from both post-consumer and pre-consumer sources. Post-consumer recycled glass sources are numerous, but the most common sources are curbside recycling and salvaged glass from demolished buildings.

Curbside recycled glass from homes and businesses is the largest source of post consumer recycled glass (California recycled 79% of its glass bottles in 2008), but salvage glass is an emerging new source. Many more state and municipal governments have begun to mandate the separation and recycling of the glass pulled from demolished buildings. Some companies are now processing locally sourced glass into aggregate to more closely meet the LEED, USGBC specifications.

Where the recycled glass countertops are made is another variable that is taken into account when determining the sustainability of a recycled glass countertop. At this time most of the companies manufacture their products in the United States, but recycled glass countertops have gained international interest. Now some foreign companies have begun to export recycled glass countertops to the USA. This practice of importing heavy (14.5 lbs/sqr ft) recycled glass countertops from overseas when there is such a large quantity of recycled glass in the USA has quickly come under scrutiny. The Leadership in Energy and Environmental Design (LEED) certification program has begun to put increasing weight on the importance of regional manufacturing to reduce the carbon footprint of building materials. This focus on materials made in the USA has the secondary affect of encouraging the support of US business in a time of economic downturn.

Several recycled glass countertop manufacturers have committed to providing green jobs to the low-income neighborhoods surrounding their manufacturing facilities. Some, but not all of the manufacturers use facilities that are solar powered and/or day-lit by south facing skylights. These manufacturers will also incorporate some degree of water recycling in their polishing process.

Cement vs. Petroleum Binders

The use of cement versus petroleum based binders has raised some questions about the sustainability of petroleum binders for a number of reasons. The use of a petroleum binder creates a superior quality product in most aspects; its performance withstanding heat is diminished though comparable to other products using resin binders (engineered quartz, and solid surface). Cement-based products are the second largest contributor of greenhouse gases.

Petroleum based binders do present the advantage that they do not need to be sealed once to twice a year like cement based recycled glass countertops, and are able to be manufactured in more usable sizes, thickness, and curved forms. The main advantage of petroleum based products is the anti staining properties over cement products.

Criticism

The potential environmental benefits of finding a secondary market for recycled glass may be outweighed by the vast amounts of greenhouse gas produced in the production of standard cement. With nearly a ton and a half of CO_2 released in the manufacturing of a single ton of cement, the binder used in these composites is among the planet's most egregious polluters.

The alkali–silica reaction (ASR), which occurs when glass aggregate is combined with cement, produces production difficulties for recycled glass surfaces that may compromise structural integrity if left unaddressed. Ongoing research at Columbia University and Argonne National Labs has ruled out glass and cement composites as a viable alternative to conventional surface materials, but new technologies are under development which may mitigate the challenges of producing these surfaces.

Glass Wool

Glass wool is an insulating material made from fibres of glass arranged using a binder into a texture similar to wool. The process traps many small pockets of air between the glass, and these small air pockets result in high thermal insulation properties.

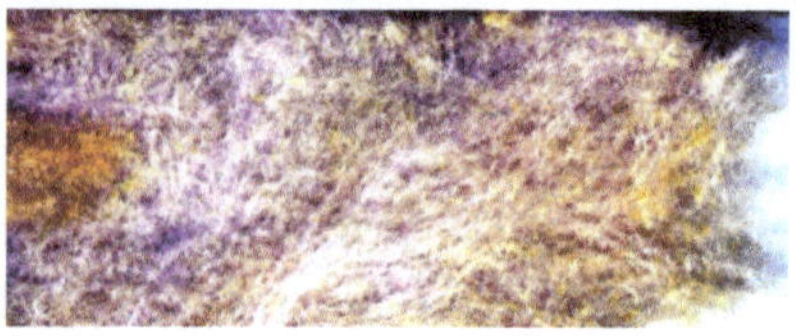

Fiberglass insulation from a ceiling tile

Glass wool batt insulation

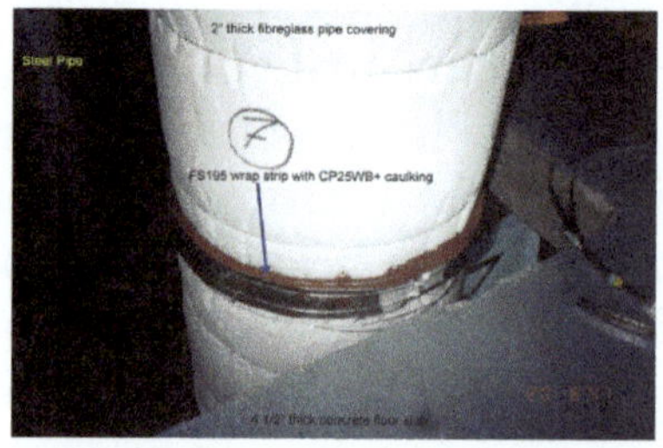

Fiberglass pipe insulation with ASJ (All Service Jacket) penetrating concrete slab opening about to be firestopped. Intumescent wrap strip is used to seal off where the fiberglass will be consumed by fire.

Glass wool is produced in rolls or in slabs, with different thermal and mechanical properties. It may also be produced as a material that can be sprayed or applied in place, on the surface to be insulated.

The modern method for producing glass wool is the invention of Games Slayter working at the Owens-Illinois Glass Co. (Toledo, Ohio). He first applied for a patent for a new process to make glass wool in 1933.

Principles of Function

Gases possess poor thermal conduction properties compared to liquids and solids, and thus makes a good insulation material if they can be trapped. In order to further augment the effectiveness of a gas (such as air) it may be disrupted into small cells which cannot effectively transfer heat by natural convection. Convection involves a larger bulk flow of gas driven by buoyancy and temperature differences, and it does not work well in small cells where there is little density difference to drive it.

In order to accomplish formation of small gas cells in man-made thermal insulation, glass and polymer materials can be used to trap air in a foam-like structure. The same principle used in glass wool is used in other man-made insulators such as rock wool, styrofoam, wet suit neoprene foam fabrics, and fabrics such as Gore-Tex and polar fleece. The air-trapping property is also the insulation principle used in nature in down feathers, and insulating hair such as natural wool.

Manufacturing Process

After the mixture of natural sand and recycled glass at 1,450 °C, the glass that is produced is converted into fibers. It is typically produced in a method similar to making cotton candy, forced through a fine mesh by centripetal force, cooling on contact with the air. The cohesion and mechanical strength of the product is obtained by the presence of a binder that "cements" the fibers together. Ideally, a drop of bonder is placed at each fiber intersection. This fiber mat is then heated to around 200 °C to polymerize the resin and is calendered to give it strength and stability. The final stage involves cutting the wool and packing it in rolls or panels under very high pressure before palletizing the finished product in order to facilitate transport and storage.

Uses

Glass wool is a thermal insulation that consists of intertwined and flexible glass fibers, which causes it to "package" air, resulting in a low density that can be varied through compression and binder content (as noted above, these air cells are the actual insulator). Glass wool can be a loose fill material, blown into attics, or, together with an active binder sprayed on the underside of structures, sheets and panels that can be used to insulate flat surfaces such as cavity wall insulation, ceiling tiles, curtain walls as well as ducting. It is also used to insulate piping and for soundproofing.

Fiberglass Batts and Blankets

Installing glass wool batts as ceiling insulation

Batts are precut, whereas blankets are available in continuous rolls. Compressing the material reduces its effectiveness. Cutting it to accommodate electrical boxes and other obstructions allows air a free path to cross through the wall cavity. One can install batts in two layers across an unfinished attic floor, perpendicular to each other, for increased effectiveness at preventing heat bridging. Blankets can cover joists and studs as well as the space between them. Batts can be challenging and unpleasant to hang under floors between joists; straps, or staple cloth or wire mesh across joists, can hold it up.

Gaps between batts (bypasses) can become sites of air infiltration or condensation (both of which reduce the effectiveness of the insulation) and requires strict attention during the installation. By the same token careful weatherization and installation of vapour barriers is required to ensure that the batts perform optimally. Air infiltration can be also reduced by adding a layer of cellulose loose-fill on top of the material.

Health Problems

Fiberglass will irritate the eyes, skin, and the respiratory system. Potential symptoms include irritation of eyes, skin, nose, throat, dyspnea (breathing difficulty); sore throat, hoarseness and cough. Scientific evidence demonstrates that fiber glass is safe to manufacture, install and use when recommended work practices are followed to reduce temporary mechanical irritation.

In the US, the National Toxicology Program ("NTP"), in June 2011, removed from its Report on Carcinogens all biosoluble glass wool used in home and building insulation and for non-insulation products. Similarly, California's Office of Environmental Health Hazard Assessment ("OEHHA"), in November 2011, published a modification to its Proposition 65 listing to include only "Glass wool fibers (inhalable and biopersistent)." The U.S. NTP and California's OEHHA action means that a cancer warning label for biosoluble fiber glass home and building insulation is no longer required under Federal or California law. All fiber glass wools commonly used for thermal and acoustical insulation were reclassified by the International Agency for Research on Cancer ("IARC") in October 2001 as Not Classifiable as to carcinogenicity to humans (Group 3).

Fiberglass is resistant to mold but often the binder is organic, therefore the binder if organic supports mould growth. In tests glass wool was quite resistant to the growth of mold but at very high relative humidity, 96% or above or if it becomes wet from saturation or indeed liquid water the growth will be moderate.

Abrasive

An abrasive is a material, often a mineral, that is used to shape or finish a workpiece through rubbing which leads to part of the workpiece being worn away by friction. While finishing a material often means polishing it to gain a smooth, reflective surface, the process can also involve roughening as in satin, matte or beaded finishes. In short, the Ceramics which are used to cut, grind and polish other softer materials are known as Abrasives.

Abrasives are extremely commonplace and are used very extensively in a wide variety of industrial, domestic, and technological applications. This gives rise to a large variation in the physical and chemical composition of abrasives as well as the shape of the abrasive. Common uses for abrasives include grinding, polishing, buffing, honing, cutting, drilling, sharpening, lapping, and sanding. (For simplicity, "mineral" in this article will be used loosely to refer to both minerals and mineral-like substances whether man-made or not.)

Files are not abrasives; they remove material not by scratching or rubbing, but by the cutting action of sharp teeth which have been cut into the surface of the file, very much like those of a saw. However, diamond files are a form of coated abrasive (as they are metal rods coated with diamond powder).

Mechanics of Abrasion

Abrasives generally rely upon a difference in hardness between the abrasive and the material being worked upon, the abrasive being the harder of the two substances. However, this is not necessary as any two solid materials that repeatedly rub against each other will tend to wear each other away; examples include, softer shoe soles wearing away wooden or stone steps over decades or centuries or glaciers abrading stone valleys.

Typically, materials used as abrasives are either hard minerals (rated at 7 or above on Mohs scale of mineral hardness) or are synthetic stones, some of which may be chemically and physically identical to naturally occurring minerals but which cannot be called minerals as they did not arise naturally. (While useful for comparative purposes, the Mohs scale is of limited value to materials engineers as it is an arbitrary, ordinal, irregular scale.) Diamond, a common abrasive, for instance occurs both naturally and is industrially produced, as is corundum which occurs naturally but which is nowadays more commonly manufactured from bauxite. However, even softer minerals like calcium carbonate are used as abrasives, such as "polishing agents" in toothpaste.

Grit size ranging from 2 mm (the large grain) (about F 10 using FEPA standards) to about 40 micrometres (about F 240 or P 360).

These minerals are either crushed or are already of a sufficiently small size (anywhere from macroscopic grains as large as about 2 mm to microscopic grains about 0.001 mm in diameter) to permit their use as an abrasive. These grains, commonly called grit, have rough edges, often terminating in points which will decrease the surface area in contact and increase the localised contact pressure. The abrasive and the material to be worked are brought into contact while in relative motion to each other. Force applied through the grains causes fragments of the worked material to break away while simultaneously smoothing the abrasive grain and/or causing the grain to work loose from the rest of the abrasive.

Some factors which will affect how quickly a substance is abraded include:

- Difference in hardness between the two substances: a much harder abrasive will cut faster and deeper
- Grain size (grit size): larger grains will cut faster as they also cut deeper
- Adhesion between grains, between grains and backing, between grains and matrix: determines how quickly grains are lost from the abrasive and how soon fresh grains, if present, are exposed
- Contact force: more force will cause faster abrasion
- Loading: worn abrasive and cast off work material tends to fill spaces between abrasive grains so reducing cutting efficiency while increasing friction
- Use of lubricant/coolant/metalworking fluid: Can carry away swarf (preventing loading), transport heat (which may affect the physical properties of the workpiece or the abrasive), decrease friction (with the substrate or matrix), suspend worn work material and abrasives allowing for a finer finish, conduct stress to the workpiece.

Abrasive Minerals

Abrasives may be classified as either natural or synthetic. When discussing sharpening stones, natural stones have long been considered superior but advances in material technology are seeing this distinction become less distinct. Many synthetic abrasives are effectively identical to a natural mineral, differing only in that the synthetic mineral has been manufactured rather than been mined. Impurities in the natural mineral may make it less effective.

Some naturally occurring abrasives are:

- Calcite (calcium carbonate)
- Emery (impure corundum)
- Diamond dust (synthetic diamonds are used extensively)
- Novaculite
- Pumice
- Rouge
- Sand
- Corundum
- Garnet
- Sandstone
- Tripoli
- Powdered Feldspar
- Staurolite

Some abrasive minerals (such as zirconia alumina) occur naturally but are sufficiently rare or sufficiently more difficult/costly to obtain such that a synthetic stone is used industrially. These and other artificial abrasives include:

- Borazon (cubic boron nitride or CBN)
- Ceramic
- Ceramic aluminium oxide
- Ceramic iron oxide
- Corundum (alumina or aluminium oxide)
- Dry ice
- Glass powder

- Steel abrasive
- Silicon carbide (carborundum)
- Zirconia alumina
- Boron carbide
- Slags

Manufactured Abrasives

Abrasives are shaped for various purposes. Natural abrasives are often sold as dressed stones, usually in the form of a rectangular block. Both natural and synthetic abrasives are commonly available in a wide variety of shapes, often coming as bonded or coated abrasives, including blocks, belts, discs, wheels, sheets, rods and loose grains.

Bonded Abrasives

Assorted grinding wheels as examples of bonded abrasives.

A grinding wheel with a reservoir to hold water as a lubricant and coolant.

A bonded abrasive is composed of an abrasive material contained within a matrix, although very fine aluminium oxide abrasive may comprise sintered material. This matrix is called a binder and is often a clay, a resin, a glass or a rubber. This mixture of binder and abrasive is typically shaped into blocks, sticks, or wheels. The most usual abrasive used is aluminium oxide. Also common are silicon carbide, tungsten carbide and garnet. Artificial sharpening stones are often a bonded abrasive and are readily available as a two sided block, each side being a different grade of grit.

Grinding wheels are cylinders that are rotated at high speed. While once worked with a foot pedal or hand crank, the introduction of electric motors has made it necessary to construct the wheel to withstand greater radial stress to prevent the wheel flying apart as it spins. Similar issues arise with cutting wheels which are often structurally reinforced with impregnated fibres. High relative speed between abrasive and workpiece often makes necessary the use of a lubricant of some kind. Traditionally they were called coolants as they were used to prevent frictional heat build up which could damage the workpiece (such as ruining the temper of a blade). Some research suggests that the heat transport property of a lubricant is less important when dealing with metals as the metal will quickly conduct heat from the work surface. More important are their effects upon lessening tensile stresses while increasing some compressive stresses and reducing "thermal and mechanical stresses during chip formation".

Various shapes are also used as heads on rotary tools used in precision work, such as scale modelling.

Bonded abrasives need to be trued and dressed after they are used. Dressing is cleaning the waste material (swarf and loose abrasive) from the surface and exposing fresh grit. Depending upon the abrasive and how it was used, dressing may involve the abrasive being simply placed under running water and brushed with a stiff brush for a soft stone or the abrasive being ground against another abrasive, such as aluminium oxide used to dress a grinding wheel.

Truing is restoring the abrasive to its original surface shape. Wheels and stones tend to wear unevenly, leaving the cutting surface no longer flat (said to be "dished out" if it is meant to be a flat stone) or no longer the same diameter across the cutting face. This will lead to uneven abrasion and other difficulties.

Coated Abrasives

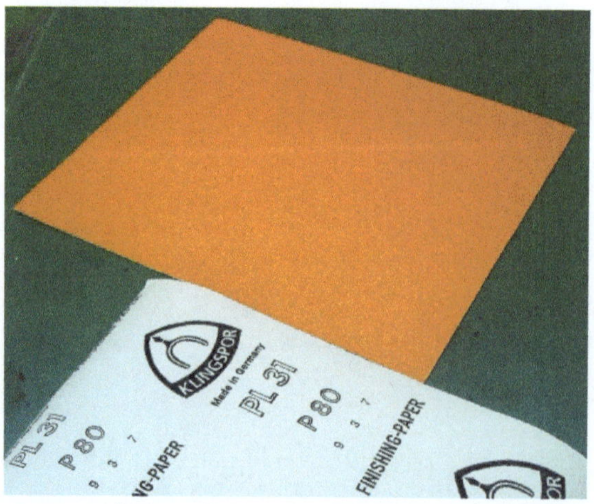

A German sandpaper showing its backing and FEPA grit size.

A *coated abrasive* comprises an abrasive fixed to a backing material such as paper, cloth, rubber, resin, polyester or even metal, many of which are flexible. Sandpaper is a very common coated abrasive. Coated abrasives are commonly the same minerals as are used for bonded abrasives. A bonding agent (often some sort of adhesive or resin) is applied to the backing to provide a flat surface to which the grit is then subsequently adhered. A woven backing may also use a filler agent (again, often a resin) to provide additional resilience.

Coated abrasives may be shaped for use in rotary and orbital sanders, for wrapping around sanding blocks, as handpads, as closed loops for use on belt grinders, as striking surfaces on matchboxes, on diamond plates and diamond steels. Diamond tools, though for cutting, are often abrasive in nature.

Other Abrasives and their Uses

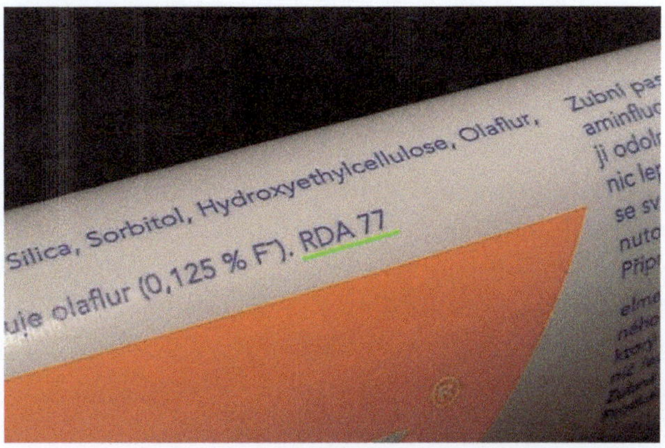

Here the abrasiveness of toothpaste is detailed by its Relative Dentin Abrasivity (RDA)

Sand, glass beads, metal pellets copper slag and dry ice may all be used for a process called sandblasting (or similar, such as the use of glass beads which is "bead blasting"). Dry ice will sublimate leaving behind no residual abrasive.

Cutting compound used on automotive paint is an example of an abrasive suspended in a liquid, paste or wax, as are some polishing liquids for silverware and optical media. The liquid, paste or wax acts as a binding agent that keeps the abrasive attached to the cloth which is used to as a backing to move the abrasive across the work piece. On cars in particular, wax may serve as both a protective agent by preventing exposure of the paint of metal to air and also act as an optical filler to make scratches less noticeable. Toothpaste contains calcium carbonate or silica as a "polishing agent" to remove plaque and other matter from teeth as the hardness of calcium carbonate is less than that of tooth enamel but more than that of the contaminating agent.

Very fine rouge powder was commonly used for grinding glass, being somewhat replaced by modern ceramics, and is still used in jewellery making for a highly reflective finish.

Cleaning products may also contain abrasives suspended in a paste or cream. They are chosen to be reasonably safe on some linoleum, tile, metal or stone surfaces. However, many laminate surfaces and ceramic topped stoves are easily damaged by these abrasive compounds. Even ceramic/pottery tableware or cookware can damage these surfaces, particularly the bottom of the tableware which is often unglazed in part or in whole and acts as simply another bonded abrasive.

Metal pots and stoves are often scoured with abrasive cleaners, typically in the form of the aforementioned cream or paste or of steel wool and non woven scouring pads which holds fine grits abrasives.

Human skin is also subjected to abrasion in the form of exfoliation. Abrasives for this can be much softer and more exotic than for other purposes and may include things like almond and oatmeal. Dermabrasion and microdermabrasion are now rather commonplace cosmetic procedures which use mineral abrasives.

Scratched compact discs and DVDs may sometimes be repaired through buffing with a very fine compound, the principle being that a multitude of small scratches will be more optically transparent than a single large scratch. However, this does take some skill and will eventually cause the protective coating of the disc to be entirely eroded (especially if the original scratch is deep), after which the data surface will be destroyed if abrasion continues.

Choice of Abrasive

The shape, size and nature of the workpiece and the desired finish will influence the choice of the abrasive used. A bonded abrasive grind wheel may be used to commercially sharpen a knife (producing a hollow grind), but an individual may then sharpen the same knife with a natural sharpening stone or an even flexible coated abrasive (like a sandpaper) stuck to a soft, non-slip surface to make achieving a convex grind easier. Similarly, a brass mirror may be cut with a bonded abrasive, have its surface flattened with a coated abrasive to achieve a basic shape, and then have finer grades of abrasive successively appied culminating in a wax paste impregnated with rouge to leave a sort of "grainless finish" called, in this case, a "mirror finish".

Also, different shapes of adhesive may make it harder to abrade certain areas of the workpiece. Health hazards can arise from any dust produced (which may be ameliorated through the use of a lubricant) which could lead to silicosis (when the abrasive or workpiece is a silicate) and the choice of any lubricant. Besides water, oils are the most common lubricants. These may present inhalation hazards, contact hazards and, as friction necessarily produces heat, flammable material hazards.

An abrasive which is too hard or too coarse can remove too much material or leave undesired scratch marks. Besides being unsightly, scratching can have other, more serious effects. Excessive abrasion or the presence of scratches may:

- diminish or destroy usefulness (as in the case of scratching optical lenses and compact discs or dulling knives);
- trap dirt, water, or other material;
- increase surface area (permitting greater chemical reactivity such as increased rusting which is also affected by matter caught in scratches);
- erode or penetrate a coating (such as a paint or a chemical or wear resistant coating);
- overly quickly cause an object to wear away (such as a blade or a gemstone);
- increase friction (as in jeweled bearings and pistons).

A finer or softer abrasive will tend to leave much finer scratch marks which may even be invisible to the naked eye (a "grainless finish"); a softer abrasive may not even significantly abrade a certain object. A softer or finer abrasive will take longer to cut as tends to cut less deeply than a coarser, harder material. Also, the softer abrasive may become less effective more quickly as the abrasive is itself abraded. This allows fine abrasives to be used in the polishing of metal and lenses where the series of increasingly fine scratches tends to take on a much more shiny or reflective appearance or greater transparency. Very fine abrasives may be used to coat the strop for a cut-throat razors, however, the purpose of stropping is not to abrade material but to straighten the burr on an edge. The final stage of sharpening Japanese swords is called polishing and may be a form of superfinishing.

Different chemical or structural modifications may be made to alter the cutting properties of the abrasive.

Other very important considerations are price and availability. Diamond, for a long time considered the hardest substance in existence, is actually softer than fullerite and even harder aggregated diamond nanorods, both of which have been synthesised in laboratories but no commercial process has yet been developed. Diamond itself is expensive due to scarcity in nature and the cost of synthesising it. Bauxite is a very common ore which, along with corundum's reasonably high hardness, contributes to corundum's status as a common, inexpensive abrasive.

Thought must be given to the desired task about using an appropriately hard abrasive. At one end, using an excessively hard abrasive wastes money by wearing it down when a cheaper, less hard abrasive would suffice. At the other end, if too soft, abrasion does not take place in a timely fashion, effectively wasting the abrasive as well as any accruing costs associated with loss of time.

Other Instances of Abrasion

Aside from the aforementioned uses of shaping and finishing, abrasives may also be used to prepare surfaces for application of some sort of paint of adhesive. An excessively smooth surface may prevent paint and adhesives from adhering as strongly as an irregular surface could allow. Inflatable tyre repair kits (which, on bicycles particularly, are actually patches for the inner tube rather than the tyre) require use of an abrasive so that the self-vulcanising cement will stick strongly.

Inadvertently, people who use knives on glass or metal cutting boards are abrading their knife blades. The pressure at the knife edge can easily create microscopic (or even macroscopic) cuts in the board. This cut is a ready source of abrasive material as well as a channel full of this abrasive through which the edge slides. For this reason—without regard for the health benefits—wooden boards are much more desirable. A similar occurrence arises with glass-cutters. Glass-cutters are have circular blades that are designed to roll not slide. They should never retrace an already effected cut.

Undesired abrasion may result from the presence of carbon in internal combustion engines. While smaller particles are readily transported by the lubrication system, larger carbon particles may abrade components with close tolerances. The carbon arises from the excessive heating of engine oil or from incomplete combustion. This soot may contain fullerenes which are noted for their extreme hardness—and small size and limited quantity which would tend to limit their effect.

Permissions

All chapters in this book are published with permission under the Creative Commons Attribution Share Alike License or equivalent. Every chapter published in this book has been scrutinized by our experts. Their significance has been extensively debated. The topics covered herein carry significant information for a comprehensive understanding. They may even be implemented as practical applications or may be referred to as a beginning point for further studies.

We would like to thank the editorial team for lending their expertise to make the book truly unique. They have played a crucial role in the development of this book. Without their invaluable contributions this book wouldn't have been possible. They have made vital efforts to compile up to date information on the varied aspects of this subject to make this book a valuable addition to the collection of many professionals and students.

This book was conceptualized with the vision of imparting up-to-date and integrated information in this field. To ensure the same, a matchless editorial board was set up. Every individual on the board went through rigorous rounds of assessment to prove their worth. After which they invested a large part of their time researching and compiling the most relevant data for our readers.

The editorial board has been involved in producing this book since its inception. They have spent rigorous hours researching and exploring the diverse topics which have resulted in the successful publishing of this book. They have passed on their knowledge of decades through this book. To expedite this challenging task, the publisher supported the team at every step. A small team of assistant editors was also appointed to further simplify the editing procedure and attain best results for the readers.

Apart from the editorial board, the designing team has also invested a significant amount of their time in understanding the subject and creating the most relevant covers. They scrutinized every image to scout for the most suitable representation of the subject and create an appropriate cover for the book.

The publishing team has been an ardent support to the editorial, designing and production team. Their endless efforts to recruit the best for this project, has resulted in the accomplishment of this book. They are a veteran in the field of academics and their pool of knowledge is as vast as their experience in printing. Their expertise and guidance has proved useful at every step. Their uncompromising quality standards have made this book an exceptional effort. Their encouragement from time to time has been an inspiration for everyone.

The publisher and the editorial board hope that this book will prove to be a valuable piece of knowledge for students, practitioners and scholars across the globe.

Index

K

L

M

N

O

P

R

S

T

V